Universitext

Falko Lorenz

Algebra

Volume I:
Fields and Galois Theory

With the collaboration of the translator, Silvio Levy

 Springer

Falko Lorenz
FB Mathematik Institute
University Münster
Münster, 48149
Germany
lorenz@math.uni-muenster.de

Mathematics Subject Classification (2000): 11-01, 12-01, 13-01

Library of Congress Control Number: 2005932557

ISBN-10: 0-387-28930-5   Printed on acid-free paper.
ISBN-13: 978-0387-28930-4

Printed in the United States of America.   (SHER)

9 8 7 6 5 4 3 2 1

springeronline.com

Foreword

The present textbook is my best effort to write a lively, problem-oriented and under-standable introduction to classical modern algebra. Besides careful exposition, my goals were to lead the reader right away to interesting subject matter and to assume no more background than that provided by a first course in linear algebra.

In keeping with these goals, the exposition is by and large geared toward certain motivating problems; relevant conceptual tools are introduced gradually as needed. This way of doing things seems more likely to hold the reader's attention than a more or less systematic stringing together of theorems and proofs. The pace is more leisurely and gentle in the beginning, later faster and less cautious, so the book lends itself to self-study.

This first volume, primarily about fields and Galois theory, in order to deal with the latter introduces just the necessary amount of group theory. It also covers basic applications to number theory, ring extensions and algebraic geometry. I have found it advantageous for various reasons to bring into play early on the notion of the algebraic closure of a field. Naturally, Galois' beautiful results on solvable groups of prime degree could not be left out, nor could Dedekind's Galois-theoretical arith-metic reduction principle. Infinite Galois extensions are not neglected either. Finally, it seemed appropriate to include the fundamentals of transcendental extensions.

At the end of the volume there is a collection of exercises, interspersed with remarks that enrich the text. The problems chosen are of widely varying degrees of difficulty, but very many of them are accompanied by hints — sometimes amounting to an outline of the solution — and in any case there are no outright riddles. These exercises are of course meant to allow readers to practice their grasp of the material, but they serve another important purpose as well: precisely because the main text was kept short and to the point, without lots of side-results, the appendix will give the reader a better idea of the wealth of consequences and applications derived from the theory.

The linear algebra facts used, when not totally elementary, are accompanied by references to my *Lineare Algebra*, now published by Spektrum Akademischer Verlag and abbreviated LA I and LA II. This has not been translated, but equivalent spots in other linear algebra textbooks are not hard to find. Theorems and lesser results are numbered within each chapter in sequence, the latter being marked F1, F2, . . . — the F is inherited from the German word Feststellung. Allusions to historical matters are made only infrequently (but certainly not at random). When a theorem or other result bears the name of a mathematician, this is sometimes a matter of tradition more than of accurate historical origination.

The first German edition of this book appeared in 1987. I thank my colleagues who, already back at the writing stage, favored it with their interest and gave me encouragement — none more than the late H.-J. Nastold, with whom I had many fruitful conversations, W. Lütkebohmert, who once remarked that there was no suitable textbook for the German Algebra I course, O. Willhöft, who suggested several good problems, and H. Schulze-Relau and H. Epkenhans, whose critical perusal of large portions of the manuscript was a great help. The second (1991) and third (1995) editions benefited from the remarks of numerous readers, to whom I am likewise thankful, in particular R. Alfes, H. Coers, H. Daldrop and R. Schopohl. The response and comments on the part of students were also highly motivating. Special thanks are due to the publisher BI-Wissenschaftsverlag (later acquired by Spektrum) and its editor H. Engesser, who got me going in the first place.

The publication of this English version gives me great pleasure. I'm grateful to Springer-Verlag New York and its mathematics editor Mark Spencer, for their support and competent handling of the project. And not least for seeing to it that the translation be done by Silvio Levy: I have observed the progress of his task with increasing appreciation and have incorporated many of the changes he suggested, in a process of collaboration that led to noticeable improvements. Further perfecting is of course possible, and readers' suggestions and criticism will continue to be welcome and relevant for future reprints.

Münster, July 2005   Falko Lorenz

Contents

1

Constructibility with Ruler and Compass

1. In school one sometimes learns to solve problems where a certain geometric figure must be constructed from given data. Such construction problems can be quite difficult and afford a real challenge to the student's intelligence and ingenuity.

If you have tried long and hard to solve a certain construction problem, to no avail, you might then wonder whether the required construction can be carried out at all. Whether a construction exists is a fundamental question: so much so that there are some construction problems that had already been entertained by the ancient Greeks, and yet remained unsolved for two thousand years and more.

For example, nobody has ever been able to state a procedure capable of dividing an *arbitrary* angle into three equal angles, using ruler and compass. Now, of course construction problems range widely in degree of difficulty: think of the similar-looking problem of dividing an arbitrary *segment* into three equal parts using ruler and compass — not totally trivial, but after some thought just about anyone can carry out the construction. The problem of constructing a *regular pentagon* is also solvable, but already somewhat more complicated. So it is certainly understandable that even a construction problem that has eluded would-be solvers for a long time should leave room for hoping that success might yet be achieved through greater ingenuity. Perhaps, then, the question of whether a particular construction with ruler and compass is *possible* is not one that comes to mind immediately.

Even if someone asks this question of principle, it is not clear *a priori* that there is a promising way to tackle it. Yet there is, as the development of algebra since Gauss (1777–1855) has shown. I would like to explain now, at the beginning of our introduction to algebra, how one can arrive at broad statements about the general *constructibility problem*, by translating this geometric problem into an algebraic one. As we elaborate on this, we will have the chance to motivate quite naturally certain fundamental algebraic concepts. Moreover the subsequent treatment of the derived algebraic problem will require many of the tools usually treated in an Algebra I course. This procedure has the advantage that one starts from a concrete and easily understood question and keeps the goal of solving it in mind as one goes along.

Let it be said, however, that the problem of constructibility with ruler and compass by no means played a central role in the development of algebra. In this regard the problem of solving algebraic equations by means of radicals was surely more significant, not to mention other motivations and stimuli coming from outside algebra — from number theory and analysis, for example. Incidentally, in due time we will make precise the problem of *solubility of equations by radicals* and keep it in view as the exposition unfolds.

2. First let's describe properly what is to be understood by constructibility with ruler and compass. For this we start from the plane $\mathbb{R}^2$ of elementary geometry. A construction problem asks whether a certain point P of the plane can be constructed with ruler and compass, starting from a given initial set M of points. Thus, let a subset M of $\mathbb{R}^2$ be given (we may as well assume it has at least two points). Then look at the set

$$\triangle M \,=\, \{P \in \mathbb{R}^2 \mid P \text{ is constructible from } M \text{ with ruler and compass}\},$$

to be defined more precisely as follows. Let

$\mathrm{Li}(M) =$ set of straight lines joining two distinct points of M,

$\mathrm{Ci}(M) =$ set of circles whose center belongs to M
and whose radius equals the distance between two points of M.

Then consider the following *elementary steps* for the construction of "new" points:

(i) intersecting two distinct lines in $\mathrm{Li}(M)$;

(ii) intersecting a line in $\mathrm{Li}(M)$ with a circle in $\mathrm{Ci}(M)$;

(iii) intersecting two distinct circles in $\mathrm{Ci}(M)$.

Let M' be the union of M with the set of points obtained by the application of one of these steps. The points of $\mathbb{R}^2$ that can be obtained by *repeated* application of steps (i)–(iii), starting from M and replacing M by M' each time, are said to be *constructible from M with ruler and compass*. They form the set $\triangle M$.

We just mention right now four well known constructibility problems that were posed already by the ancient Greeks.

α: Trisection of the angle.

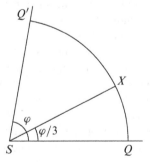

Given an angle of measure φ, construct an angle of measure $\varphi/3$ with ruler and compass. We regard the given angle as determined by its vertex S and points Q, Q' on each of its sides; one may as well assume that Q and Q' are equidistant from S. Let X be the point indicated in the figure. The question then is whether $X \in \triangle\{S, Q, Q'\}$.

β: Doubling of the cube (Delian problem).

Given a cube of side length a, find a cube of twice the volume. The side length x of the desired cube satisfies $x^3 = 2a^3$, so $x = a \sqrt[3]{2}$. Thus let P, Q, X be points on the real line such that $\overline{PQ} = a$ and $\overline{PX} = a\sqrt[3]{2}$; the question is whether $X \in \triangle\{P, Q\}$.

γ: Quadrature of the circle.

Given a circle, construct a square of the same surface area. The given circle of radius r is determined by points P and Q a distance $r = \overline{PQ}$ apart. The side length x of the desired square must satisfy $x^2 = \pi r^2$, so $x = r\sqrt{\pi}$. What must be decided, then, is whether some point X such that $\overline{PX} = r\sqrt{\pi}$ belongs to $\mathbb{A}\{P, Q\}$.

δ: Construction of a regular n-gon (n-section of the circle).

As before, we think of the circle as being given by two points P and Q, the center and a point on the circumference. Let X be the point shown on the right. For what natural numbers n does X lie in $\mathbb{A}\{P, Q\}$?

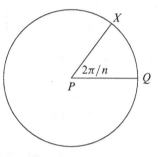

This is the case, for example, for $n = 6$ (and therefore for all numbers of the form $n = 3 \cdot 2^m$): it is enough to draw a circle of radius $\overline{PQ}$ with center Q. By successively repeating the procedure with the newly found points one gets the well known rosette:

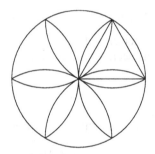

3. To make the problem of constructibility with ruler and compass accessible, one must first "algebraize" it. To that end it is useful to employ the identification

$$\mathbb{R}^2 = \mathbb{C},$$

that is, to regard points in the plane as complex numbers, and so take advantage of the possibility not only of (vector) addition but also of *multiplication*. Assuming the basic properties of the field $\mathbb{C}$ of complex numbers, the problem of dividing the circle into n parts (see δ above) amounts to the following question: Is it the case that

$$e^{2\pi i/n} \in \mathbb{A}\{0, 1\} \ ?$$

The next statement points out that the fundamental algebraic operations of $\mathbb{C}$ can be described constructively.

F1. *Let M be any subset of $\mathbb{C}$ containing the numbers 0 and 1. Then:*

(1) $i \in \mathbb{A}M$;

(2) $z \in \mathbb{A}M \Rightarrow \bar{z} \in \mathbb{A}M$;

(3) $z \in \mathbb{A}M \Rightarrow \operatorname{Re} z, \operatorname{Im} z \in \mathbb{A}M$;

(4) $z \in \mathbb{A}M \Rightarrow -z \in \mathbb{A}M$;

(5) $z_1, z_2 \in \mathbb{A}M \Rightarrow z_1 + z_2 \in \mathbb{A}M$;

(6) $z_1, z_2 \in \mathbb{A}M \Rightarrow z_1 z_2 \in \mathbb{A}M$;

(7) $z \in \mathbb{A}M, z \neq 0 \Rightarrow 1/z \in \mathbb{A}M$.

Proof. (1) The line connecting 0 and 1, that is, the real line $\mathbb{R}$, belongs to Li(M) by definition. Intersecting $\mathbb{R}$ with the unit circle, which belongs to Ci(M), we see that $-1 \in \mathbb{A}M$. If we now construct the perpendicular bisector of the interval $[-1, 1]$ in the well-known way and intersect it with the unit circle, we obtain $i \in \mathbb{A}M$.

(2) Drop a perpendicular from z to $\mathbb{R}$. From the foot of this perpendicular, say a, draw a circle whose radius is the distance from a to z. Its second intersection with the straight line through z and a gives $\bar{z} \in \mathbb{A}M$.

(3) As just verified, we have $a = \operatorname{Re} z \in \mathbb{A}M$. To obtain $b = \operatorname{Im} z$, draw the perpendicular to the imaginary axis through z, and then transfer to $\mathbb{R}$ the absolute value of the foot bi of the perpendicular.

(4) Intersect the line through 0 and z with the circle of radius $|z|$ and center 0.

(5) Intersect the circle of center z_1 and radius $|z_2|$ with the circle of center z_2 and radius $|z_1|$. One of the intersections is the vertex $z_1 + z_2$ of the parallelogram determined by z_1, z_2.

(6) If $z_1 = a_1 + ib_1$ and $z_2 = a_2 + ib_2$ we have

$$z_1 z_2 = (a_1 a_2 - b_1 b_2) + (a_1 b_2 + a_2 b_1)i.$$

Now $z_1, z_2 \in \mathbb{A}M$ implies $a_1, b_1, a_2, b_2 \in \mathbb{A}M$, by (3). If we assume the claim is true *for real numbers*, it will also be true for arbitrary complex numbers, because of (4) and (5). Therefore we must prove that given real numbers r_1 and r_2,

$$r_1, r_2 \in \mathbb{A}M \Rightarrow r_1 r_2 \in \mathbb{A}M.$$

Clearly one can assume $r_1, r_2 > 0$. To complete the proof, consider this diagram:

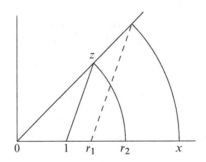

Algebraization of the problem 5

Here z is the appropriate intersection of the line through 0 and $1+i$ with the circle of radius r_2 and center 0, and the dashed line indicates a parallel to the line through 1 and z. By similarity of triangles we have $x:r_2 = r_1:1$, and therefore $x = r_1 r_2$. Since x lies in $\triangle M$, this proves the claim.

(7) Since $z^{-1} = \bar{z} \cdot (z\bar{z})^{-1}$, it suffices in view of the earlier parts to show that if $r > 0$ lies in $\triangle M$, so does r^{-1}. To do this we refer to the following diagram:

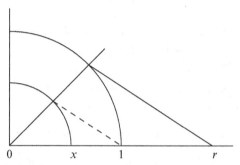

By similarity of triangles, $r:1 = 1:x$, and this proves the claim since $x \in \triangle M$.

$\square$

As a consequence of F1 we will explicitly state again:

F2. *Let M be a subset of $\mathbb{C}$ containing the points 0 and 1. Then $\triangle M$ is a **subfield** of $\mathbb{C}$. It is called the field of numbers constructible from M.*

In particular, $\mathbb{Q} \subseteq \triangle M$, since $\mathbb{Q}$ is the smallest subfield of $\mathbb{C}$. Also the set $\{a + bi \mid a, b \in \mathbb{Q}\}$ — which incidentally is also a subfield of $\triangle M$ — is contained in $\triangle M$. But the field $\triangle M$ is substantially larger:

F3. *The field $\triangle M$ is **quadratically closed**, that is, for every $z \in \mathbb{C}$ we have*

(8) $$z \in \triangle M \implies \sqrt{z} \in \triangle M,$$

where $\sqrt{z}$ represents any complex number w with $w^2 = z$.

Proof. Suppose $w^2 = z = re^{i\varphi}$. Letting $\sqrt{r}$ be the positive square root of $r \in \mathbb{R}$, we have $w = \pm\sqrt{r}\, e^{i\varphi/2}$. Since it is always possible to bisect an angle with ruler and compass, it is enough in order to prove (8) to show that for any $r > 0$ in $\triangle M$, the square root $\sqrt{r}$ is also in $\triangle M$. To do this we raise the perpendicular to the segment $[-1, r]$ through 0 and intersect it with the semicircle constructed over the same segment, to obtain a point v:

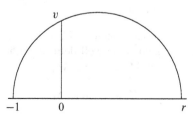

Then $x = |v|$ belongs to $\triangle M$. By *Thales' Theorem* the triangle with vertices $-1, v, r$ has a right angle; applying the formula for the altitude of a right triangle we get $x^2 = 1 \cdot r$, and so $x = \sqrt{r}$. $\qquad\square$

4. Now let's return to the statement of F2, which says that $\triangle M$ is a field containing $\mathbb{Q}$ as a subfield. Trivially,

$$(9) \qquad\qquad M \subseteq \triangle M.$$

Thus the field $E := \triangle M$ also contains all numbers that can be obtained from $k := \mathbb{Q}$ and from M by means of arithmetic operations in E. This serves as the motivation for a simple but fundamental definition:

Definition 1. Let E be a field and k a subfield of E. We say that E is an *extension* of k. Let A be any subset of E. Set

$\qquad k(A) :=$ intersection of all subfields F of E such that $k \subseteq F$ and $A \subseteq F$.

We call $k(A)$ the subfield of E *generated by A over k*, and we also say that $k(A)$ arises from k by *adjoining* to k the elements of A. (The corresponding noun is *adjunction*.) *Clearly, $k(A)$ is the smallest subfield of E containing k and A.* In the case of a finite set $A = \{\alpha_1, \ldots, \alpha_m\}$ we also denote the field $k(A)$ by

$$k(\alpha_1, \alpha_2, \ldots, \alpha_m).$$

Example. Take $E = \mathbb{C}$, $k = \mathbb{Q}$, $A = \{i\}$. We claim that

$$\mathbb{Q}(i) = \{a + bi \mid a, b \in \mathbb{Q}\}.$$

Proof. Let F_0 be the set on the right-hand side. Then $\mathbb{Q} \subseteq F_0$ and $\{i\} \subseteq F_0$ no matter what. Since $\mathbb{Q}(i)$ is a subfield of E, we also have $F_0 \subseteq \mathbb{Q}(i)$. To prove the claim we must show that F_0 is a subfield of $\mathbb{C}$. Clearly, F_0 is closed under addition and multiplication. There remains to show that if $z = a + bi \neq 0$ lies in F_0, so does z^{-1}. But

$$z^{-1} = \bar{z}(z\bar{z})^{-1} = \frac{a - ib}{a^2 + b^2} = \frac{a}{a^2 + b^2} - i\frac{b}{a^2 + b^2},$$

so z^{-1} does lie in F_0. $\qquad\square$

Warning. In the situation of Definition 1 it is not generally true that

$$k(\alpha) = \{a_0 + a_1\alpha + \cdots + a_n\alpha^n \mid a_i \in k, \, n \geq 0\}.$$

We will return to this point in Chapter 3.

Now let's return to the earlier situation. We know that $\triangle M$ is an extension of $\mathbb{Q}$ containing M. Thus it contains the well-defined subfield $\mathbb{Q}(M)$ obtained from $\mathbb{Q}$ by adjunction of M. We set $\overline{M} = \{\bar{z} \mid z \in M\}$ and consider the subfield

$$(10) \qquad\qquad K := \mathbb{Q}(M \cup \overline{M})$$

of $\mathbb{A} M$ obtained from $\mathbb{Q}$ by adjunction of the set $A := M \cup \overline{M}$ (because of (2) and (9), we know that $A \subset \mathbb{A} M$). Since $M \subseteq K \subseteq \mathbb{A} M$ we obviously have

$$(11) \qquad\qquad \mathbb{A} M = \mathbb{A} K.$$

In other words, when considering the set of numbers constructible from a set, the initial set M can always be replaced by the field given by (10). By the way, the field K is mapped to itself under complex conjugation:

$$(12) \qquad\qquad \overline{K} = K.$$

This is clear from (10), but we will justify it in detail for pedagogical purposes: By definition, $K = \mathbb{Q}(A)$, and so $\mathbb{Q} = \overline{\mathbb{Q}} \subseteq \overline{K}$ and $A = \overline{A} \subseteq \overline{K}$. By Definition 1, then, $\mathbb{Q}(A) \subseteq \overline{K}$, that is, $K \subseteq \overline{K}$. By complex conjugation we then get $\overline{K} \subseteq \overline{\overline{K}} = K$. (Where have we used the fact that the complex conjugate of a subfield of $\mathbb{C}$ is also one?)

F4. *Let K be any subfield of $\mathbb{C}$ with $K = \overline{K}$.*

(a) *If z is the intersection of two distinct lines in $\mathrm{Li}(K)$, then $z \in K$.*

(b) *If z is an intersection of a line in $\mathrm{Li}(K)$ with a circle in $\mathrm{Ci}(K)$, then*

$$(*) \qquad\qquad \text{there exists } w \in \mathbb{C} \text{ with } w^2 \in K \text{ and } z \in K(w).$$

(c) *If z is an intersection of two distinct circles in $\mathrm{Ci}(K)$, condition $(*)$ again holds.*

These statements simply reflect the well-known fact that the analytic counterpart of the elementary construction steps (i)–(iii) can only lead to solving *linear* or *quadratic* equations. Therefore we postpone the proof of F4 and instead derive from F4 certain consequences of great import to the constructibility problem. First we equip ourselves with appropriate terminology:

Definition 2. Let E be an extension of the field K.

(a) We say that E arises from K by *adjoining a square root* if there exists $w \in E$ such that
$$w^2 \in K \quad \text{and} \quad E = K(w).$$
We call w a *square root* of the element $v := w^2$ of K, and we write $w = \sqrt{v}$.

(b) We say that E arises from K by *successively adjoining square roots* if there is a chain $K = K_0 \subseteq K_1 \subseteq \cdots \subseteq K_m = E$ of subfields K_i of E where each K_i is obtained from K_{i-1} by adjoining a square root.

Examples.   1. $E = \mathbb{Q}(\sqrt{2})$ is obtained from $\mathbb{Q}$ by adjoining a square root.

2. $E = \mathbb{Q}(z)$, where $z = e^{2\pi i/3}$, is obtained from $\mathbb{Q}$ by adjoining a square root. For since $z = -\frac{1}{2} + \frac{1}{2}i\sqrt{3}$, we have $\mathbb{Q}(z) = \mathbb{Q}(\sqrt{-3})$.

3. $E = \mathbb{Q}(e^{2\pi i/5})$ is obtained from $\mathbb{Q}$ by successively adjoining square roots. This is not immediately obvious, but can be seen as follows: The complex number $w = e^{2\pi i/5}$ satisfies $w^4 + w^3 + w^2 + w + 1 = 0$, since $w^5 = 1$ and $w \neq 1$. Dividing by w^2 yields

(13) $$w^2 + w^1 + 1 + w^{-1} + w^{-2} = 0.$$

Now set $z := w + w^{-1}$; then (13) becomes, via $(w + w^{-1})^2 = w^2 + 2 + w^{-2}$,

(14) $$z^2 + z - 1 = 0.$$

This equation has the solutions $z = -\frac{1}{2} \pm \frac{1}{2}\sqrt{5}$. The field $K_1 := \mathbb{Q}(z)$ therefore satisfies

$$K_1 = \mathbb{Q}(\sqrt{5}).$$

But $E = K_1(w)$, and w obviously satisfies the quadratic equation

(15) $$w^2 - zw + 1 = 0,$$

whose coefficients lie in K_1. Thus, as can be seen from the quadratic formula, E is obtained from K_1 by adjoining a square root of $z^2 - 4$, which lies in K_1.

The following theorem fulfills our first goal, the reduction of the geometric problem of constructibility with ruler and compass to a purely *algebraic* problem.

Theorem 1. *Suppose* $M \subseteq \mathbb{C}$ *contains* 0 *and* 1. *Set*

$$K := \mathbb{Q}(M \cup \overline{M}).$$

For a given $z \in \mathbb{C}$, *the following statements are equivalent:*

(i) $z \in \triangle M$, *that is*, z *is constructible from* M *with ruler and compass.*

(ii) z *lies in a subfield* E *of* $\mathbb{C}$ *obtainable from* K *by successively adjoining square roots.*

Proof. (ii) $\Rightarrow$ (i): By assumption, there exists a finite chain of subfields of $\mathbb{C}$, say

$$K = K_0 \subseteq K_1 \subseteq \cdots \subseteq K_m = E,$$

satisfying $K_i = K_{i-1}(w_i)$ with $w_i^2 \in K_{i-1}$ for each i, and also $z \in E$. We also know that $K_0 = K \subseteq \triangle M$. Now consider the field $K_1 = K(w_1)$, with $w_1^2 \in K$. By F3, w_1 lies in $\triangle M$ because w_1^2 does. Since $\triangle M$ is a field, we have $K(w_1) \subseteq \triangle M$, that is, $K_1 \subseteq \triangle M$. Analogously we get $K_2 \subseteq \triangle M$, and so on until we finally get $E = K_m \subseteq \triangle M$. Since $z \in E$ we have $z \in \triangle M$.

(i) $\Rightarrow$ (ii): We first consider a $z \in \mathbb{C}$ arising from M by applying only one of the elementary construction steps (i), (ii), (iii). Now we make use of F4. We conclude that $z \in K$ in the case of step (i); in cases (ii) and (iii) we get $z \in K(w)$, where $w \in \mathbb{C}$ is such that w^2 lies in K. We claim that in each case z lies in a subfield

K' of $\mathbb{C}$ obtained from K by successively adjoining square roots and satisfying the condition

(16) $$\overline{K'} = K'.$$

In case (i) this is clear; see (12). In cases (ii) and (iii) consider the field

(17) $$K' = K(w, \overline{w}) = K(w)(\overline{w}).$$

It clearly satisfies condition (16). Since $w^2 \in K$, we have $\overline{w}^2 = \overline{w^2} \in \overline{K} = K \subseteq K(w)$, so by (17) K' is indeed obtained from K by successively adjoining square roots.

For an arbitrary $z \in \triangle M$ the assertion follows by induction on the number of elementary steps needed to construct z. $\qquad\qquad\square$

Now that Theorem 1 has accomplished the desired algebraization of our problem, the four classical constructibility problems listed earlier can also be reformulated algebraically:

Doubling of the cube: Does $\sqrt[3]{2}$ lie in a subfield of $\mathbb{C}$ obtainable from $\mathbb{Q}$ through the successive adjunction of square roots?

Quadrature of the circle: Does the number π lie in a subfield of $\mathbb{C}$ that can be obtained from $\mathbb{Q}$ by successively adjoining square roots?

Construction of a regular n-gon: For what natural numbers n is the complex number $e^{2\pi i/n}$ contained in a subfield of $\mathbb{C}$ obtainable from $\mathbb{Q}$ by the successive adjunction of square roots? (By Example 3 after Definition 2, this is certainly the case for $n = 5$: thus a regular pentagon is constructible with ruler and compass. You are encouraged to derive a practical construction from the calculation in Example 3; it is not hard to do.)

Angle trisection: Let φ be any real number. Is it the case that the complex number $e^{i\varphi/3}$ always lies in a subfield of $\mathbb{C}$ obtained from the field

(18) $$K = \mathbb{Q}(e^{i\varphi})$$

by successively adjoining square roots? (Note: For $z := e^{i\varphi}$ we have $\overline{z} = e^{-i\varphi} = z^{-1}$, so $\mathbb{Q}(z, \overline{z}) = \mathbb{Q}(z)$, and (18) does indeed represent the right ground field for the purposes of Theorem 1.)

5. We now carry out the proof of F4. We start with an arbitrary subfield K of $\mathbb{C}$ satisfying $\overline{K} = K$.

(a) An arbitrary line $\mathfrak{g}$ in $\mathbb{R}^2 = \mathbb{C}$ is given by an equation

(19) $$\mathfrak{g} = \{z_0 + t z_1 \mid t \in \mathbb{R}\},$$

where $z_0, z_1 \in \mathbb{C}$ and $z_1 \neq 0$. If $\mathfrak{g} \in \mathrm{Li}(K)$ we can assume that $z_0, z_1 \in K$. Now suppose that $\mathfrak{g}' = \{z_0' + t' z_1' \mid t' \in \mathbb{R}\}$, with $z_0', z_1' \in K$, is another line in $\mathrm{Li}(K)$, distinct from $\mathfrak{g}$, and that $z \in \mathfrak{g} \cap \mathfrak{g}'$. There exist uniquely determined real numbers

t, t' such that $z = z_0 + t z_1 = z_0' + t' z_1'$. Looking separately at real and imaginary parts, we see that (t, t') is the *unique* solution of the system of linear equations

$$t x_1 - t' x_1' = x_0' - x_0,$$
$$t y_1 - t' y_1' = y_0' - y_0.$$

Multiplying the second equation by i we get a system of linear equations over K. It follows that $t, t' \in K$ and so $z \in K$.

(b) An arbitrary circle $\mathfrak{c}$ in $\mathbb{R}^2 = \mathbb{C}$ is given by

$$(20) \qquad \mathfrak{c} = \{ z \mid (z - a)\overline{(z - a)} = r^2 \},$$

with $a \in \mathbb{C}$ and $r > 0$ real. Suppose $\mathfrak{c} \in \mathrm{Ci}(K)$. Then $a \in K$. Since $\mathfrak{c}$ also contains an element b from K, and since $r^2 = (b - a)(\overline{b} - \overline{a})$ and $\overline{K} = K$, we have $r^2 \in K$. Now let $\mathfrak{g} \in \mathrm{Li}(K)$ be given by (19) and suppose $z \in \mathfrak{g} \cap \mathfrak{c}$. Then there exists $t \in \mathbb{R}$ with $z = z_0 + t z_1$ and, in view of (20),

$$(z_0 + z_1 t - a)(\overline{z}_0 + \overline{z}_1 t - \overline{a}) = r^2.$$

Multiplying out and dividing by $|z_1|^2$, we get an equation

$$t^2 + pt + q = 0,$$

with p and q in K. Then $w := t + \frac{1}{2} p$ satisfies $w^2 \in K$, and since $z = z_0 + t z_1$ and $K(t) = K(w)$ we have $z \in K(w)$.

(c) Now let two distinct circles $\mathfrak{c}_1$ and $\mathfrak{c}_2$ be given and suppose $z \in \mathfrak{c}_1 \cap \mathfrak{c}_2$. Then z satisfies a system of equations of the form

$$(21) \qquad \begin{aligned} (z - a)(\overline{z} - \overline{a}) &= r^2, \\ (z - b)(\overline{z} - \overline{b}) &= s^2, \end{aligned}$$

with $a, b, r^2, s^2 \in K$ and $a \neq b$. Subtracting one equation from the other yields

$$(22) \qquad z(\overline{b} - \overline{a}) + \overline{z}(b - a) = c,$$

with $c = r^2 - s^2 - a\overline{a} + b\overline{b} \in K$. Solving equation (22) for $\overline{z}$ and substituting the resulting value into the first line of (21), we get for z a quadratic equation with coefficients in K. The assertion follows immediately.   □

6. The algebraic translation of the constructibility problem (Theorem 1) thus leads us to a more detailed study of the extensions of a given field K. In this forthcoming investigation the following statement is both simple and fundamental:

F5 (Dedekind). *Let K be a field and E an extension of K. Then E can be regarded as a vector space over K.*

Proof. This is clear: we consider on E the existing addition operation and a scalar multiplication $K \times E \to E$ defined simply by restricting the existing multiplication map $E \times E \to E$ of the field E. With these operations E of course obeys all the axioms of a K-vector space. (Incidentally, by considering on the K-vector space E thus obtained the original multiplication operation, we can make E into a K-algebra; for this concept see LA I, p. 87.)   □

By regarding extensions of a field K as K-vector spaces, one gains access to the powerful methods of *linear algebra*, which have demonstrated their fruitfulness over and over in many different areas of mathematics and applications.

Definition 3. If E is an extension of the field K, we denote by

$$[E:K]$$

the *dimension* of the K-vector space E. Instead of $[E:K]$ we can also write $E:K$. This number is called the *degree of E over K*.

Examples. (1) As an $\mathbb{R}$-vector space, $\mathbb{C} = \mathbb{R}^2$, so $\mathbb{C}:\mathbb{R} = 2$.

(2) We have $\mathbb{Q}(i):\mathbb{Q} = 2$; see the example after Definition 1 and observe that $i \notin \mathbb{Q}$.

(3) We will see in Chapter 2 that $\mathbb{Q}(\sqrt[3]{2}):\mathbb{Q} = 3$.

(4) Because $\mathbb{R}$ is uncountable, the degree $\mathbb{R}:\mathbb{Q}$ cannot be finite.

The usefulness of the viewpoint introduced in F5 becomes apparent already from the next statement:

F6. *Let E be an extension of the field K, and suppose that $1 + 1 \neq 0$ in K. Then these two statements are equivalent*:

(i) $E:K = 2$.

(ii) E *is obtained from K by adjoining a square root that is not already in K.*

Proof. Suppose (i) holds, and let α be an element of E not belonging to K. Since $E:K = 2$, the set $\{1, \alpha\}$ is necessarily a K-basis of E. In particular there is a relation of the form

$$\alpha^2 + p\alpha + q1 = 0, \quad \text{with } p, q \in K.$$

For $w := \alpha + \tfrac{1}{2}p$ we then have $w^2 = \tfrac{1}{4}p^2 - q \in K$. Since $E = K(\alpha) = K(w)$ this implies (ii).

Suppose, conversely, that $E = K(w)$ with $w^2 =: d \in K$ and $w \notin K$. Clearly $E' := \{a + bw \mid a, b \in K\}$ is a *subring* of E containing K. To prove (i) therefore we just have to show that for every $a + bw \neq 0$ in E' the inverse $(a + bw)^{-1}$ also belongs to E'. This follows from

$$(a + bw)(a - bw) = a^2 - b^2 d \in K,$$

because we know (from $w = \sqrt{d} \notin K$) that $a^2 - b^2 d \neq 0$.   □

Thanks to F6 we can recast Theorem 1 as follows:

Theorem 1'. *As before, suppose $M \subseteq \mathbb{C}$ contains 0 and 1, and set $K := \mathbb{Q}(M \cup \overline{M})$. Then these two statements are equivalent:*

  (i) $z \in \triangle M$.

  (ii) *There is a finite chain $K = K_0 \subseteq K_1 \subseteq \cdots \subseteq K_m$ of subfields of $\mathbb{C}$ such that $z \in K_m$ and*

$$K_i : K_{i-1} = 2 \quad \text{for } 1 \le i \le m.$$

This result suggests that we should study the relationship between the degrees of the extensions in the diagram

(23)

whose meaning is that F is an extension of K and E is an extension of F.

F7 (Degree formula). *Let E be an extension of K and let F be a subfield of E containing K. Then*

(24)
$$[E : K] = [E : F] \cdot [F : K].$$

Proof. If $E : K$ is finite, so are $E : F$ and $F : K$, of course. Now assume

(25)
$$F : K = m \quad \text{and} \quad E : F = n,$$

with m and n natural numbers. We show that $E : K$ is also finite and satisfies equation (24). Indeed, by (25) there is an isomorphism $F \simeq K^m$ of K-vector spaces and an isomorphism $E \simeq F^n$ of F-vector spaces. This results in an isomorphism

$$E \simeq F^n \simeq (K^m)^n = K^{mn}$$

of K-vector spaces. It follows that $E : K = mn$, so equation (24) holds. The essential content of (24) is thus proved. (Incidentally, it is clear how to modify the argument in case any of the degrees are infinite, so as to prove (24) regarded as an equality between cardinals.) But in addition we establish the following: If $\alpha_1, \ldots, \alpha_m$ form a basis of F over K and $\beta_1, \ldots, \beta_n$ form a basis of E over F, the elements

(26)
$$(\alpha_i \beta_j)_{1 \le i \le m, \, 1 \le j \le n}$$

form a basis of the K-vector space E. For any $\alpha \in E$ can be written in the form $\alpha = \sum_j b_j \beta_j$ with coefficients $b_j \in F$, which in turn can be written as $b_j = \sum_i a_{ij} \alpha_i$ with $a_{ij} \in K$; it follows that $\alpha = \sum_j \left(\sum_i a_{ij} \alpha_i \right) \beta_j = \sum_{i,j} a_{ij} \alpha_i \beta_j$, so the elements in (26) span the K-vector space E. Since $E : K = mn$ they must form a basis. $\qquad\square$

We immediately get, as a consequence of F7 (and F6):

F8. *If E is obtained from K by successively adjoining square roots,*

$$E : K = 2^m \quad \text{for some integer } m \geq 0.$$

Remark. The converse of F8 is unfortunately not true; see Appendix, §5.7.

At any rate, F8 gives us a *necessary* condition for a number to be constructible:

F9. *Let K be a subfield of* $\mathbb{C}$ *with* $\overline{K} = K$. *If* $z \in \mathbb{C}$ *is constructible from K,*

(27) $$K(z) : K \quad \text{is a power of 2.}$$

Proof. Take $z \in \triangle K$. By Theorem 1, z lies in an extension E of K that can be obtained from K by successively adjoining square roots. By F8, $E : K = 2^m$ is a power of 2. Since $z \in E$ we have $K(z) \subseteq E$. Because of the degree formula $[E : K] = [E : K(z)] \cdot [K(z) : K]$, the integer $K(z) : K$ is a power of 2, since it divides $E : K$. □

As remarked, the converse of F9 is not generally true. Only in Chapter 11 will we be able to explain how condition (27) can be modified to give a *necessary and sufficient* condition for the constructibility of a number.

Regarding the four classical constructibility problems listed near the beginning of this chapter, F9 tells us that we should be investigating the following questions:

α) $\mathbb{Q}(e^{i\varphi/3}) : \mathbb{Q}(e^{i\varphi}) = ?$

β) $\mathbb{Q}(\sqrt[3]{2}) : \mathbb{Q} = ?$

γ) $\mathbb{Q}(\pi) : \mathbb{Q} = ?$

δ) $\mathbb{Q}(e^{2\pi i/n}) : \mathbb{Q} = ?$

If we can show, for example, that $\mathbb{Q}(\sqrt[3]{2}) : \mathbb{Q} = 3$, this would prove that the problem of the doubling of the cube is *insoluble* with ruler and compass.

2

Algebraic Extensions

1. Let K be a field and E an extension of K. One writes this assumption in short as

$$\text{Let } E/K \text{ be a field extension,}$$

and the word "field" is often omitted when it can be inferred from the context.

An element α of E is called *algebraic over K* if there exists a polynomial $f(X) \neq 0$ in $K[X]$ such that

$$f(\alpha) = 0.$$

If α is not algebraic over K, we say that α is *transcendental over K*.

Remarks. (a) If $K = \mathbb{Q}$ and $E = \mathbb{C}$, the elements of E algebraic over K are called simply *algebraic numbers*, and the elements of E transcendental over K are called *transcendental numbers*. Example: $\alpha := \sqrt[3]{2}$ is an algebraic number, since α is a root of the polynomial $X^3 - 2 \in \mathbb{Q}[X]$.

(b) The set of algebraic numbers is countable (since $\mathbb{Q}[X]$ is countable and any nonzero polynomial in $\mathbb{Q}[X]$ has finitely many roots in $\mathbb{C}$). Therefore the set of transcendental numbers must be uncountable. To actually be able to exhibit a transcendental number is a different (and much harder) matter.

Theorem 1. *Let M be a subset of $\mathbb{C}$ containing 0 and 1. Any point $z \in \triangle M$ is algebraic over $K := \mathbb{Q}(M \cup \overline{M})$.*

The proof will be given later in this chapter. But first we quote a famous result:

Theorem 2 (Lindemann 1882). *The number π is transcendental.*

Corollary. *The quadrature of the circle with ruler and compass is impossible.*

Proof. If it were possible, we would have $\pi \in \triangle \mathbb{Q}$; by Theorem 1 then π would be algebraic, which by Lindemann's Theorem is not the case. $\qquad \square$

Lindemann's Theorem can be proved using relatively elementary algebraic and analytic arguments, but the proof is on the whole quite intricate. We will go into it later on (Chapter 17).

2. Now we start our study of field theory with the following statement:

F1. *Let E/K be a field extension. If $\alpha \in E$ is algebraic over K, then*

$$K(\alpha):K < \infty.$$

Proof. Suppose there exists a nonzero polynomial

$$(1) \qquad f(X) = X^n + a_{n-1}X^{n-1} + \cdots + a_0 \in K[X]$$

such that $f(\alpha) = 0$; we have assumed without loss of generality that f is normalized (has leading coefficient 1). There exists a unique homomorphism of K-algebras φ from the polynomial ring $K[X]$ into E such that $\varphi(X) = \alpha$ (see page 21); its image

$$R = \operatorname{im}\varphi \subset E$$

consists precisely of those elements of E that can be written as polynomial expressions $g(\alpha)$ in α with coefficients in K. But in writing such an expression we immediately see from the relation

$$(2) \qquad \alpha^n = -(a_{n-1}\alpha^{n-1} + \cdots + a_1\alpha + a_0)$$

that only terms of degree less than n are needed, so in fact

$$(3) \qquad R = \{c_0 + c_1\alpha + \cdots + c_{n-1}\alpha^{n-1} \mid c_i \in K\}.$$

Thus, as a vector space over K, the dimension of R is at most n. Since R, being a subring of E, has no zero-divisors, a simple argument (given a bit further down) shows that R is actually a field. It follows that $K(\alpha) \subseteq R$ (using the definition of $K(\alpha)$), and therefore that $R = K(\alpha)$. From (3) we then get

$$(4) \qquad K(\alpha) = \{c_0 + c_1\alpha + \cdots + c_{n-1}\alpha^{n-1} \mid c_i \in K\}.$$

In particular,

$$(5) \qquad K(\alpha):K \leq n. \qquad \square$$

F2. *Let R be an integral domain (that is, a commutative ring with no zero divisors and with $1 \neq 0$), and let K be a subfield of R. If R is finite-dimensional as a K-vector space, R is a field.*

Proof. For a given $a \neq 0$ in R, consider the map $h : R \to R$ given by multiplication by a, namely, $h(x) = ax$ for all x in R. Then h is an endomorphism (linear map) of the K-vector space R. Since R has no zero-divisors, h is injective. Because R is assumed finite-dimensional over K, it is also surjective. In particular, there exists $b \in R$ such that $ab = 1$. $\qquad \square$

Remark. It can be proved in an analogous way that *an integral domain that has finite cardinality is a field.*

3. Let E/K be a field extension, and let $\alpha \in E$ be algebraic over K. Consider on the K-vector space $K(\alpha)$ the endomorphism h defined by multiplication by α. The minimal polynomial of h is called the *minimal polynomial of α over K*, and we denote it by

$$\mathrm{MiPo}_K(\alpha).$$

This is the lowest-degree normalized polynomial in $K[X]$ that has α as a zero. (That there can be only one such polynomial is clear: if f, g are both normalized and of degree n, the degree of $f - g$ is less than n.) The degree of $f = \mathrm{MiPo}_\alpha(K)$ is also called the *degree of α over K*, and is denoted by $[\alpha : K]$.

Example. Consider $E = \mathbb{C}$, $K = \mathbb{Q}$ and $\alpha = e^{2\pi i/3}$. Then α is a root of $X^3 - 1$. But $X^3 - 1 = (X - 1)g(X)$, with $g(X) = X^2 + X + 1$; since $\alpha \neq 1$, we have $g(\alpha) = 0$. Let $f = \mathrm{MiPo}_K(\alpha)$; we claim that $f = g$. Otherwise necessarily $\deg f < \deg g$, so f could only be of the form $f(X) = X - \alpha$, which is impossible since $\alpha \notin \mathbb{R}$.

F3. *Let E/K be a field extension and let $\alpha \in E$ be algebraic over K, of degree $n := [\alpha : K]$. The elements*

$$(6) \qquad\qquad 1, \alpha, \alpha^2, \ldots, \alpha^{n-1}$$

of E form a basis of $K(\alpha)$ over K. In particular,

$$(7) \qquad\qquad K(\alpha) : K = [\alpha : K] = \deg \mathrm{MiPo}_K(\alpha).$$

Proof. Let $f(X) = X^n + \cdots + a_1 X + a_0$ the minimal polynomial of α over K. We know that

$$K(\alpha) : K \leq n;$$

see (5) in the proof of F1. There remains to show that $1, \alpha, \alpha^2, \ldots, \alpha^{n-1}$ are linearly independent over K. Suppose there is a relation

$$(8) \qquad\qquad \sum_{i=0}^{n-1} c_i \alpha^i = 0 \quad \text{with } c_i \in K.$$

Set $g(X) := \sum_{i=0}^{n-1} c_i X^i$. If some c_i in (8) were nonzero, $g(X)$ would be a nonzero polynomial in $K[X]$ of degree less than n and vanishing at α. Contradiction! $\qquad\square$

4. Let E/K be a field extension and assume $\alpha \in E$ is algebraic over K. Is it the case that any $\beta \in K(\alpha)$ is also algebraic over K?

Definition. An extension E/K is called *algebraic* if every element of E is algebraic over K. An extension E/K is called *finite* if $E : K < \infty$.

Remarks. $\mathbb{C}/\mathbb{R}$ is a finite extension, since $\mathbb{C} : \mathbb{R} = 2$. The extension $\mathbb{R}/\mathbb{Q}$ is not algebraic; see Remark (b) in Section 2.1.

An extension E/K is called *transcendental* if it is not algebraic.

F4. *If an extension E/K is finite, it is also algebraic; for each $\beta \in E$ the degree $[\beta : K]$ is a divisor of $E : K$.*

Proof. Let E/K be finite of degree n. Given $\beta \in E$, the $n + 1$ elements $1, \beta$, $\beta^2, \ldots, \beta^n$ of the n-dimensional K-vector space E are linearly dependent. Therefore there exist $a_0, a_1, \ldots, a_n \in K$, not all zero, such that

$$a_0 1 + a_1 \beta + \cdots + a_n \beta^n = 0.$$

Thus β is algebraic over K. By F3, $[\beta : K] = K(\beta) : K$, and $K(\beta) : K$ is a divisor of $E : K$ by the degree formula (Chapter 1, F7). □

We now can easily answer in the affirmative the question asked at the beginning of this section.

F5. *Let E/K be a field extension. If $\alpha \in E$ is algebraic over K, the extension $K(\alpha)/K$ is algebraic.*

Proof. If α is algebraic over K, we know from F1 that $K(\alpha)/K$ is finite. But every finite field extension is algebraic, by F4. □

Together, F1 and F4 afford the following criterion:

F6. *Let E/K be a field extension. An element α of E is algebraic over K if and only if $K(\alpha)/K$ is finite.*

Now it is a cinch to prove Theorem 1, which we can reformulate as follows:

Theorem 1. *Let M be a subset of $\mathbb{C}$ containing 0 and 1. Let $K = \mathbb{Q}(M \cup \overline{M})$. The field extension $\triangle M/K$ is algebraic.*

Proof. Take $z \in \triangle M$. From F9 of Chapter 1 we know that $K(z) : K < \infty$. Then F6 says that z is algebraic over K. □

Remark. The converse of F4 is not true: Not every algebraic extension is finite. This will soon become obvious. In fact a counterexample comes up naturally in our context: If $E = \triangle\{0, 1\}$ is the field of all numbers constructible from $\{0, 1\}$ with ruler and compass, the field extension $E/\mathbb{Q}$ is algebraic but not finite. (With what we know so far this is not very easy to prove, but it's worth thinking about; see §2.5 in the Appendix.)

Among algebraic extensions, finite extensions can be characterized thus:

F7. *Let E/K be a field extension. The following conditions are equivalent:*

   (i) *There are elements $\alpha_1, \ldots, \alpha_m$ of E, finite in number and algebraic over K, such that $E = K(\alpha_1, \ldots, \alpha_m)$.*

   (ii) *E/K is finite.*

Proof. (ii) ⇒ (i) is clear; all we need to do is choose a basis $\alpha_1, \ldots, \alpha_m$ for E/K. Then we actually have $E = K\alpha_1 + \cdots + K\alpha_m$, and by F4 all the α_i are algebraic over K.

To show (i) ⇒ (ii) we use induction over m. For $m = 0$ there is nothing to prove. Assume that (i) holds for some $m \geq 1$ and set

$$K' = K(\alpha_1, \ldots, \alpha_{m-1}).$$

Then $E = K'(\alpha_m)$. Since α_m is algebraic over K, it is *a fortiori* algebraic over the larger field K'. By F1 this implies $E : K' < \infty$. But by the induction hypothesis, K'/K is finite. The degree formula (Chapter 1, F7) then implies that E/K is finite. □

5. Let E/K be a field extension. A subfield L of E containing K is called an *intermediate field* of the extension E/K.

F8. *Let E/K be a field extension. The subset*

$$F = \{\alpha \in E \mid \alpha \text{ is algebraic over } K\}$$

is an intermediate field of E/K. It is called the algebraic closure of K in E. In particular, the set of all algebraic numbers is a subfield of $\mathbb{C}$.

Proof. Take $\alpha, \beta \in F$. Consider the subfield $K(\alpha, \beta)$ of E. By F7 the extension $K(\alpha, \beta)/K$ is finite (prove this again for practice). Now apply F4; all elements of $K(\alpha, \beta)$ are algebraic over K, so

$$K(\alpha, \beta) \subseteq F.$$

The elements $\alpha + \beta$, $\alpha - \beta$, $\alpha\beta$ and $1/\alpha$ (if $\alpha \neq 0$) lie in $K(\alpha, \beta)$, and thus also in F. So F really is a subfield of E. Clearly $K \subseteq F$, since any $\alpha \in K$ is a zero of a polynomial $X - \alpha \in K[X]$ and therefore algebraic over K. This completes the proof. □

This proof qualifies as easy, but it's only easy because we have the right notions at our disposal. Otherwise, would you be able to write down, at the drop of a hat, a nontrivial rational polynomial that vanishes at the sum of two numbers, given only rational polynomials vanishing at one and the other number respectively?

F9 (Transitivity of algebraicness). *Let L be an intermediate field of the extension E/K. If E/L and L/K are algebraic, so is E/K (and vice versa).*

Proof. Take $\beta \in E$. By assumption β is algebraic over L. Let $\alpha_0, \alpha_1, \ldots, \alpha_{n-1}$ be the coefficients of $\mathrm{MiPo}_L(\beta)$; then β is also algebraic over the subfield $F := K(\alpha_0, \alpha_1, \ldots, \alpha_{n-1})$. By assumption all the α_i are algebraic over K. Therefore we can apply F7 to conclude that $F : K$ is finite. But $F(\beta) : F$ is also finite, by F6; therefore the degree formula gives

$$F(\beta) : K < \infty.$$

Using F4 we see in particular that β is algebraic over K. □

F10. *Let E/K be a field extension and A a subset of E. If all elements of A are algebraic over K, the extension $K(A)/K$ is algebraic.*

Proof. Clearly $K(A)$ is the union of all subfields of the form $K(M)$, where M ranges over *finite* subsets of A. By F7, each $K(M)/K$ is finite and therefore also algebraic. Thus $K(A)$ contains only elements algebraic over K. (Of course F10 also follows directly from F8.) □

F11. *Let E/K be a field extension, and L_1, L_2 intermediate fields of E/K. The field*

$$(9) \qquad\qquad L_1 L_2 := L_1(L_2) = L_2(L_1)$$

is called the **composite** *of L_1 and L_2 in E.*

  (a) *If L_1/K is algebraic, so is $L_1 L_2/L_2$.*
  (b) *If L_1/K is finite, so is $L_1 L_2/L_2$; moreover $L_1 L_2 : L_2 \leq L_1 : K$.*
  (c) *If L_1/K and L_2/K are algebraic, so is $L_1 L_2/K$.*
  (d) *If L_1/K and L_2/K are finite, so is $L_1 L_2/K$; if, moreover, the extension degrees $n_1 = L_1 : K$ and $n_2 = L_2 : K$ are relatively prime, we have $L_1 L_2 : K = n_1 n_2$.*

Proof. Part (a) follows from F10, taking (9) into account. Part (c) therefore also follows, thanks to F9. Let L_1/K and L_2/K be finite. Assuming (b) already proved, we see from the degree formula that

$$(10) \qquad L_1 L_2 : K = (L_1 L_2 : L_2)(L_2 : K) \leq (L_1 : K)(L_2 : K),$$

which is the first part of (d). Again from the degree formula we obtain that $L_1 L_2 : K$ is divisible by n_1 and by n_2. If n_1, n_2 are relatively prime, $L_1 L_2 : K$ is divisible by $n_1 n_2$, which together with (10) gives the second part of (d).

There remains to prove (b). Consider the set R of all finite sums of products ab with $a \in L_1, b \in L_2$. Clearly R is a subring of E containing L_1 and L_2. It is also clear that any basis of L_1/K generates R as an L_2-vector space R, so in particular $R : L_2 \leq L_1 : K$. If $L_1 : K < \infty$, this implies that R is a field (see F2). It follows that $R = L_1 L_2$, which concludes the proof. □

3

Simple Extensions

1. We have seen that in considering constructibility questions one must investigate certain algebraic field extensions, for example $\mathbb{Q}(\sqrt[3]{2})/\mathbb{Q}$ in the case of the doubling of the cube and $\mathbb{Q}(e^{2\pi i/n})/\mathbb{Q}$ in the case of the construction of a regular n-gon. How can such extensions be described? What can be said about their degree?

More generally, let E/K be a field extension and take $\alpha \in E$. How can $K(\alpha)/K$ be described?

Definition. A field extension L/K is called *simple* if there exists an element α of L such that $L = K(\alpha)$. Such an α is called a *primitive element* of L/K.

Let K be a field. The polynomial ring $K[X]$ is an *algebra over K*, or *K-algebra* (for the definition of this notion see LA I, p. 87). Now, if E/K is a field extension, E can also be regarded as a K-algebra. Let α be an element of E. Because of the *universal property* of the polynomial ring there exists a unique homomorphism of K-algebras

$$\text{(1)} \qquad \varphi : K[X] \longrightarrow E \quad \text{with } \varphi(X) = \alpha,$$

namely the *substitution homomorphism* given by

$$\text{(2)} \qquad g(X) = \sum a_i X^i \longmapsto \sum a_i \alpha^i =: g(\alpha).$$

We will denote the image of φ by $K[\alpha]$. We have

$$\text{(3)} \qquad K[\alpha] = \{g(\alpha) \mid g \in K[X]\}.$$

$K[\alpha]$ is a subring of E, and indeed a subalgebra of the K-algebra E.

F1. *Let E/K be a field extension and take $\alpha \in E$. The following statements are equivalent:* (i) α *is algebraic;* (ii) $K(\alpha) = K[\alpha]$; (iii) $K[\alpha]$ *is a field.*

Proof. (i) $\Rightarrow$ (ii): If α is algebraic over K, the field $K(\alpha)$ can be described as in equation (4) of Chapter 2, so $K(\alpha) = K[\alpha]$. The implication (ii) $\Rightarrow$ (iii) is trivial. Finally, if (iii) is true, α (if nonzero) has an inverse in $K[\alpha]$; that is,

$$\alpha(a_0 + a_1\alpha + \cdots + a_m\alpha^m) = 1 \quad \text{for appropriate } a_i \in K.$$

Hence α is algebraic over K.   □

2. Let E/K be a field extension and α any element of E. In connection with the description of $K[\alpha]$ in (3), we compare $K[\alpha]$ with the polynomial ring $K[X]$. In $K[X]$,

(4) $$\sum a_i X^i = \sum b_i X^i \quad \text{implies} \quad a_i = b_i \text{ for all } i.$$

In contrast with X in $K[X]$, the element α may satisfy a nontrivial relation in $K[\alpha]$, so that in general we cannot deduce from $\sum a_i \alpha^i = \sum b_i \alpha^i$ that $a_i = b_i$ for all i. For $g_1, g_2 \in K[X]$ the equation $g_1(\alpha) = g_2(\alpha)$ is equivalent to $(g_1 - g_2)(\alpha) = 0$, so we must study the kernel

(5) $$I = I_\alpha := \{ g \in K[X] \mid g(\alpha) = 0 \}$$

of the homomorphism φ of (1) and (2).

Definition. Let R be any ring with unity. A nonempty subset I of R is called a (two-sided) *ideal* of R if

  (i) $a, b \in I \;\Rightarrow\; a + b \in I$, and

  (ii) $a \in I,\ x \in R \;\Rightarrow\; xa,\ ax \in I$.

Thus a subset of R is an ideal of R if and only if it is a subgroup of the additive group of R and is mapped into itself by multiplication, whether on the left or on the right, with any element of R.

The set I_α in (5) is an example of an ideal, called the *ideal of relations of α*. In general:

F2. *Suppose $\varphi : R \to R'$ is a ring homomorphism (of rings with unity, so $\varphi 1 = 1$). Then*

$$\ker \varphi = \{ a \in R \mid \varphi a = 0 \}$$

is an ideal of R. Conversely, if I is an ideal in a ring R, there is a (canonical) surjective homomorphism of rings $\pi : R \to \bar{R}$ such that $\ker \pi = I$.

Proof. The first assertion is clear. (But incidentally, the image of φ is generally not an ideal of R'.) The proof of the second assertion results from the following construction:

The quotient modulo I. Consider the relation $\sim$ on R defined as follows: $a \sim b$ means $a - b \in I$. Clearly this is an equivalence relation on R. Instead of $a \sim b$ one generally writes (following Gauss)

(6) $$a \equiv b \ \mathrm{mod}\ I,$$

read "a and b are congruent modulo I". This terminology is felicitous (among other reasons because it stresses that the relation $\sim$ depends on the ideal I). Now consider the equivalence classes determined by $\sim$, for which we use the notations

$$\bar{a} = \{ a' \in R \mid a' \sim a \} = a + I = \{ a + y \mid y \in I \}.$$

Thus in the notation of (6) we have

(7) $\bar{a} = \bar{b} \iff a \equiv b \text{ mod } I.$

The set $\bar{a}$ is called the *residue class* of a mod I. We denote the set of all residue classes modulo I by

$$R/I,$$

read "R modulo I" or "R quotient I". Now take the map

(8)
$$\pi : R \to R/I$$
$$a \mapsto \bar{a},$$

which assigns to each a in R the residue class of a mod I (that is, the $\sim$-equivalence class containing a).

We claim that R/I is a ring in a natural way; more precisely, *there is a unique ring structure on R/I that makes π into a ring homomorphism.*

Proof: Given $\bar{a}, \bar{b} \in R/I$, we have no choice but to specify

(9) $\bar{a} + \bar{b} := \overline{a+b}, \quad \bar{a}\bar{b} := \overline{ab}.$

But we have to check that this sum and product are well defined — in other words, that $a' \equiv a$ mod I and $b' \equiv b$ mod I necessarily imply that $a' + b' \equiv a + b$ mod I and $a'b' \equiv ab$ mod I. So suppose that $a' = a + x$ and $b' = b + y$, with $x, y \in I$. Then

$$a' + b' = a + b + (x + y), \quad a'b' = ab + (ay + bx + xy).$$

Since I is an ideal of R, the expressions in parentheses both lie in I, which proves the claim.

To prove F2 we now just have to show that $I = \ker \pi$. But this is clear, because $\pi(a) = \bar{0} \iff \bar{a} = \bar{0} \iff a \equiv 0 \text{ mod } I \iff a \in I.$ □

We call R/I the *quotient ring of R by I*, or *modulo I*. ("Residue-class ring" is an alternative name for "quotient ring".) When no misunderstanding is likely, one can simply write $\bar{R}$ instead of R/I. The map π in (8) is called the *quotient homomorphism* under I or the *canonical map* from R onto R/I.

The role of the quotient ring in the description of ring homomorphisms is the following:

F3 (Fundamental Homomorphism Theorem). *Let $\varphi : R \to R'$ be a homomorphism of rings. There exists a unique ring isomorphism*

$$\psi : R/\ker \varphi \longrightarrow \text{im } \varphi$$

such that the diagram

(10)

R — φ → *R'*, π, ι, $R/\ker\varphi$ $\xrightarrow{\psi}$ $\text{im } \varphi$

commutes. Here π denotes the quotient homomorphism and ι the inclusion of im φ
in R'. In particular, there is an isomorphism

$$(11) \qquad\qquad R/\ker\varphi \simeq \operatorname{im}\varphi.$$

*Analogous statements hold regarding (instead of rings) K-algebras, modules over
rings, abelian groups, arbitrary groups, etc.*

Proof. We have no choice but to define ψ by setting $\psi(\bar{a}) = \varphi(a)$. Is ψ then well
defined? Yes, because $\bar{a} = \bar{b}$ implies the existence of $x \in \ker\varphi$ such that $a = b + x$,
whence we get the desired equality $\varphi(a) = \varphi(b + x) = \varphi(b) + \varphi(x) = \varphi(b)$.

Clearly ψ is surjective. We still have to prove that $\ker\psi = 0$, that is, ψ is
injective: but from $\psi(\bar{a}) = 0$ we get $\varphi(a) = 0$, hence $a \in \ker\varphi$, so $\bar{a} = 0$ as desired.

This reasoning applies wholly analogously to any type of algebraic structure;
in most cases it is clear which subsets actually occur as kernels of the type of
homomorphism in question, so the construction of the quotient structure carries
over. In the case of a group G kernels are subgroups U of G that satisfy $Ux = xU$
for every $x \in G$; these are called *normal subgroups* of G. $\qquad\qquad\square$

Applying the Fundamental Homomorphism Theorem to the situation considered
at the beginning of the chapter leads to the isomorphism of K-algebras

$$(12) \qquad\qquad K[\alpha] \simeq K[X]/I_\alpha.$$

F4. *Let E/K be a field extension. Given $\alpha \in E$ the following statements are equivalent:*

(i) *α satisfies no algebraic relation, that is, $f(\alpha) = 0$ with $f \in K[X]$ implies
$f = 0$.*

(ii) *α is transcendental over K.*

(iii) *$K[\alpha] \simeq K[X]$ as K-algebras.*

(iv) *$K[\alpha]$ is not a field.*

Proof. (i) $\Longleftrightarrow$ (ii) is clear. Suppose (i) holds. Then $I_\alpha = \{0\}$, so $K[\alpha] \simeq$
$K[X]/\{0\} \simeq K[X]$ by (12), which implies (iii). Using a degree argument we see
that the group of units of $K[X]$ is $K^\times$, the multiplicative group of K; therefore (iii)
$\Rightarrow$ (iv). The implication (iv) $\Rightarrow$ (ii) is already contained in F1. $\qquad\square$

We now look at the case where α is *algebraic* over K. The ideal I_α of relations
of α is then nontrivial; in particular, it contains $f := \operatorname{MiPo}_K(\alpha)$. For an arbitrary
$g \in K[X]$,

$$(13) \qquad\qquad g(\alpha) = 0 \quad\Longrightarrow\quad f \text{ divides } g \text{ in } K[X].$$

We recall the proof of this well-known fact. Division with remainder yields

$$(14) \qquad\qquad g = qf + r \quad \text{with } q, r \in K[X], \ \deg r < \deg f.$$

So if $g(\alpha) = 0$, one concludes by substituting α in (14) that $r(\alpha) = 0$, which (since $\deg r < \deg f$ and $f = \text{MiPo}_K(\alpha)$) can only happen if $r = 0$. Therefore $g = qf$.

Thus, according to (13) the ideal I_α has the form

$$(15) \qquad I_\alpha = \{qf \mid q \in K[X]\} = K[X]f.$$

This motivates the next definition:

Definition. Let R be a commutative ring with unity. Given $a \in R$, denote by (a) the ideal $Ra = \{ca \mid c \in R\}$ in R. This is the *principal ideal* of R generated by a. Instead of $x \equiv y \bmod (a)$, it is common to write $x \equiv y \bmod a$. The quotient $R/(a)$ is also denoted by R/a. The ideal $\{0\} = (0)$ is often simply written 0.

We summarize the work so far:

Theorem 1. *Let L/K be a simple algebraic field extension and α a primitive element of L/K. Put $f = \text{MiPo}_K(\alpha)$. The substitution homomorphism corresponding to α,*

$$K[X] \longrightarrow L = K(\alpha) = K[\alpha],$$

gives rise to an isomorphism (of K-algebras)

$$K[X]/f \longrightarrow L = K(\alpha).$$

In particular, given $g \in K[X]$, we have $g(\alpha) = 0 \iff f \mid g$.

In the situation of Theorem 1, the isomorphism

$$(16) \qquad K(\alpha) \simeq K[X]/f$$

gives a good description of a simple algebraic field extension: it all boils down to computing in $K[X]$ modulo f. Crucially, this description also provides a hint for how to generate simple algebraic extensions of a given field. We address this question now.

3. So let K be an arbitrary field and f a polynomial in $K[X]$ of degree $n \geq 1$. Consider the quotient algebra

$$(17) \qquad K_f := K[X]/f$$

over K and denote by $\pi : g \mapsto \bar{g}$ the corresponding quotient homomorphism. Since $\deg f \geq 1$, the homomorphism of K-algebras

$$(18) \qquad K \to K_f \text{ with } a \mapsto \bar{a}$$

is clearly injective. Through this homomorphism we can regard K as a subfield of K_f: $K \subseteq K_f$. Now let $\alpha := \pi(X) = \overline{X}$ be the residue class of $X \bmod f$. Then π is also the unique algebra homomorphism $K[X] \to K_f$ mapping X to α. Therefore $\pi(g) = g(\alpha)$ for all $g \in K[X]$. Thus, since $\pi(f) = 0$,

$$(19) \qquad f(\alpha) = 0.$$

Thus f has α as a root in K_f. Even more: given any $g \in K[X]$,

(20) $$g(\alpha) = 0 \iff f \mid g \text{ in } K[X].$$

Also

(21) $$K_f = K[\alpha] = \{g(\alpha) \mid g \in K[X]\}.$$

More precisely, we claim that $1, \alpha, \ldots, \alpha^{n-1}$ *form a basis of the K-vector space* K_f, *so this space has dimension* n. Proof: Any element of K_f is of the form $g(\alpha)$ with $g \in K[X]$. Division with remainder lets us write $g = qf + r$, and substituting α we get $g(\alpha) = r(\alpha)$ with $\deg r \leq n - 1$. There remains to show that $1, \alpha, \ldots, \alpha^{n-1}$ are linearly independent. A linear relation $c_0 + c_1\alpha + \cdots + c_{n-1}\alpha^{n-1} = 0$ with $c_i \in K$ results in $h(\alpha) = 0$ with $h(X) = c_0 + c_1 X + \cdots + c_{n-1}X^{n-1} \in K[X]$, which is only possible if $h = 0$, because f divides h (look at the degrees).

Is K_f a field? We will show that this is so if and only if f is irreducible. Recall that a polynomial $f \in K[X]$ is *irreducible* if $\deg f \geq 1$ and any factorization $f = f_1 f_2$ with $f_1, f_2 \in K[X]$ implies that $f_1 \in K$ or $f_2 \in K$. An irreducible polynomial is also called a *prime polynomial*. The following result is fundamental and well-known:

Theorem 2. *Let* f *be an irreducible polynomial in* $K[X]$. *If* f *divides* gh *for* $g, h \in K[X]$, *then* $f \mid g$ *or* $f \mid h$.

Proof. This assertion, which you're surely familiar with, will follow from general considerations in Chapter 4. Here we give an *ad hoc* argument: Suppose that f, g, h contradict the theorem. Division with remainder gives $g = qf + r$. Since $f \mid gh$ we have $f \mid rh$, so we might as well assume that $\deg g < \deg f$ to begin with. Among all triples f, g, h contradicting the theorem choose one where $\deg g$ is minimal. Since $f = qg + r$ with $\deg r < \deg g$ we first get $f \mid rh$, and since the degree of g is minimal and less than that of f we next get $r = 0$. Because f is irreducible it follows that g is a unit — a contradiction. □

Theorem 2 was first formulated by Simon Stevin in 1585; the analogous statement for the ring $\mathbb{Z}$ is already in the works of Euclid (ca. -330).

F5. $K_f = K[X]/f$ *is a field if and only if* f *is irreducible in* $K[X]$.

Proof. Let K_f be a field. If $f = f_1 f_2$ in $K[X]$ we have $f_1(\alpha) f_2(\alpha) = f(\alpha) = 0$ and therefore $f_1(\alpha) = 0$ or $f_2(\alpha) = 0$. Because of (20), either f_2 or f_1 lies in K, so f is irreducible.

Conversely, assume that f is irreducible. We already know that K_f is finite-dimensional over K; keeping in mind Chapter 2, F2, we then just have to show that K_f is an integral domain. So let $\bar{g}\bar{h} = 0$. Since $\bar{g}\bar{h} = \overline{gh} = 0$ we get $f \mid gh$; by Theorem 2 this implies $f \mid g$ or $f \mid h$, which is to say $\bar{g} = 0$ or $\bar{h} = 0$. □

F6. *Let* E/K *be a field extension and suppose* $\alpha \in E$ *is algebraic over* K. *Then* $f := \mathrm{MiPo}_K(\alpha)$ *is an irreducible polynomial in* $K[X]$. *Conversely, a normalized irreducible polynomial in* $K[X]$ *that vanishes at* α *must equal* $\mathrm{MiPo}_K(\alpha)$.

Proof. (i) Because of (16) we have $K[X]/f \simeq K(\alpha)$, so f is irreducible by F5.

(ii) Any polynomial g such that $g(\alpha) = 0$ is divisible by f. If g is irreducible and normalized, it must equal f because f is also a normalized polynomial of degree at least 1. □

Example. Let $K = \mathbb{Q}$, $E = \mathbb{R}$, $\alpha = \sqrt[3]{2}$, $g(X) = X^3 - 2$. We wish to show that $g = \mathrm{MiPo}_{\mathbb{Q}}(\alpha)$. Since $g(\alpha) = 0$ all we have to do, thanks to F6, is show that g is irreducible. Suppose $g = g_1 g_2$ were a nontrivial factorization of g in $\mathbb{Q}[X]$. Then one or the other factor, say g_1, has degree 1. Being linear, g_1 has a zero β in $\mathbb{Q}$. It follows that $g(\beta) = 0$, so $\beta^3 = 2$. This contradicts the fact that $\sqrt[3]{2}$ does not lie in $\mathbb{Q}$. (This is something we assume known; later we will be able to show that $X^3 - 2$ is irreducible without resorting to this fact, but rather as an immediate consequence of Chapter 5, F10.)

Taking into account equation (7) of Chapter 2, we immediately get the corollary

$$\mathbb{Q}(\sqrt[3]{2}) : \mathbb{Q} = 3.$$

As a first fruit of our algebraic study of the constructibility problem, we obtain from this and from Chapter 1, F9:

Theorem 3. *$\sqrt[3]{2}$ is not constructible from $\{0, 1\}$ with ruler and compass. Consequently, the Delian problem of the doubling of the cube is also not soluble.*

Here is an important field-theoretical application of the results from this chapter:

Theorem 4 (Kronecker). *Every nonconstant polynomial $f(X)$ over a field K has a root in some appropriate extension of K.*

Proof. Since $\deg f \geq 1$, there must be an irreducible polynomial g dividing f (consider all nonconstant factors of f and take one of least degree). If an extension of K contains a root of g it will also serve for f; therefore we assume without loss of generality that f is irreducible. Then $K_f = K[X]/f$ is a field, by F5. Up to isomorphism K_f is an extension of K, and the image α of X is a zero of f; see (18) and (19). □

Kronecker's Theorem is unsurprising from the point of view of modern algebra, and its proof is simple. Nonetheless it does remove one of the criticisms leveled by Gauss at earlier justifications of the Fundamental Theorem of Algebra, which was that Euler and Lagrange simply started off from the premise that a nonconstant polynomial always has roots (somewhere) and then sought to prove that these roots must be in $\mathbb{C}$. Gauss wrote: "How these magnitudes, which we cannot even begin to visualize — mere shadows of shadows — are to be added or multiplied is something that surely cannot be grasped with the clarity that mathematics always demands."

4. We now consider a *simple transcendental* extension L/K. Let α be a primitive element of L/K. Then α must be transcendental (by F5 in Chapter 2), and so F4 yields

$$K[\alpha] \simeq K[X].$$

What can we say about $L = K(\alpha)$ itself? (By F4 we know $K[\alpha]$ is not a field, so $K(\alpha) \neq K[\alpha]$.)

For brevity we set $R = K[\alpha]$. We claim that

(22) $$L = \{\beta/\gamma \mid \beta, \gamma \in R, \ \gamma \neq 0\}.$$

Proof: Let Q be the subset of L defined by the right-hand side of (22). Clearly L is a *subfield* of L, and $K[\alpha] = R \subseteq Q$. This implies $K(\alpha) \subseteq Q$, so we get $L = Q$.

Definition. Let E be a field and R a subring of E. The *field of fractions of R in E* is the intersection L of all subfields of E that contain R. One can express L exactly as in (22), the justification being the same as above.

Now, is any integral domain R a subring of a field?

F7 (Fraction field). *Let R be a integral domain. There exists a field F and an injective ring homomorphism $\iota : R \to F$ with the following property: If $\kappa : R \to E$ is any injective ring homomorphism from R into a field E, there is a unique ring homomorphism $\lambda : F \to E$ such that $\lambda \circ \iota = \kappa$ — in other words, making the following diagram commutative:*

(23)

$$
\begin{array}{ccc}
F & \xrightarrow{\ \lambda\ } & E \\
\iota \uparrow & \nearrow_{\kappa} & \\
R & &
\end{array}
$$

Such a field F is called a fraction field of R. It is uniquely determined up to isomorphism: more precisely, if F' is another fraction field and $\iota' : R \to F'$ the corresponding map, there exists a unique isomorphism $\lambda : F \to F'$ such that

(24)

$$
\begin{array}{ccc}
F & \xrightarrow{\ \lambda\ } & F' \\
\iota \uparrow & & \uparrow \iota' \\
R & \xrightarrow{\ \text{id}\ } & R
\end{array}
$$

commutes. Moreover, F is the field of fractions of ιR in F, in the sense of the preceding definition.

(Another name for "fraction field" is "field of quotients". This use of "quotient" is not the same as in the expression "quotient ring" defined earlier; in a field of quotients the elements of the field are themselves the quotients.)

Remark. In view of the uniqueness statement in F7, we talk from now on about *the* fraction field of R; we denote it by Frac R. For simplicity we will generally assume that $R \subseteq$ Frac R, which entails no loss of generality. We then have

$$\text{Frac } R = \{a/b \mid a, b \in R, \ b \neq 0\}.$$

The reason we were so punctilious in the statement of F7 is that this is a key example of *solving a universal problem* of the kind that one often comes across in algebra (and elsewhere).

Before proving F7, we state one more result:

F8. *Let R be a integral domain. If $\iota : R \to K$ is an injective ring homomorphism from R into a field K, the field F of fractions of ιR in K is a fraction field of R.*

Proof. Let $\kappa : R \to E$ be an injective ring homomorphism of R into a field E. We define $\lambda : F \to E$ by setting

$$\lambda(\iota a / \iota b) = \kappa a / \kappa b.$$

It is easy to check that λ is well defined. It is also clear that λ is a ring homomorphism, and that in fact it's the only one for which diagram (23) commutes. $\square$

Proof of F7. We first show the uniqueness statement. By assumption there exist homomorphisms $\lambda : F \to F'$ and $\lambda' : F' \to F$ with $\lambda \circ \iota = \iota'$ and $\lambda' \circ \iota' = \iota$. It follows that $\lambda' \circ \lambda \circ \iota = \lambda' \circ \iota' = \iota$, and thus, because of the uniqueness requirement, $\lambda' \circ \lambda = \mathrm{id}_F$; analogously we have $\lambda \circ \lambda' = \mathrm{id}_{F'}$. Therefore λ is an isomorphism.

In view of F8 what is left to show is that there is a field K and an injective ring homomorphism $\iota : R \to K$. For this consider the set $M = \{(a,b) \mid a,b \in R, \ b \neq 0\}$ with the relation $\sim$ defined by

$$(a,b) \sim (c,d) \quad \text{means} \quad ad = bc.$$

It is easy to prove that this is an equivalence relation; let $K = M/\sim$ the set of equivalence classes. Denote the class of $(a,b) \in M$ by $[a/b]$. Define addition and multiplication on K as follows:

$$[a/b] + [c/d] = [(ad + bc)/bd], \quad [a/b] \cdot [c/d] = [ac/bd].$$

Checking that these operations are well defined is left to the reader. It is easy to see that with these operations K becomes a commutative ring with unity; the zero element is $[0/1]$ and the unity is $[1/1]$. The map $\iota : R \to K$ defined by $\iota(a) = [a/1]$ is a homomorphism. By definition, $[a/b] = 0 = [0/1]$ if and only if $a = 0$. In particular, ι is injective. In addition, every $[a/b] \neq 0$ in K has a multiplicative inverse, namely $[b/a]$. Therefore K is a field. $\square$

The classical example of the construction above is the field of rational numbers

$$\mathbb{Q} = \mathrm{Frac}\,\mathbb{Z}.$$

Other key examples arise as follows:

Definition. Let K be a field and $K[X]$ the polynomial ring over K. The field

$$K(X) := \mathrm{Frac}\,K[X]$$

is called the *field of rational functions in one variable over K*. It satisfies

$$K(X) = \left\{ \frac{f(X)}{g(X)} \ \middle|\ f, g \in K[X], \ g \neq 0 \right\};$$

thus every "rational function"—which is to say, every element of $K(X)$—is a quotient of polynomials. (If K is infinite, the elements of $K(X)$ can really be seen as rational *functions* in the usual sense of sending each point in the domain of definition—here K minus some points—to its image.)

F9. *Let E/K be a field extension and take $\alpha \in E$. If α is transcendental over K, there is a natural isomorphism*

$$(25) \qquad\qquad K(\alpha) \simeq K(X)$$

of fields (and of K-algebras). Conversely, if (25) holds, α is transcendental over K.

Proof. If α is transcendental over K, the homomorphism $\varphi : K[X] \to K[\alpha]$ given by $\varphi X = \alpha$ is an isomorphism and so can be uniquely extended to an isomorphism $\tilde{\varphi} : K(X) \to K(\alpha)$ of the corresponding fraction fields (see F7).

The converse part of F9 follows for instance from the fact that $K[X]$—and therefore also $K(X)$—is infinite-dimensional over K.   □

Thus the simple transcendental extensions of a given K are all of *one type*, represented by $K(X)/K$.

5. This is a good place for one more essential remark about fields. Let K be a field. For each $n \in \mathbb{Z}$, consider the n-th multiple $n_K = n1_K$ of the unity 1_K in K. If $n_K \neq 0$ for all $n \neq 0$, we say that K is a field of *characteristic zero*, and write

$$(26) \qquad\qquad \text{char } K = 0.$$

If, on the contrary, there is a natural number n such that $n_K = 0$, and if p is the smallest such number, p is called the *characteristic of K* and we write

$$(27) \qquad\qquad \text{char } K = p.$$

Because $(mn)_K = m_K n_K$, this p must be *prime*. For the moment, denote by $\mathbb{Z}_K$ the subring of K consisting of all n_K, for $n \in \mathbb{Z}$. Consider the uniquely defined homomorphism

$$(28) \qquad\qquad \varphi : \mathbb{Z} \to K \quad \text{such that} \quad \varphi 1 = 1_K.$$

Two cases can be distinguished:

Case A: $\ker \varphi \neq 0$. Then we are in situation (27) above. If $n \in \ker \varphi$, division by p with remainder shows that $n \in p\mathbb{Z}$. It follows that $\ker \varphi = p\mathbb{Z}$, so the Fundamental Homomorphism Theorem applied to φ yields an isomorphism

$$(29) \qquad\qquad \mathbb{Z}_K \simeq \mathbb{Z}/p\mathbb{Z}.$$

In particular, $\mathbb{Z}_K$ has exactly p elements. Being a finite integral domain, $\mathbb{Z}_K$ is a field! (See Chapter 2, Remark after F2.)

Case B: $\ker \varphi = 0$. This occurs if and only if char $K = 0$, and the Homomorphism Theorem applied to (28) then gives an isomorphism

$$(30) \qquad\qquad \mathbb{Z}_K \simeq \mathbb{Z}.$$

Definition. A field is called a *prime field* if it has no proper subfields.

F10. (a) *Any field K has exactly one prime field as a subfield. (This is called the prime field of K.)*

   (b) *Any prime field K is isomorphic either to $\mathbb{Q}$ or to some $\mathbb{Z}/p\mathbb{Z}$ for p prime (depending on whether* char $K = 0$ *or* char $K = p > 0$*).*

Proof. (a) The intersection of all subfields of K is a subfield of K. It is the smallest subfield of K, hence a prime field.

(b) Let K be any field and K_0 its prime field. Clearly, $\mathbb{Z}_K \subseteq K_0$. Now, in case A above, $\mathbb{Z}_K$ is itself already a subfield of K, so $K_0 = \mathbb{Z}_K \simeq \mathbb{Z}/p\mathbb{Z}$. In case B we have $K_0 = \operatorname{Frac} \mathbb{Z}_K \simeq \operatorname{Frac} \mathbb{Z} = \mathbb{Q}$.   □

Remarks. (a) It is customary to write just n instead of n_K, and we will do so. But you should keep an eye open in each case for whether the n represents an integer or an element of K.

(b) Clearly $\mathbb{Q}$ is a prime field (indeed, up to isomorphism, the only prime field of characteristic 0). For any prime number p,

$$(31) \qquad \mathbb{F}_p := \mathbb{Z}/p\mathbb{Z}$$

is a field (see Chapter 2, Remark after F2; naturally, to show that $\mathbb{Z}/p\mathbb{Z}$ has no zero-divisors, it is necessary to use the well-known Euclidean result: *if p is a prime dividing ab, then p divides a or b*; see also Chapter 4). For a given p, the field $\mathbb{F}_p$ is, up to isomorphism, the only prime field of characteristic p. As an example of a nonfinite field of characteristic p consider the field of rational functions $\mathbb{F}_p(X)$ over $\mathbb{F}_p$.

(c) It's good to keep in mind the following trivial fact: If K is a subfield of E, then char $E = $ char K.

6. To conclude this chapter we will go into another interesting characterization of simple algebraic field extensions. First we prove:

F11. *Let E/K be a **simple** algebraic extension with primitive element α. Let L be an intermediate field of E/K, and denote by*

$$g(X) = X^m + \beta_{m-1} X^{m-1} + \cdots + \beta_1 X + \beta_0 \in L[X]$$

*the minimal polynomial of α **over L**. Then*

$$L = K(\beta_0, \beta_1, \ldots, \beta_{m-1}).$$

Proof. Set $F = K(\beta_0, \beta_1, \ldots, \beta_{m-1})$. Trivially, $F \subseteq L$. Since $g \in F[X]$ we see that g is the minimal polynomial of α *also over F*. Consequently,

$$(32) \qquad F(\alpha) : F = L(\alpha) : L$$

(see Chapter 2, F3). But since $E = K(\alpha)$ we get $F(\alpha) = L(\alpha) = E$, so (32) says simply that $E : F = E : L$. By the degree formula this means $F : K = L : K$, which (since $F \subseteq L$) demands that $F = L$.   □

Theorem 5. *Suppose E/K is an **algebraic** extension. Then E/K is simple if and only if it possesses only finitely many intermediate fields.*

Proof. Denote by $\mathfrak{L}$ be the set of all intermediate fields of E/K.

(i) Assume $E = K(\alpha)$, and set $f = \text{MiPo}_K(\alpha)$. To prove the finiteness of $\mathfrak{L}$, consider the set

$$\mathfrak{D} = \{g \in E[X] \mid g \text{ is normalized and divides } f \text{ in } E[X]\}.$$

Now, it is well known that $E[X]$ enjoys unique factorization into prime factors (see for example LA II, p. 142, or the next chapter in this book). Therefore f has only finitely many normalized factors in $E[X]$, and thus $\mathfrak{D}$ is finite. Now consider the map

(33) $$\mathfrak{D} \to \mathfrak{L}$$

that takes each $g(X) = X^m + \beta_{m-1} X^{m-1} + \cdots + \beta_0$ in $\mathfrak{D}$ to the intermediate field $K(\beta_0, \ldots, \beta_{m-1})$. Given $L \in \mathfrak{L}$, the element $g = \text{MiPo}_L(\alpha)$ is a factor of f in $L[X]$, therefore also in $E[X]$. Thus g lies in $\mathfrak{D}$. By F11, L is the image of g under the map (33). The map (33) is thus surjective, and since $\mathfrak{D}$ is finite, so is $\mathfrak{L}$.

(ii) The converse will be proved here only in the case where K has infinitely many elements. Suppose that $\mathfrak{L}$ is finite. Then $E = K(\alpha_1, \ldots, \alpha_n)$ with *finitely many* elements α_i; otherwise there would be an infinite chain of intermediate fields obtained by adjoining ever more elements.

Now, to start an induction, we assume that $E = K(\alpha, \beta)$. Since $\mathfrak{L}$ is finite but K is infinite, there exist distinct $\lambda_1, \lambda_2 \in K$ such that

$$K(\lambda_1 \alpha + \beta) = K(\lambda_2 \alpha + \beta) =: L.$$

Then $(\lambda_1 \alpha + \beta) - (\lambda_2 \alpha + \beta) = (\lambda_1 - \lambda_2)\alpha$ lies in L, and therefore so does α, and likewise β. It follows that $E = L = K(\lambda_1 \alpha + \beta)$, so that E/K is simple (with $\gamma = \lambda_1 \alpha + \beta$ as a primitive element). To prove the case $E = K(\alpha_1, \ldots, \alpha_n)$, apply the induction hypothesis to write $K(\alpha_1, \ldots, \alpha_{n-1}) = K(\alpha)$, so $E = K(\alpha, \alpha_n)$.

For K a finite field the assertion follows from the fundamental theorem of the theory of finite fields, which we will study later (Theorem 2 in Chapter 9). $\quad\square$

4

Fundamentals of Divisibility

Throughout this chapter,

> *R stands for a commutative ring with unity.*

Much of the content of this chapter is probably familiar to you from earlier courses. We nonetheless lay it out here because of its fundamental importance; in connection with the problems pursued up to now, we will be particularly interested in the question of irreducibility of polynomials.

1. Given elements a, b in R, we say that a is a *divisor* of b (or *divides* b, or that b *is divisible* by a) if there exists c in R such that $b = ca$. In this case we write

$$\text{(1)} \qquad\qquad\qquad a \mid b.$$

The negation of (1) is denoted by

$$\text{(2)} \qquad\qquad\qquad a \nmid b.$$

The divisibility relation satisfies some obvious rules:

(3) $a \mid a$ (reflexivity);

(4) $a \mid b$ and $b \mid c \implies a \mid c$ (transitivity);

(5) $1 \mid a,\ a \mid 0$;

(6) $a \mid b$ and $c \mid d \implies ac \mid bd$.

Item (5) says that 1 is a minimal and 0 is a maximal element for the divisibility relation. Divisibility is compatible with addition in the following sense:

(7) $a \mid b$ and $a \mid c \implies a \mid b + c$.

If R is an integral domain,

(8) $ac \mid bc \implies a \mid b$ for $c \neq 0$.

Elements of $R^\times$, that is, *units* or *invertible elements* of R, can be characterized thus:

(9) $$\varepsilon \mid 1 \iff \varepsilon \in R^\times.$$

Definition 1. If $a \mid b$ and $b \mid a$, we say that a is *associated* to b and write $a \stackrel{\wedge}{=} b$. Clearly $\stackrel{\wedge}{=}$ is an equivalence relation on R.

F1. *In an integral domain R we have $a \stackrel{\wedge}{=} b$ if and only if there is a unit ε of R such that $b = \varepsilon a$.*

The simple proof is left to the reader.

In the ring $\mathbb{Z}$ we have $a \stackrel{\wedge}{=} b$ if and only if $a = \pm b$. But in general an integral domain has more units than just 1 and -1:

Examples. (a) If R is a integral domain, so is the polynomial ring $R[X]$, and $R[X]^\times = R^\times$.

(b) For the subring $R = \mathbb{Z}[i]$ of $\mathbb{C}$ we have $R^\times = \{\pm 1, \pm i\}$.

(c) For a field K we of course have $K^\times = K \smallsetminus \{0\}$.

(d) For the subring $R = \mathbb{Z}[\sqrt{2}]$ of $\mathbb{R}$ we have $R^\times = \{\pm(1 + \sqrt{2})^j \mid j \in \mathbb{Z}\}$. (The proof of this is not totally straightforward and is left to the reader as a more challenging exercise.)

Definition 2. Let $a_1, \ldots, a_n$ be given in R. An element $d \in R$ is called a *greatest common divisor* (gcd) of $a_1, \ldots, a_n$ if the following conditions are satisfied:

(i) d is a common divisor of $a_1, \ldots, a_n$.

(ii) Every common divisor of $a_1, \ldots, a_n$ also divides d.

We say that the elements $a_1, \ldots, a_n$ are *relatively prime* if 1 is a gcd of $a_1, \ldots, a_n$. The notion of the *least common multiple* (lcm) of $a_1, \ldots, a_n$ is defined analogously.

F2. *Any two gcd's of $a_1, \ldots, a_n$ are associated to one another. Likewise for any two lcm's of $a_1, \ldots, a_n$.*

This follows immediately from the definitions.

But how about the *existence* of a gcd or lcm for given elements of R?

2. In investigating divisibility questions it is relevant to consider in connection with an element a of R the set of its multiples, i.e., the *principal ideal* generated by a:

$$(a) = Ra = \{xa \mid x \in R\}.$$

Clearly,

(10) $$a \mid b \iff (b) \subseteq (a).$$

This translation of the divisibility relation into a simple inclusion relation is very fruitful. We have, for example,

(11) $$a \stackrel{\wedge}{=} b \iff (a) = (b).$$

Moreover: v is a common multiple of a and b if and only if $(v) \subseteq (a) \cap (b)$. From this we deduce easily that

(12) m is an lcm of $a, b \iff (a) \cap (b) = (m)$.

Thus there exists a lowest common multiple for a, b if and only if the ideal $(a) \cap (b)$ is a principal ideal.

F3. *Let I_1 and I_2 be ideals of R. Then $I_1 \cap I_2$ and $I_1 + I_2 := \{\alpha_1 + \alpha_2 \mid \alpha_i \in I_i\}$ are also ideals of R; and in fact $I_1 \cap I_2$ is the largest ideal of R contained in I_1 and I_2, and $I_1 + I_2$ is the smallest ideal of R containing I_1 and I_2.*

Of course similar statements hold for the intersection $I_1 \cap \cdots \cap I_n$ and the sum $I_1 + \cdots + I_n$ of more than two ideals. In the case of principal ideals we also use the notation

(13) $(a_1, \ldots, a_n) := (a_1) + \cdots + (a_n).$

This set consists of all R-linear combinations of $a_1, \ldots, a_n$.

So t is a common divisor of a, b if and only if $(a) + (b) \subseteq (t)$. Now, (12) does not have a complete analog for the gcd (why not?); but if $(a) + (b)$ is a *principal ideal*, say $(a) + (b) = (d)$, then d *is* a gcd of a, b. (We already know that d is a common divisor of a and b; if t is another, then $(d) \subseteq (t)$, so $t \mid d$.)

Definition 3. An integral domain R is called a *principal ideal domain*, or PID, if every ideal of R is a principal ideal.

F4. *In a principal ideal domain R any tuple of elements $a_1, \ldots, a_n$ of R has a gcd. If d is a gcd of $a_1, \ldots, a_n$, it can be represented as*

(14) $d = x_1 a_1 + \cdots + x_n a_n$

for appropriate $x_i \in R$.

Proof. Given $a_1, \ldots, a_n$, we use the assumption that R is a PID to find d such that

(15) $(a_1) + \cdots + (a_n) = (d).$

This means d is a gcd of the a_i, by the argument preceding Definition 3; moreover, d clearly has a representation of the desired from. And so does any other gcd d' of the a_i, since $(d') = (d)$, by F2 and (11).   □

Thus in a principal ideal domain not only is the existence of a gcd for any $a_1, \ldots, a_n$ assured, but it's true to boot that any gcd has an *additive* representation of the form (14), which is astonishing. But none of this would help if we could not prove the existence of interesting principal ideal domains...

F5. *The ring $\mathbb{Z}$ of integers is a principal ideal domain.*

Proof. Let I be an ideal of $\mathbb{Z}$; we may except the trivial case $I = (0)$. Among all nonzero elements of I, let a be one with smallest absolute value $|a|$. We claim that $I = (a)$. Obviously $(a) \subseteq I$. Now let $b \in I$. By considering division with remainder we see that there exist $q, r \in \mathbb{Z}$ such that

$$b = qa + r \quad \text{and} \quad |r| < |a|$$

(we can even demand that $0 \leq r < |a|$ or alternatively that $-\frac{1}{2}|a| < r \leq \frac{1}{2}|a|$). Because $r = b - qa$, this r is an element of I. If r were nonzero we'd have a contradiction with our choice of a, because $|r| < |a|$. It follows that $b = qa$, and therefore $b \in (a)$. $\qquad\square$

A study of this proof leads to the following generalization:

Definition 4. An integral domain R is called a *Euclidean domain* if there exists a map $v : R \to \mathbb{N} \cup \{0\}$ such that $v(0) = 0$ and that, for every $a, b \in R$ with $a \neq 0$, there exist $q, r \in R$ with

$$b = qa + r \quad \text{and} \quad v(r) < v(a).$$

Such a map v is called a *Euclidean valuation* on R.

Examples. (i) $R = \mathbb{Z}$ with $v(a) = |a|$.

(ii) If K is a field, $R = K[X]$ is a Euclidean domain, with valuation v defined by setting $v(0) = 0$ and $v(g) = (\deg g) + 1$ for $g \neq 0$.

F6. *Every Euclidean domain R is a principal ideal domain.*

This is proved exactly like the case $R = \mathbb{Z}$ of F5.

3. We now generalize the familiar notion of prime numbers in $\mathbb{Z}$ and irreducible polynomials in $K[X]$.

Definition 5. An element π of R is called *irreducible* if $\pi \notin R^\times$ and

$$(16) \qquad\qquad \pi = ab \implies a \in R^\times \text{ or } b \in R^\times.$$

Remarks. (1) The irreducible elements of $\mathbb{Z}$ are precisely the prime numbers p and their negatives $-p$.

(2) By the Fundamental Theorem of Algebra (see for example LA I, p. 191), we know that the only irreducible polynomials in $\mathbb{C}[X]$ are the linear polynomials. As an exercise, deduce the following: Apart from linear polynomials, the only irreducible polynomials in $\mathbb{R}[X]$ are those of the form $f = aX^2 + bX + c$ with $b^2 - 4ac < 0$.

(3) A divisor a of b is called *proper* if it is neither a unit of R nor an element associated to b. Thus a nonzero element π is irreducible if and only if it is not a unit and has no proper divisors (here we're assuming that R is an integral domain).

Definition 6. We say that $a \in R$ can be *decomposed into irreducible factors* if it has an expression of the form

(17) $a = \varepsilon\pi_1\pi_2 \ldots \pi_r$ with $\varepsilon \in R^\times$ and each π_i irreducible.

(Here we allow $r = 0$, in which case (17) is to be read as saying that $a = \varepsilon 1 = \varepsilon$). An integral domain where every $a \neq 0$ has a decomposition into irreducible factors is called a *factorization domain*.

We say that a has a *unique* decomposition into irreducible factors if it has a decomposition into irreducible factors and the following uniqueness condition holds: If in addition to (17) we have another such decomposition

(18) $$a = \varepsilon'\pi_1'\pi_2' \ldots \pi_{r'}',$$

then $r' = r$ and, after a permutation, $\pi_i' \stackrel{\wedge}{=} \pi_i$ for $1 \leq i \leq r$. An integral domain where every $a \neq 0$ has a **unique** decomposition into irreducible factors is called a **unique factorization domain** (UFD).

F7. *For a factorization domain R, the following conditions are equivalent:*

  (i) *R is a **unique factorization domain**.*

  (ii) *For any irreducible element π of R we have*

(19) $$\pi \mid ab \;\Rightarrow\; \pi \mid a \text{ or } \pi \mid b.$$

Proof. (i) $\Rightarrow$ (ii): We may as well assume $a, b \neq 0$. Given factorizations $a = \varepsilon\pi_1 \ldots \pi_r$ and $b = \tilde{\varepsilon}\tilde{\pi}_1 \ldots \tilde{\pi}_s$ of a and b into irreducible factors we get for ab the factorization $ab = \varepsilon\tilde{\varepsilon}\pi_1 \ldots \pi_r\tilde{\pi}_1 \ldots \tilde{\pi}_s$. Now, if $\pi \mid ab$, there is a decomposition of ab into irreducible factors where π appears. From the assumption it follows that π is associated with one of the elements $\pi_1, \ldots, \pi_r, \tilde{\pi}_1, \ldots, \tilde{\pi}_s$. Therefore π is a divisor of a or b.

(ii) $\Rightarrow$ (i): Assume that (17) and (18) are true and that $r \geq 1$, the case $r = 0$ being trivial. Now, π_1 is always a divisor of the product on the right-hand side of (18). Assumption (ii) then implies that π_1 must divide one of the π_i' — let's say π_1'. Then there is a unit η such that $\pi_1' = \eta\pi_1$. By cancellation of π_1 we then get $\varepsilon\pi_2 \ldots \pi_r = \varepsilon'\eta\pi_2' \ldots \pi_{r'}'$. The assertion follows by induction.   □

Definition 7. An element π in R is called *prime in R*, or *a prime of R*, if it is not a unit and it satisfies (19).

Remarks. (1) Clearly, in an integral domain every nonzero prime is irreducible.

(2) F7 suggests a question: Under what circumstances are the irreducible elements of an integral domain necessarily prime? That this is not always the case can be seen from the example of $R = \mathbb{Z}[\sqrt{-5}]$, where 2 is irreducible but not prime. (Prove this as an exercise; notice that $6 = 2 \cdot 3 = (1 + \sqrt{-5})(1 - \sqrt{-5})$ in R.)

F8. *An integral domain R is a unique factorization domain if and only if the following two conditions are satisfied:*

(i) *Every chain* $(a_1) \subseteq (a_2) \subseteq \cdots \subseteq (a_n) \subseteq (a_{n+1}) \subseteq \cdots$ *of principal ideals is stationary, that is,* $(a_j) = (a_n)$ *for some n and all* $j \geq n$. (This is the *ascending chain condition* for principal ideals.)

(ii) *Every irreducible element of R is prime.*

Proof. We show first that (i) implies that any nonzero a in R can be decomposed into irreducible factors. Let M be the set of all ideals $(a) \neq 0$ such that a has no such decomposition, and assume that $M \neq \varnothing$. Then M has a maximal element; otherwise there would be a nonstationary chain $(a_1) \subsetneq (a_2) \subsetneq \cdots \subsetneq (a_n) \subsetneq (a_n) \subsetneq (a_{n+1}) \subsetneq \cdots$, in contradiction with assumption (i). So let (a) be maximal in M. The generator a can be neither irreducible nor a unit. Thus $a = bc$ with $(a) \subsetneq (b)$ and $(a) \subsetneq (c)$. Because (a) is maximal, both b and c have decompositions into irreducible factors. But then the same is true of $a = bc$, contradicting the assumption that $(a) \in M$.

If condition (ii) is satisfied as well as (i), we see from F7 that R is a UFD.

Conversely, assume that R is a UFD. Then (ii) is immediately true, by F7. Let a and t be elements of R with $(0) \neq (a) \subseteq (t) \neq (a)$, and suppose that a satisfies (17). From the uniqueness of the decomposition into irreducible factors we conclude that, since t is a proper divisor of a, it has (after reordering the π_i) a decomposition of the form $t = \varepsilon' \pi_1 \ldots \pi_s$, with $s < r$. From this one easily concludes (i).   □

F9. *Every principal ideal domain is a unique factorization domain.*

Proof. Let R be a PID. We will use the characterization of UFDs in F8. Consider a chain $(a_1) \subseteq (a_2) \subseteq \cdots$ of principal ideals. Let I be the union of all the (a_j). It's easy to check that I is an *ideal* of R. By assumption, it is a *principal ideal*, $I = (a)$. By the definition of I, there exists n such that $a \in (a_n)$. Then $(a_j) \subseteq I = (a) \subseteq (a_n) \subseteq (a_j)$ for every $j \geq n$, so the chain is stationary.

Now let π be any irreducible element of R, and let a be an element of R not divisible by π. Since π is irreducible, π and a are relatively prime. By F4 we have $1 = x\pi + ya$, for appropriate x, y in R. Multiplying by an arbitrary $b \in R$ we get

$$(20) \qquad\qquad b = (xb)\pi + y(ab).$$

This says that if π divides ab, it divides b. Therefore π is prime.   □

Remarks. (1) The converse of F9 is not true. For example, the polynomial ring $\mathbb{Z}[X]$ over $\mathbb{Z}$ is a unique factorization domain, by a theorem of Gauss (see next chapter), but it is not a principal ideal domain (again see next chapter; but this is easy to see directly — for instance, the ideal $(2) + (X)$ cannot be principal in $\mathbb{Z}[X]$.)

(2) *Euclidean domains* are *principal ideal domains* (F6), and principal ideal domains are *unique factorization domains*. In particular, $\mathbb{Z}$ is a UFD (Euclid, ca. -330), and so is any polynomial ring $K[X]$ over a field K (Stevin, 1585). In a Euclidean domain, though, there are additional benefits arising from the Euclidean valuation v. For example, a gcd of two given elements $a \neq 0$ and b can be computed step-

by-step through the *Euclidean algorithm*:

$$b = q_0 a + r_1 \quad \text{with } v(r_1) < v(a),$$
$$a = q_1 r_1 + r_2 \quad \text{with } v(r_2) < v(r_1),$$
$$\vdots$$
$$r_{i-1} = q_i r_i + r_{i+1} \quad \text{with } v(r_{i+1}) < v(r_i),$$
$$\vdots$$
$$r_{n-1} = q_n r_n + 0.$$

Then r_n is a gcd of a, b and these equations even provide, by recursion, elements x, y such that $r_n = xa + yb$. In the case $R = K[X]$, with v as in Example (ii) following Definition 4, the elements q_i, r_i are uniquely determined by a, b. The same is true for $R = \mathbb{Z}$, if we demand that each r_i be nonnegative.

4. We now wish to inspect more closely the situation in *unique factorization domains*. By taking a factorization of the form (17) and grouping together irreducible factors that are associated to one another, we arrive at a representation of the form

$$(21) \qquad a = \eta \pi_1^{e_1} \pi_2^{e_2} \dots \pi_m^{e_m} \quad \text{with } \eta \in R^\times, \ e_i \in \mathbb{N},$$

where π_i is not associated to π_j if $i \neq j$. If the ring in question is a UFD, this representation is essentially unique; indeed, if besides (21) there were another such decomposition $a = \varepsilon \rho_1^{f_1} \dots \rho_n^{f_n}$, we would have $m = n$, and (after renumbering) $\rho_i \hat{=} \pi_i$ and $e_i = f_i$ for all $i = 1, 2, \dots, m$. In this sense (21) is called **the** prime factorization of a.

It turns out to be useful to extend our terminology a little in a formal sense. Toward this goal we first fix a *directory of primes* $\mathcal{P}$ of R specifying a representative for each class of associated primes; that is, $\mathcal{P}$ is a set of nonzero prime elements of R such that every nonzero prime of R is associated with one and only one element of $\mathcal{P}$. (Such a $\mathcal{P}$ exists by the *axiom of choice*.) In many cases there is a canonical choice for $\mathcal{P}$ — for example, in $R = \mathbb{Z}$ the set of *natural prime numbers* stands out, and in the polynomial ring $K[X]$ over a field K we can take for $\mathcal{P}$ the set of all *normalized prime polynomials*. In any case we have:

F10. *Let R be a unique factorization domain and $\mathcal{P}$ a directory of primes of R. Every nonzero $a \in R$ possesses a **unique** representation of the form*

$$(22) \qquad a = \varepsilon \prod_{\pi \in \mathcal{P}} \pi^{e_\pi},$$

where ε is a unit of R and the e_π are nonnegative integers with $e_\pi = 0$ for almost all $\pi \in \mathcal{P}$ (that is, all but finitely many $\pi \in \mathcal{P}$).

There is also a sort of converse to this statement:

F11. *Let R be a integral domain and $\mathcal{P}$ a subset of $R \smallsetminus \{0\}$. If every nonzero $a \in R$ can be **uniquely** represented in the form (22) above, then R is a unique factorization domain and $\mathcal{P}$ is a directory of primes of R.*

Proof. Each $\pi \in \mathcal{P}$ is of course irreducible, by assumption. Now let π be any irreducible element of R. Again from the assumption we have $\pi = \varepsilon \pi'$, with ε a unit and $\pi' \in \mathcal{P}$, both being uniquely determined. Thus π is associated to exactly one $\pi' \in \mathcal{P}$. Overall, the assumption implies that every $a \neq 0$ has a unique decomposition into irreducible factors — that is, R is a UFD. Since in a UFD being an irreducible element is the same as being a nonzero prime, the proof is complete.   □

Let R be a *unique factorization domain* and π any irreducible element of R. For every nonzero $a \in R$ we denote by $w_\pi(a)$ the (highest) exponent with which π appears in a. Thus we have a unique representation of the form

(23) $$a = \pi^{w_\pi(a)} a' \quad \text{with } \pi \nmid a'.$$

We also set $w_\pi(0) = \infty$. Thus we obtain a map

(24) $$w_\pi : R \to \mathbb{Z} \cup \{\infty\},$$

which obviously enjoys the following properties:

(25) $$w_\pi(ab) = w_\pi(a) + w_\pi(b),$$

(26) $$w_\pi(a + b) \geq \min\big(w_\pi(a), w_\pi(b)\big).$$

It should be stressed that both the definition of w_π and property (25) depend on the assumption that R is a UFD.

If K is the fraction field of R, we can extend w_π to a map

(27) $$w_\pi : K \to \mathbb{Z} \cup \{\infty\},$$

by setting

(28) $$w_\pi(a/b) = w_\pi(a) - w_\pi(b).$$

Because of (25), this w_π is well defined. Moreover now (25) and (26) hold for all $a, b \in K$. We call w_π the π-*adic valuation on* K. The ability to extend arithmetic considerations to fraction fields has certain advantages.

F12. *Let R be a unique factorization domain with fraction field $K = \text{Frac } R$ and fix a directory of primes $\mathcal{P}$.*

(i) *Every nonzero element $x \in K$ has a representation*

(29) $$x = \varepsilon \prod_{\pi \in \mathcal{P}} \pi^{w_\pi(x)} \quad \text{with } \varepsilon \in R^\times,$$

where $w_\pi(x) = 0$ for almost all $\pi \in \mathcal{P}$.

(ii) *An element $x \in K$ is in R if and only if $w_\pi(x) \geq 0$ for all $\pi \in \mathcal{P}$.*

(iii) *For $a, b \in R$ we have $a \mid b$ if and only if $w_\pi(a) \leq w_\pi(b)$ for all $\pi \in \mathcal{P}$.*

(iv) *Given arbitrary* $a_1, \ldots, a_n \in R$, *there exist a gcd and an lcm for* $a_1, \ldots, a_n$, *and in fact*

$$d := \prod_{\pi \in \mathscr{P}} \pi^{\min(w_\pi(a_1), \ldots, w_\pi(a_n))} \quad \text{is a gcd}$$

and

$$m := \prod_{\pi \in \mathscr{P}} \pi^{\max(w_\pi(a_1), \ldots, w_\pi(a_n))} \quad \text{is an lcm}$$

of $a_1, \ldots, a_n$ *(where* π^∞ *is to be understood as* 0 *if it occurs).*

Proof. Part (i) follows easily from F10 with the help of (28). If $w_\pi(x) \geq 0$ for all π, then $x \in R$, by (29). The converse is clear, so (ii) is established. Since $a \mid b$ is equivalent to $b/a \in R$, part (iii) follows using (28). Part (iv) now is an automatic consequence of (iii). □

5. The foregoing sections have dealt with little more than the general foundations of elementary arithmetic. We now wish to introduce some ring-theoretical concepts connected with our discussion in Section 3.2.

Definition 8. Let R be a (not necessarily commutative) ring with unity $1 \neq 0$. We call R *simple* if every homomorphism $R \to R'$ into an arbitrary ring R' is either injective or the zero map. Clearly (see Section 3.2) a ring R (with $1 \neq 0$) is simple if and only if $\{0\}$ and R are the only ideals of R. An ideal $I \neq R$ of R is a *maximal ideal* of R if there is no ideal of R distinct from I and R and containing I.

F13. *I is a maximal ideal of R if and only if the quotient ring R/I is simple.*

Proof. Ideals of R containing I are in one-to-one correspondence, via the quotient map $\pi : R \to R/I$, with ideals of R/I. □

Definition 9. Let R be a commutative ring with unity. An ideal I of R is called a *prime ideal* of R if R/I is an integral domain. This condition is equivalent to saying that $I \neq R$ and

(30) $$ab \in I \implies a \in I \text{ or } b \in I.$$

Thus a *principal ideal* (π) of R is prime if and only if π is a prime element of R.

F14. *Let R be a commutative ring with unity. R is simple if and only if R is a field. Therefore an ideal of R is maximal if and only if R/I is a field. Moreover, every maximal ideal of R is also a prime ideal.*

Proof. Only the first assertion needs to be proved. Let R be a field and $\varphi : R \to R'$ a ring homomorphism. If the kernel of φ contains a nonzero element a, it contains *every* element x of R, because $x = (xa^{-1})a$; thus φ is the zero map. Conversely, assume R is simple and take a nonzero $a \in R$. Then $(a) = R$, so there exists $x \in R$ such that $ax = 1$. □

We single out a special case:

F15. *If E, F are fields and $\varphi : E \to F$ is a homomorphism of rings with unity (meaning that $\varphi 1_E = 1_F$), then φ is injective, and so provides an isomorphism between E and a subfield E' of F.*

Definition 10. Let R be a commutative ring with unity. Two ideals I_1, I_2 of R are *relatively prime* if $I_1 + I_2 = R$; in other words, when there exists $a \in I_1$ and $b \in I_2$ such that $a + b = 1$. The *product* $I_1 I_2$ of two ideals I_1, I_2 of R is the ideal of R generated by all products xy, where $x \in I_1$ and $y \in I_2$; thus is consists of all finite sums of such products. Clearly $I_1 I_2 \subseteq I_1 \cap I_2$.

Lemma.   (a) *For I_1, I_2 relatively prime ideals of R we have $I_1 I_2 = I_1 \cap I_2$.*

   (b) *If an ideal I_1 of R is relatively prime to each of the ideals $I_2, I_3, \ldots, I_n$ of R, it is also relatively prime to the product $I_2 I_3 \ldots I_n$.*

Proof. (a) From $1 = a + b$ with $a \in I_1$ and $b \in I_2$ we conclude by multiplying with an arbitrary $c \in I_1 \cap I_2$ that $c = ca + cb \in I_1 I_2$.

(b) By assumption there exists for each $i = 2, 3, \ldots, n$ an element $a_i \in I_1$ and a $b_i \in I_i$ such that $1 = a_i + b_i$. It follows that

$$1 = \prod_i (a_i + b_i) \in I_1 + I_2 I_3 \ldots I_n. \qquad \square$$

F16 (Chinese Remainder Theorem). *Let $I_1, I_2, \ldots, I_n$ be pairwise relatively prime ideals of a commutative ring R with unity. The natural ring homomorphism*

$$(31) \qquad\qquad R \to R/I_1 \times R/I_2 \times \cdots \times R/I_n$$

is surjective, that is, given any elements $x_1, x_2, \ldots, x_n$ of R there exists $x \in R$ such that

$$(32) \qquad\qquad x \equiv x_i \bmod I_i \quad \text{for } i = 1, 2, \ldots, n.$$

The kernel of the map (31) *is the ideal*

$$(33) \qquad\qquad I_1 \cap I_2 \cap \cdots \cap I_n = I_1 I_2 \ldots I_n,$$

so the element x in (32) *is uniquely determined modulo the ideal* (33).

Proof. Consider first the case $n = 2$. By assumption there exist $e_1 \in I_1$ and $e_2 \in I_2$ such that $e_1 + e_2 = 1$. For arbitrary $x_1, x_2 \in R$, the element

$$x = x_2 e_1 + x_1 e_2$$

is then a solution of the system (32). Now let $n \geq 2$ be arbitrary. By induction we can assume that there exists $x' \in R$ such that

$$x' \equiv x_i \bmod I_i \quad \text{for } i = 2, \ldots, n.$$

By part (b) of the preceding lemma, I_1 is relatively prime to the product $I_2 I_3 \ldots I_n$; thus, thanks to the previously settled case $n = 2$, there exists $x \in R$ such that

$$x \equiv x_1 \bmod I_1 \quad \text{and} \quad x \equiv x' \bmod I_2 \ldots I_n.$$

Then x clearly satisfies all the congruences in (32). As for the kernel of (31), it obviously equals the intersection of the I_i. But by the preceding lemma one easily concludes by induction that

$$I_1 \cap I_2 \cap \cdots \cap I_n = I_1 \cap (I_2 \ldots I_n) = I_1 I_2 \ldots I_n. \qquad \square$$

5

Prime Factorization in Polynomial Rings. Gauss's Theorem

1. Let $\alpha \in \mathbb{C}$ be an algebraic number. When looking into whether α is constructible from $\{0, 1\}$ with ruler and compass, we were led to investigate in particular the degree of the field extension $\mathbb{Q}(\alpha)/\mathbb{Q}$ (Chapter 1, F9). Now, $\mathbb{Q}(\alpha) : \mathbb{Q}$ equals the degree of the minimal polynomial of α over K (Chapter 2, F3). Thus, if we already know a (normalized) polynomial $f \in \mathbb{Q}[X]$ such that $f(\alpha) = 0$, our task is to determine whether f is irreducible. If so, we have found our desired minimal polynomial — it is f (Chapter 3, F6). If not, we must continue the search by looking for irreducible factors of f.

Example. Consider $\alpha = e^{2\pi i/n}$, for $n > 1$ a natural number. Since

$$X^n - 1 = (X - 1)(1 + X + \cdots + X^{n-1}),$$

the polynomial $f(X) = X^{n-1} + \cdots + X + 1$ satisfies $f(\alpha) = 0$. Is f irreducible? If n has a proper divisor d, surely not, since in this case $(X^d - 1)/(X - 1)$ is a divisor of f. But if n is prime we will see in F11 that f is indeed irreducible.

This is not the place for a comprehensive study of the problem mentioned in the first paragraph. Nonetheless, we would like to shed some light on certain theoretical aspects of the issue. Two very natural questions will guide us: (i) Is the ring $\mathbb{Z}[X]$ a UFD? (ii) Is every polynomial $f \in \mathbb{Z}[X]$ that is irreducible in $\mathbb{Z}[X]$ also irreducible in $\mathbb{Q}[X]$? These questions are intimately connected and were both answered by Gauss in the affirmative.

2. In the remainder of this chapter,

R will always be an integral domain.

Question (i) above can be generalized to read: When is $R[X]$ a UFD? For practice with basic algebraic constructs, we start by establishing the following:

F1. *$R[X]$ is a principal ideal domain if and only if R is a field.*

Proof. If $R = K$ is a field, $K[X]$ is a Euclidean domain and therefore a principal ideal domain (Chapter 4, F6).

Conversely, assume $R[X]$ is a principal ideal domain. Consider the kernel I of the substitution homomorphism $\varphi : R[X] \to R$ with $\varphi(X) = 0$. Since φ sends each polynomial in $R[X]$ to its constant term, $I = (X)$. Thus

(1) $$R[X]/X = R[X]/I \simeq R$$

is an *integral domain*, that is, X is *prime* in $R[X]$, by Definition 9 of Chapter 4. But then the next statement shows that $R[X]/X$ is actually a field, and therefore by (1) R is also a field. □

F2. *If A is a principal ideal domain and π is irreducible in A, the quotient A/π is a field.*

Proof. According to F14 in Chapter 4, we have to show that (π) is a *maximal ideal* of A. Let I be a proper ideal of A containing (π). By assumption, $I = (a)$ for some a. Then $a \mid \pi$ and $a \notin A^\times$, so since π is irreducible we have $(\pi) = (a) = I$. (Incidentally this shows again that every irreducible element of a principal ideal domain is prime.) □

F3. *If $R[X]$ is a UFD, so is R.*

Proof. Take a nonzero $a \in R$. Since $R[X]$ is a UFD, we can write

$$a = \varepsilon p_1(X) p_2(X) \ldots p_r(X)$$

with $\varepsilon \in R[X]^\times = R^\times$ and $p_1(X), \ldots, p_r(X)$ all prime in $R[X]$. But then all the factors have degree zero, so define $\pi_i := p_i(X) \in R$. For elements of R, divisibility in $R[X]$ coincides with divisibility in R, so the π_i are prime in R. Thus every nonzero $a \in R$ can be represented as

(2) $$a = \varepsilon \pi_1 \pi_2 \ldots \pi_r$$

with $\varepsilon \in R^\times$ and the π_i prime in R. If $a \in R$ is *irreducible*, we must have $r = 1$ in this representation, so a is associated to π_1 and therefore also prime. This shows that R is a UFD, by F7 in Chapter 4. □

3. We now prove the converse:

Gauss's Theorem. *If R is a UFD, so is the polynomial ring $R[X]$.*

We begin with some preliminary observations, which are of interest in and of themselves. Every ring homomorphism $\varphi : R \to R'$ can be naturally extended to a homomorphism $R[X] \to R'[X]$ between the corresponding polynomial rings, by setting

(3) $$\sum a_i X^i \mapsto \sum \varphi(a_i) X^i;$$

we will denote the extension by φ as well. (By the way, if we regard $R'[X]$ as an R-algebra via φ, the map (3) can be thought of as a substitution homomorphism.)

Given $a \in R$, we consider in particular the quotient map $R \to R/a$ and its natural extension

(4) $$R[X] \to (R/a)[X].$$

F4. (i) *The homomorphism (4) yields a natural isomorphism (of R-algebras)*

$$R[X]/a \to (R/a)[X].$$

(ii) *An element $a \in R$ is prime in R if and only if it is prime in $R[X]$.*

Proof. Part (i) follows from the Fundamental Homomorphism Theorem, since the kernel of (4) is clearly $I = aR[X]$. Part (ii): since $R[X]/a \simeq (R/a)[X]$, we have a prime in $R \iff R/a$ is an integral domain $\iff (R/a)[X]$ is an integral domain $\iff R[X]/a$ is an integral domain $\iff a$ is prime in $R[X]$. $\qquad\square$

From now on we assume R is a *unique factorization domain*. We denote by

$$K = \mathrm{Frac}\, R$$

the *fraction field* of R. Now let π be a given nonzero prime of R. The corresponding π-adic valuation $w_\pi : K \to \mathbb{Z} \cup \{\infty\}$ in R (page 40) can be extended to a map

(5) $$w_\pi : K[X] \to \mathbb{Z} \cup \{\infty\}$$

as follows: set

(6) $$w_\pi\left(\sum a_i X^i\right) = \min\left\{w_\pi(a_i) \mid i \geq 0\right\}.$$

Thus, for $f \in R[X]$, the value of $w_\pi(f)$ is the exponent of the highest π-power that fits in all coefficients of f. By equation (25) in Chapter 4,

(7) $$w_\pi(cf) = w_\pi(c) + w_\pi(f) \quad \text{for } c \in K, \ f \in K[X].$$

The springboard for the proof of Gauss's Theorem is provided by the next result:

F5. *Let R be a unique factorization domain and $\pi \neq 0$ a prime in R. With the notations introduced above, we have, for all $g, h \in K[X]$,*

(8) $$w_\pi(gh) = w_\pi(g) + w_\pi(h).$$

Proof. Clearly, for every $f \in K[X]$ there exists $c \in R$ such that $cf \in R[X]$. So taking (7) into account, we can assume without loss of generality that $g, h \in R[X]$. For simplicity we set $w = w_\pi$. By the definition of w we then have $g = \pi^{w(g)} g_1$ and $h = \pi^{w(h)} h_1$, where $g_1, h_1 \in R[X]$ are polynomials satisfying

(9) $$w(g_1) = 0, \quad w(h_1) = 0.$$

We obtain $gh = \pi^{w(g)+w(h)} g_1 h_1$, which together with (7) implies $w(gh) = w(g) + w(h) + w(g_1 h_1)$. Thus we must show that

$$w(g_1 h_1) = 0.$$

Suppose to the contrary that $w(g_1 h_1) > 0$, that is, $\pi \mid g_1 h_1$. By F4(ii) we then have $\pi \mid g_1$ or $\pi \mid h_1$, that is, $w(g_1) > 0$ or $w(h_1) > 0$. But this contradicts (9). $\qquad\square$

Definition. A nonconstant polynomial $f \in R[X]$ (that is, one whose degree is at least 1) is called *primitive* if the gcd of the coefficients of f is 1.

Thus a normalized polynomial in $R[X]$ is trivially primitive. If R is a UFD, every nonconstant polynomial $g \in R[X]$ can be represented as

$$g = a g_1, \quad \text{with } a \in R \smallsetminus \{0\} \text{ and } g_1 \in R[X] \text{ primitive.}$$

Also, a is determined up to associatedness, being the gcd of the coefficients of g. We call a the *content* of g; more precisely, the uniquely determined principal ideal (a) is called that. As can be proved easily from F5,

$$(10) \qquad\qquad \text{Content}(gh) = \text{Content}(g) \cdot \text{Content}(h). \qquad\qquad \square$$

We now formulate Gauss's result a bit more precisely:

Theorem 1 (Gauss). *Let R be a UFD with fraction field K. Let $\mathcal{P}_1$ be a directory of primes for R and $\mathcal{P}_2$ a directory of primes for $K[X]$ containing only primitive polynomials of $R[X]$. Then $R[X]$ is a UFD and $\mathcal{P}_1 \cup \mathcal{P}_2$ is a directory of primes for $R[X]$.*

(It is clear that there exists a $\mathcal{P}_2$ with the required properties.)

Proof. Take any nonzero $g \in R[X]$. Since $K[X]$ is a UFD, there is a unique factorization

$$(11) \qquad g = a \prod_{f \in \mathcal{P}_2} f^{e_f} \quad \text{with } a \in K^\times = K[X]^\times \text{ and integers } e_f \geq 0,$$

where $e_f = 0$ for almost all $f \in \mathcal{P}_2$. Now, for any $\pi \in \mathcal{P}_1$ we have (see F5) $w_\pi(g) = w_\pi(a) + \sum e_f w_\pi(f) = w_\pi(a)$, the latter equation because the f's are primitive. It follows that $w_\pi(a) \geq 0$ for all π, and thus also $a \in R$ (see F12 in Chapter 4). Now let

$$(12) \qquad\qquad\qquad a = \varepsilon \prod_{\pi \in \mathcal{P}_1} \pi^{e_\pi}$$

be the prime factorization of a in R. Together, (11) and (12) yield

$$(13) \qquad\qquad\qquad g = \varepsilon \prod_{\pi \in \mathcal{P}_1} \pi^{e_\pi} \prod_{f \in \mathcal{P}_2} f^{e_f}.$$

This representation is unique, that is, ε, the e_π and the e_f are uniquely determined by g. For if a representation of the form (13) is given, a comparison with (11) immediately yields (12), since $K[X]$ is a UFD; but now since R too is a UFD, the representation (13) is completely fixed. Now keeping in mind Chapter 4, F11, the proof is complete. $\qquad\qquad \square$

F6. *Let R be a UFD with $K = \text{Frac } R$, and let $g \in R[X]$ be nonconstant. If g is irreducible in $R[X]$, it is irreducible in $K[X]$.*

Proof. By Theorem 1 we have (with $\mathscr{P}_2$ as in the theorem's statement)

$$g = \varepsilon f \quad \text{with } \varepsilon \in R^{\times} \text{ and } f \in \mathscr{P}_2.$$

Thus g is irreducible in $K[X]$. □

Conversely, if $g \in R[X]$ is irreducible in $K[X]$, then g is irreducible in $R[X]$ if and only if g is primitive.

F7 (Gauss's Lemma). *Let R be a UFD and $K = \operatorname{Frac} R$. If $f(X) \in R[X]$ can be expressed as*

$$f(X) = g(X)h(X) \quad \text{with normalized } g, h \in K[X],$$

all the coefficients of g and h lie in R.

Proof. For any prime $\pi \neq 0$ of R we have

$$w_\pi(f) \geq 0 \qquad\qquad \text{since } f \in R[X],$$
$$w_\pi(g), w_\pi(h) \leq w_\pi(1) = 0 \quad \text{since } g, h \text{ are normalized.}$$

But $w_\pi(f) = w_\pi(g) + w_\pi(h)$, so all three integers vanish. Since π was arbitrary, all the coefficients of g and h belong to R (see Chapter 4, F12). □

F8. *Let R be a UFD and $K = \operatorname{Frac} R$. Let $f \in R[X]$ be a **normalized** polynomial with coefficients in R. Then any root α of f that lies in K actually lies in R, and moreover divides the constant term of f.*

Proof. Take a factorization $f(X) = (X - \alpha)h(X)$ in $K[X]$. Since f is normalized, so is h. By Gauss's Lemma (F7), all the coefficients of $X - \alpha$ and $h(X)$ are in R. In particular, $\alpha \in R$, and since $a_0 := f(0) = (-\alpha)h(0)$ we have $\alpha \mid a_0$. □

Application. Consider the special case $R = \mathbb{Z}$, $K = \mathbb{Q}$. We prove that $f(X) = X^3 - 2$ is irreducible. Suppose f were reducible. Since it has degree 3, it would have a linear factor, and therefore a root α in $\mathbb{Q}$. By F8, $X^3 - 2$ would also have a root in $\mathbb{Z}$, which is clearly not the case.

From the irreducibility of $X^3 - 2$ it follows that $\sqrt[3]{2}$ is not a rational number. Using similar arguments one can easily derive from F8 the irrationality of numbers such as $\sqrt[5]{3}$, $\sqrt[6]{72}$, $\sqrt[12]{27}$. (Note that although $\sqrt[12]{27}$ is irrational, $X^{12} - 27$ is not irreducible. As an exercise prove that $X^6 - 72$ *is* irreducible — compare F11(d) in Chapter 2.)

4. The next statement suggests a fundamental principle by means of which one can investigate the irreducibility of polynomials:

F9. *Let R be an integral domain and let $a \mapsto \bar{a}$ be a homomorphism of R into an integral domain $\bar{R}$; extend this to a homomorphism $R[X] \to \bar{R}[X]$ of polynomial rings in the usual way:*

$$f = \sum a_i X^i \mapsto \bar{f} = \sum \bar{a}_i X^i.$$

*Let $f(X) = a_n X^n + \cdots + a_0$ be a primitive polynomial of $R[X]$ with $\bar{a}_n \neq 0$. If $\bar{f}$
is irreducible in $\bar{R}[X]$, then f is irreducible in $R[X]$.*

Proof. Assume the opposite. Then $f = gh$, where g, h are nonconstant polynomials
in $R[X]$ (nonconstant because f is primitive). Taking the image in $\bar{R}[X]$ gives
$\bar{f} = \bar{g}\bar{h}$. Since $\bar{a}_n$ is nonzero, we must have $\deg \bar{g} = \deg g \geq 1$ and $\deg \bar{h} = \deg h \geq 1$.
Since $\bar{R}$ is an integral domain, we get a contradiction with the assumption that $\bar{f}$
is irreducible. □

In applying F9, one is usually dealing with a unique factorization domain R,
because apart from the fact that otherwise one has hardly any control over the
primitivity of f, it is also not permissible in the general case to deduce that f is
irreducible over $K = \text{Frac } R$. On the other hand, the train of thought that leads to
F9 *can* be useful even if we don't know ahead of time that $\bar{f}$ is irreducible, but
rather we know something about the possible factorizations of $\bar{f}$ in $\bar{R}[X]$. Here is
the best known application of this approach:

F10 (Eisenstein irreducibility criterion). *Let R be an integral domain, and let*

$$f(X) = a X^n + a_{n-1} X^{n-1} + \cdots + a_1 X + a_0$$

*be a **primitive** polynomial in $R[X]$. If there exists a prime π of R such that (i) $\pi \nmid a$,
(ii) $\pi \mid a_i$ for $0 \leq i \leq n-1$, and (iii) $\pi^2 \nmid a_0$, then f is irreducible in $R[X]$. If R is a
UFD, f is also irreducible in $K[X]$, where K is the fraction field of R.*

Proof. Since π was assumed prime in R, the quotient $\bar{R} := R/\pi$ is an integral
domain. We now work as in the proof of F9. Suppose f has a nontrivial factorization
in $R[X]$:

$$f = gh, \quad \text{with } r = \deg g \geq 1 \text{ and } s = \deg h \geq 1.$$

Taking the image in $\bar{R}[X]$ leads to, as before,

$$\bar{f} = \bar{g}\bar{h}, \quad \text{with } \deg \bar{g} = r \text{ and } \deg \bar{h} = s.$$

To be sure, $\bar{f}$ is not irreducible, but because of condition (ii) it has the form $\bar{f} = \bar{a} X^n$, so that in $\bar{R}[X]$ the equation

(14) $$\bar{a} X^n = \bar{g}\bar{h}$$

holds. Set $k = \text{Frac } \bar{R}$. Since $k[X]$ is a UFD and $\bar{a} \neq 0$, it follows from (14) that
$\bar{g}$ and $\bar{h}$ have the form $\bar{g} = \beta X^r$ and $\bar{h} = \gamma X^s$, with $\beta, \gamma \in k$. But $r, s \geq 1$, so
in particular $\bar{g}(0) = \bar{h}(0) = 0$, that is, $\pi \mid g(0)$ and $\pi \mid h(0)$. This implies that π^2
divides $g(0)h(0) = f(0) = a_0$, contradicting condition (iii). □

Here is an immediate consequence of Eisenstein's irreducibility criterion: All
polynomials of the form

$$X^n - a, \quad \text{with } a \in \mathbb{Z} \smallsetminus \{1, -1\} \text{ square-free}$$

are irreducible in $\mathbb{Z}[X]$ and therefore also in $\mathbb{Q}[X]$. (An integer $a \in \mathbb{Z}$ is called
square-free if it is not divisible by the square of any prime.)

F11. *If p is a prime number, the polynomial*

$$F_p(X) = X^{p-1} + X^{p-2} + \cdots + X + 1$$

is irreducible in $\mathbb{Q}[X]$.

Proof. We show that Eisenstein's criterion can be applied after a simple change of variables. Instead of $F_p(X)$, consider the polynomial

$$f(X) = F_p(X + 1).$$

This f is irreducible in $\mathbb{Q}[X]$ if and only if F_p is, because the two polynomials are mapped to one another by the automorphism of $\mathbb{Q}[X]$ coming from the invertible substitution $X \mapsto X + 1$. Since $F_p(X)(X - 1) = X^p - 1$ we get $f(X)X = (X + 1)^p - 1$, that is,

$$f(X) = \sum_{k=1}^{p} \binom{p}{k} X^{k-1} = p + \binom{p}{2} X + \cdots + \binom{p}{p-1} X^{p-2} + X^{p-1}.$$

Relative to the prime p of $\mathbb{Z}$, this is an *Eisenstein polynomial* of $\mathbb{Z}[X]$—that is, it satisfies all the conditions required for the application of the Eisenstein criterion, as a result of the fact that p divides

$$(15) \qquad \binom{p}{k} = \frac{p(p-1)\dots(p-k+1)}{1\cdot 2\cdot\dots\cdot k}$$

for $0 < k < p$. (This is seen as follows: For $0 < k < p$, the numerator, but not the denominator, of the fraction in (15) is divisible by p; since we already known — on combinatorial grounds, for instance — that $\binom{p}{k}$ is an integer, the divisibility claim is proved.) $\qquad\square$

F11 has immediate repercussions for the constructibility of regular polygons:

F12. *Let p be a prime. A construction of the regular p-gon with ruler and compass is impossible unless* $p - 1$ *is a power of 2.*

Proof. If $\zeta := e^{2\pi i/p}$ lies in $\mathbb{A}\mathbb{Q}$, the degree $\mathbb{Q}(\zeta) : \mathbb{Q}$ is a power of 2, by Chapter 1, F9. But by F11 we always have $\mathbb{Q}(\zeta):\mathbb{Q} = p-1$ (review Chapter 2, F3 and Chapter 3, F6). $\qquad\square$

Thus a regular heptagon (7-gon) is not constructible with ruler and compass, nor is an 11-gon, a 13-gon, a 14-gon, a 19-gon, and so on. And neither is a 9-gon, as can be seen from the following generalization of F11:

F13. *We keep the notations of F11. If* $n = p^r$ *is a prime power, the minimal polynomial of* $\zeta := e^{2\pi i/n}$ *over* $\mathbb{Q}$ *is*

$$(16) \qquad F_{p^r}(X) := 1 + X^{p^{r-1}} + X^{2p^{r-1}} + \cdots + X^{(p-1)p^{r-1}} = F_p(X^{p^{r-1}}).$$

In particular, $\mathbb{Q}(\zeta):\mathbb{Q} = p^{r-1}(p-1)$.

Proof. Since $X^{p^r} - 1 = (X^{p^{r-1}} - 1)F_{p^r}(X)$, our ζ is certainly a root of F_{p^r}. It suffices to show that the polynomial $f(X) := F_{p^r}(X+1)$ in $\mathbb{Z}[X]$ is an Eisenstein polynomial with respect to p. We work in $\mathbb{Z}[X]$ modulo p. As we saw in the proof of F11, $F_p(X+1) \equiv X^{p-1}$ mod p. Thus

$$f(X) = F_{p^r}(X+1) = F_p((X+1)^{p^{r-1}}) \equiv F_p(X^{p^{r-1}} + 1) \equiv (X^{p^{r-1}})^{p-1} \text{ mod } p.$$

In addition, $f(0) = F_{p^r}(1) = p$ and f is normalized, so f is indeed an Eisenstein polynomial with respect to p. $\qquad\square$

Until now we have left open the question whether for primes of the form $p = 2^m + 1$ it is indeed always possible to subdivide the circle into p parts with ruler and compass. If this is so, a moment's thought shows that the constructibility problem for arbitrary n has been completely solved (apart of course from the number-theoretical question of which primes have the form $2^m + 1$). It turns out that the division of the circle into p parts with ruler and compass *is* possible for all primes of the form $2^m + 1$, but we will only substantiate this fact after we have developed our conceptual machinery some more and acquired more powerful tools. (The reader interested in learning more about Gauss's more direct approach to this problem should consult his *Disquisitiones Arithmeticae*.) Incidentally, Gauss at the age of eighteen had already discovered a way to construct a regular 17-gon, before he had a thorough proof of the impossibility of constructing, say, a regular heptagon.

We now address the problem of trisecting an angle:

F14. *Consider an angle φ with $0 \leq \varphi < 2\pi$. If $e^{i\varphi}$ is transcendental, φ cannot be trisected with ruler and compass.*

Remarks. (1) The condition in F14 is satisfied for uncountably many φ. Indeed, the function $\varphi \mapsto e^{i\varphi}$ provides a bijection between the interval $[0, 2\pi)$ and the unit circle in $\mathbb{C}$; therefore there can be only countably many $\varphi \in [0, 2\pi)$ for which $e^{i\varphi}$ is algebraic.

(2) Even when $e^{i\varphi}$ is algebraic, the trisection of φ is by no means necessarily possible. Consider for example $\varphi = 2\pi/3$. Trisecting φ amounts to constructing a 9-gon, which as we know from F13 is impossible. Actually it is also easy to prove directly that $\mathbb{Q}(e^{i\varphi/3}) : \mathbb{Q}(e^{i\varphi}) = 3$.

(3) Again in connection with the transcendence condition in F14, the famous *Hermite–Lindemann Theorem* says that if z is any nonzero *algebraic* complex number, e^z is transcendental — in particular, $e^{i\varphi}$ is transcendental for any algebraic value of the angle φ. Since $e^{i\pi} = -1$, the theorem also implies that π is transcendental. We will give a proof of the Hermite–Lindemann Theorem in Chapter 17.

Proof of F14. Let $K = \mathbb{Q}(e^{i\varphi})$, and suppose $t = e^{i\varphi}$ is transcendental. We must show that $z = e^{i\varphi/3}$ does not belong to $\triangle K$. This will be done if we prove that

$$(17) \qquad\qquad K(z) : K = 3$$

(see Chapter 1, F9). Now, z is certainly a root of the polynomial $X^3 - t$ over K. To prove (17) therefore we have to show that

(18) $$X^3 - t \text{ is irreducible in } K[X].$$

Since t is transcendental, $K = \mathbb{Q}(t)$ is the field of rational functions in t over $\mathbb{Q}$ (Section 3.4), which is to say, the fraction field of the polynomial ring $R := \mathbb{Q}[t]$. Since R is a UFD and t is prime in R, an application of the Eisenstein criterion to the polynomial $X^3 - t$ proves (18). $\qquad\square$

6

Polynomial Splitting Fields

1. Still bearing in mind the initial problem of Chapter 1, our task now consists in the study of (finite) field extensions E/K. One fundamental question concerns the possible intermediate fields of E/K (see Chapter 1, Theorem $1'$). Our subsequent discussion will benefit from the introduction of a convenient shorthand:

Definition 1. Let E_1 and E_2 be extensions of a field K, which we regard as algebras over K. A homomorphism of K-algebras $\sigma : E_1 \rightarrow E_2$ is called a *K-homomorphism from E_1 to E_2*. We also say that σ is a *homomorphism from the extension E_1/K to the extension E_2/K* and write

(1) $$\sigma : E_1/K \rightarrow E_2/K.$$

If σ is an isomorphism, we say that the extensions E_1/K and E_2/K are isomorphic.

Remark. A field homomorphism $\sigma : E_1 \rightarrow E_2$ satisfies $\sigma(1) = 1$ by definition; therefore it is always *injective* and so gives rise to an *isomorphism* of E_1 with a subfield of E_2. If E_1 and E_2 are both extensions of a field K, a field homomorphism σ is a K-homomorphism if and only if

(2) $$\sigma(c) = c \quad \text{for all } c \in K.$$

From Definition 1 there is a steep but well-traveled path to *Galois theory* (opened largely by Dedekind and E. Artin; see the latter's *Galois Theory*). Here we will take the more leisurely and scenic route. The following result is simple but far-reaching:

F1. *Let E/K and E'/K' be field extensions and $\sigma : K \rightarrow K'$ a field homomorphism. There is a natural extension of σ to a ring homomorphism $K[X] \rightarrow K'[X]$; we still call it σ, but for $f \in K[X]$ we often write f^σ instead of $\sigma(f)$. Let $f \in K[X]$.*

(a) *Every homomorphism $\tau : E \rightarrow E'$ extending σ maps any root of f in E to a root of f^σ in E'.*

(b) *Assume $\sigma : K \rightarrow K'$ is an isomorphism. Let α be a root of f in E and α' a root of f^σ in E'. If f is **irreducible** over K, there is an isomorphism $\tau : K(\alpha) \rightarrow K'(\alpha')$ extending σ and such that $\tau(\alpha) = \alpha'$.*

Remark. As an important special case, take $K' = K$ and $\sigma = \mathrm{id}_K$. Then a τ as in part (a) is a K-homomorphism from E to E', and a τ as in part (b) is a K-homomorphism from $K(\alpha)$ onto $K(\alpha')$. By the way, the irreducibility assumption in (b) cannot be dispensed with.

Proof. (a) Suppose $f(X) = a_0 + a_1 X + \cdots + a_n X^n$. If $0 = f(\alpha) = a_0 + a_1\alpha + \cdots + a_n\alpha^n$ then $0 = \tau(f(\alpha)) = \sigma(a_0) + \sigma(a_1)\tau(\alpha) + \cdots + \sigma(a_n)\tau(\alpha)^n = f^\sigma(\tau\alpha)$.

(b) Let $f \in K[X]$ be irreducible; we may as well assume it normalized. Then f is the minimal polynomial of α over K. We define τ by setting

$$\tau(g(\alpha)) = g^\sigma(\alpha') \quad \text{for } g \in K[X].$$

Is τ well-defined? If $g_1(\alpha) = g_2(\alpha)$ we get $(g_1 - g_2)(\alpha) = 0$, so $g_1 - g_2 = hf$ with $h \in K[X]$; then $g_1^\sigma - g_2^\sigma = h^\sigma f^\sigma$ and therefore $g_1^\sigma(\alpha') - g_2^\sigma(\alpha') = h^\sigma(\alpha') f^\sigma(\alpha') = 0$. It is clear that τ is a surjective homomorphism from $K(\alpha)$ to $K'(\alpha')$ extending σ. □

In order to have some room to maneuver, we quote now a result whose proof — in spite of the statement's spartan simplicity — requires further preliminaries and is postponed to the end of the chapter. The general construction principles laid down in preparation for the proof (Sections 6.2 and 6.3) will also be important in other contexts.

Theorem 1. *Let $(E_i)_{i \in I}$ be an arbitrary family of extensions E_i of a field K. There exists an extension E of K and homomorphisms $\tau_i : E_i/K \to E/K$ such that E is obtained from K by adjoining the union of the sets $\tau_i E_i$, for $i \in I$.*

Definition 2. A field C is *algebraically closed* if every nonconstant polynomial $f(X) \in C[X]$ has a root in C.

Remark. The field $\mathbb{C}$ of complex number is algebraic closed; this is proved through analysis, function theory or algebra (see Volume II for the latter).

F2. *The following statements about a field C are equivalent:*

 (i) *C is algebraically closed.*

 (ii) *Every irreducible polynomial in $C[X]$ is linear (that is, of degree 1).*

 (iii) *Every nonconstant polynomial in $C[X]$ is completely decomposable into linear factors.*

 (iv) *If E/C is an **algebraic** field extension, $E = C$.*

Proof. (i) ⇒ (ii): Let $f \in C[X]$ be irreducible. By (i) there exists $\alpha \in C$ with $f(\alpha) = 0$. Then f is divisible in $C[X]$ by $X - \alpha$, that is, $f = \gamma(X - \alpha)$, necessarily with $\gamma \in C^\times$.

(ii) ⇒ (iii): By (ii), only linear polynomials can appear in the prime factorization of $f \in C[X]$.

(iii) ⇒ (iv): Let E/C be algebraic. The minimal polynomial $f = \mathrm{MiPo}_C(\alpha)$ of any $\alpha \in E$ is irreducible, hence linear, by (iii). Therefore $\alpha \in C$.

(iv) $\Rightarrow$ (i): Suppose $f \in C[X]$ is nonconstant. By Chapter 3, Theorem 4, there is an extension E/C and an $\alpha \in E$ such that $f(\alpha) = 0$. But $C(\alpha)/C$ is algebraic, so by (iv) we have $C(\alpha) = C$, that is, $\alpha \in C$. $\qquad\qquad\qquad\qquad\qquad\qquad$ $\square$

Theorem 2 (Steinitz). *Let K be a field.*

(I) *There exists an extension C of K with the following properties*:
  (i) *C is algebraically closed.*
  (ii) *C/K is algebraic.*
  Such a field is called an **algebraic closure** *of K.*

(II) *If C_1 and C_2 are algebraic closures of K, the extensions C_1/K and C_2/K are isomorphic.*

Proof of part (I). Let $K[X_n, \, n \in \mathbb{N}]$ be the polynomial ring in countably many indeterminates $X_1, X_2, \ldots$ over K. Consider the set I of all subsets $M \subseteq K[X_n, \, n \in \mathbb{N}]$ such that

$$M \text{ is a maximal ideal of } K[X_1, \ldots, X_m] \text{ for some } m \in \mathbb{N}.$$

For each such M, let $E_M := K[X_1, \ldots, X_m]/M$ be the corresponding quotient field. We regard E_M as an extension of K. Now apply Theorem 1 to the family $(E_M)_{M \in I}$, to conclude that there exists a field extension E/K and K-homomorphisms

$$\sigma_M : E_M \to E$$

for each $M \in I$. We claim that *for every finite field extension L/K there exists a K-homomorphism from L into E.* Indeed, if $L = K(\alpha_1, \ldots, \alpha_m)$ with each α_i algebraic over K, consider the homomorphism of K-algebras $\varphi : K[X_1, \ldots, X_m] \to L$ defined by $\varphi(X_i) = \alpha_i$. Let M be its kernel. Then φ yields an isomorphism $K[X_1, \ldots, K_m]/M \to L$. Hence M is a maximal ideal of $K[X_1, \ldots, X_m]$, and the claim is proved.

Now let C be the algebraic closure of K in E (see Chapter 2, F8). The extension C/K is certainly algebraic; we show that C is also *algebraically closed*. Suppose otherwise. Then there is an algebraic extension F/C with $F \neq C$. Take $\alpha \in F \smallsetminus C$, and let f be the minimal polynomial of α *over K*. (Note: α is algebraic over K because it is algebraic over C and C/K is an algebraic extension.) Suppose f has exactly n distinct roots $\beta_1, \ldots, \beta_n$ in C, and form the subfield $L = K(\alpha, \beta_1, \ldots, \beta_n)$ of F. Then L/K is *finite* (see Chapter 2, F7). Therefore, by the italicized statement in the previous paragraph, there exists a K-homomorphism $\varphi : K(\alpha, \beta_1, \ldots, \beta_n) \to C$. But then $\varphi(\alpha), \varphi(\beta_1), \ldots, \varphi(\beta_n)$ are $n+1$ distinct roots of f in C. Contradiction! $\qquad\qquad\qquad\qquad\qquad\qquad\qquad\qquad$ $\square$

Remark. The extensions E_M/K are all algebraic (so E/K itself can be assumed algebraic); but this is harder to prove and we don't need it here.

The proof of part (II) of Theorem 2, the uniqueness part, ensues from:

Theorem 3. *Let $\sigma : K \to K'$ be an isomorphism of fields and let L/K be an **algebraic** field extension. If C is an **algebraically closed** extension of K', the map σ can be extended to a homomorphism $\tau : L \to C$.*

Proof. (a) We deal first with the case $K' = K$ and $\sigma = \mathrm{id}_K$. An application of Theorem 1 to $E_1 := L$ and $E_2 := C$ shows there exists an extension E/K and K-homomorphisms $\sigma_i : E_i \to E$ such that $E = C'(L')$, where $L' := \sigma_1 L$ and $C' := \sigma_2 C$. The extension $C'(L')/C'$ is algebraic and C' is algebraically closed. It follows that $C' = C'(L') = E$, so $\sigma_2 : C \to E$ is an *isomorphism*. Then we can look at the K-homomorphism $\sigma_2^{-1} \circ \sigma_1 : L \to C$; this homomorphism is an extension τ of $\sigma = \mathrm{id}_K$ as desired.

(b) Now let $\sigma : K \to K'$ be any isomorphism. As can easily be checked, there is an extension L' of K' and a homomorphism $\rho : L \to L'$ that agrees with σ on K (replacing the elements of K in L by elements of K'). By (a) there exists a K'-homomorphism $\tau' : L' \to C$. Then $\tau = \tau' \circ \rho : L \to C$ is an extension of σ. $\square$

Proof of part (II) *of Theorem* 2.   By Theorem 3, there is a homomorphism $\tau : C_1/K \to C_2/K$. But $C_2/\tau C_1$ is algebraic (because C_2/K is), and τC_1 is algebraically closed (like C_1). It follows that $C_2 = \tau C_1$, so τ is an isomorphism. $\square$

F3. *Every endomorphism of an **algebraic** field extension E/K is an automorphism.*

Proof. Let $\sigma : E/K \to E/K$ be a homomorphism of field extensions. We must show that $\sigma E = E$. Take $\alpha \in E$ and set $f = \mathrm{MiPo}_K(\alpha)$. Denote by N_α the set of all roots of f in E. Then σ effects a permutation of N_α, because σ takes roots of f to roots of f (see F1) and σ is injective. Thus, since $\alpha \in N_\alpha$, there exists $\beta \in N_\alpha$ such that $\sigma(\beta) = \alpha$. $\square$

Definition 3. Let K be a field and $f \in K[X]$ a nonconstant polynomial. An extension E of K is called a *splitting field of f over K* if there exist $\alpha_1, \alpha_2, \ldots, \alpha_n \in E$ such that $f(X) = \gamma(X - \alpha_1)(X - \alpha_2) \ldots (X - \alpha_n)$ and $E = K(\alpha_1, \alpha_2, \ldots, \alpha_n)$.

The name echoes the expression "to split into linear factors", which means the same as "to have a complete decomposition into linear factors".

F4. *Every nonconstant $f \in K[X]$ has a splitting field over K. If E, E' are splitting fields of f over K, the extensions E'/K and E/K are isomorphic.*

Proof. Existence: Let C be an algebraic closure of K, which exists by Theorem 2. In $C[X]$ we have

$$f(X) = \gamma(X - \alpha_1)(X - \alpha_2) \ldots (X - \alpha_n)$$

with $\alpha_i \in C$ (and $\gamma \in K$ the leading coefficient of f). Therefore the subfield $K(\alpha_1, \alpha_2, \ldots, \alpha_n)$ of C is a splitting field of f over K. (Actually the existence of a splitting field also follows easily from Kronecker's Theorem, given as Theorem 4 in Chapter 3; see §3.3 in the Appendix.)

Uniqueness: Let C and C' be algebraic closures of E and E' (and therefore also of K and K'). By Theorem 3 there exists a K-homomorphism $\tau : E \to C'$. Clearly, since E is a splitting field of f over K, so is τE; but since the splitting fields τE and E' of f over K are both subfields of C', it follows that $\tau E = E'$. Thus τ yields a K-isomorphism between E and E'.   □

Examples.   (a) One splitting field of $X^4 - 2$ over $\mathbb{Q}$ is the subfield $\mathbb{Q}(\sqrt[4]{2}, i)$
   of $\mathbb{C}$. Indeed, $X^4 - 2 = (X - \sqrt[4]{2})(X - i\sqrt[4]{2})(X + \sqrt[4]{2})(X + i\sqrt[4]{2})$ and
   $\mathbb{Q}(\sqrt[4]{2}, i\sqrt[4]{2}) = \mathbb{Q}(\sqrt[4]{2}, i)$.

   (b) One splitting field of $X^n - 1$ over $\mathbb{Q}$ is the subfield $E := \mathbb{Q}(\zeta)$ of $\mathbb{C}$, where $\zeta = e^{2\pi i/n}$. Indeed, $X^n - 1 = \prod_{j=1}^{n}(X - \zeta^j)$ in $E[X]$, because $1, \zeta, \zeta^2, \ldots, \zeta^{n-1}$
   are all distinct.

Definition 4. An *algebraic* field extension E/K is called *normal* if every *irreducible* polynomial $f \in K[X]$ that has *some* root in E splits into linear factors over E (in other words, E contains a splitting field of f over K).

Theorem 4. *Let E/K be an **algebraic** field extension, and let C be an algebraic closure of E (and therefore also of K). The following statements are equivalent:*

   (i) *E/K is normal.*

   (ii) *For every homomorphism $\sigma : E/K \to C/K$ we have $\sigma E = E$ (that is, σ can be regarded as an automorphism of E/K).*

   (ii') *Every automorphism of C/K restricts to an automorphism of E/K.*

   (iii) *E is a splitting field over K; that is, there exists a set $M \subseteq K[X]$ of nonconstant polynomials such that $E = K(N)$, where N denotes the set of all roots of polynomials $f \in M$ in C.*

Proof. (iii) $\Rightarrow$ (ii'): Let $\sigma : C/K \to C/K$ be an isomorphism. Then $\sigma(N) \subseteq N$, so $\sigma(K(N)) \subseteq K(N)$, that is, $\sigma E \subseteq E$. Now F3 shows that $\sigma E = E$.

(ii') $\Rightarrow$ (ii): By Theorem 3, a homomorphism $\sigma : E/K \to C/K$ can be extended to a homomorphism $\tau : C/K \to C/K$. By F3, τ is an automorphism of C/K.

(ii) $\Rightarrow$ (i): Let $f \in K[X]$ be irreducible and suppose $f(\alpha) = 0$ for some $\alpha \in E$. We must show that all roots of f in C already lie in E. So suppose $f(\beta) = 0$ for some $\beta \in C$. By F1 there is a K-isomorphism $\sigma : K(\alpha) \to K(\beta)$ taking α to β. By Theorem 3, σ can be extended to a homomorphism $\tau : E \to C$. By (ii) we have $\tau E = E$. In particular $\beta = \sigma\alpha = \tau\alpha$ is an element of E.

(i) $\Rightarrow$ (iii): Set $M = \{\mathrm{MiPo}_K(\alpha) \mid \alpha \in E\}$ and let N be the set of $\beta \in C$ that are roots of polynomials $f \in M$. By definition, $E \subseteq N$. If E/K is normal, we have $N \subseteq E$. Putting it all together we get $E = N$, so $E = K(N)$.   ·□

Remark. If E/K is a *finite normal* extension, we can obviously take $M = \{f\}$ in (iii), with an appropriate choice of $f \in K[X]$. (However in general it cannot be stipulated at the same time that f be irreducible.)

As a consequence of implication (iii) $\Rightarrow$ (i) of Theorem 4, we can state explicitly:

F5. *Let $f \in K[X]$ be a nonconstant polynomial and let E be the splitting field of f over K. Then E/K is normal.*

The proof of the next statement is left to the reader as an exercise.

F6. *Let E/K be an algebraic extension. There is an extension E' of E with these properties:*

(i) *E'/K is normal.*

(ii) *If L is an intermediate field of E'/E and L/K is normal, $L = E'$.*

*Such an E' is called a **normal closure** of E/K. Any two normal closures of E/K are isomorphic as K-extensions. If E/K is finite, so is E'/K.*

2. We now equip ourselves with an important algebraic tool, which we will use to prove Theorem 1 among other things. Fix a field K. If M is a nonempty set, define

$$KM := K^{(M)} = \{ f : M \to K \mid f(\mu) = 0 \text{ for almost all } \mu \in M \}.$$

There is a natural K-vector space structure on KM, with a canonical basis $(e_\mu)_{\mu \in M}$ populated by the characteristic functions e_μ of one-point sets $\{\mu\} \subset M$ (defined by $e_\mu(\nu) = 1$ if $\mu = \nu$ and $e_\mu(\nu) = 0$ otherwise). With the identification $e_\mu = \mu$, every $f \in KM$ has a unique representation of the form

$$(3) \qquad\qquad f = \sum_{\mu \in M} c_\mu \mu \quad \text{with } c_\mu \in K,$$

where $c_\mu = 0$ for almost all $\mu \in M$.

Now give M a *monoid* structure, that is, an associative operation (written multiplicatively) having an identity element. Then KM acquires a natural *K-algebra structure*, whereby the multiplication $(\mu, \nu) \mapsto \mu\nu$ is extended distributively to all of KM. (When $M = G$ is a group we call KG the *group algebra* of G over K.)

We consider an application. Let $(A_i)_{i \in I}$ be a family of K-algebras with unity, where $I \neq \varnothing$. Set

$$M = \{ (a_i)_{i \in I} \mid a_i \in A_i, \ a_i = 1_{A_i} \text{ for almost all } i \}.$$

By setting $(a_i)_i (b_i)_i = (a_i b_i)_i$ we make M into a *monoid*, whose corresponding *monoid algebra* we denote by KM. Elements of KM have unique representations of the form

$$\sum_{\alpha = (a_i) \in M} c_\alpha \alpha, \quad c_\alpha \in K,$$

with $c_\alpha = 0$ for almost all $\alpha \in M$. We now wish to construct a *quotient algebra* of KM where certain relations are obeyed. Take the K-subspace U of KM generated by all elements of the form

(4) $(a_i) + (b_i) - (s_i)$, where $a_j + b_j = s_j$ for some $j \in I$ and $a_i = b_i = s_i$ for all other i's and those of the form

(5) $(a_i) - c(b_i)$, where $a_j = cb_j$ for some $j \in I$ and $a_i = b_i$ for all other i's.

It is easy to see that U is in fact an *ideal* of KM, so we can take the *quotient* of KM modulo U. We denote this quotient by

(6)
$$\bigotimes_{i \in I} A_i = KM/U.$$

If π is the quotient map, we set

$$\bigotimes_i a_i := \pi(\alpha) \quad \text{for } \alpha = (a_i).$$

If $I = \{1, 2, \ldots, n\}$, we also write

$$A_1 \otimes \cdots \otimes A_n := \bigotimes_{i \in I} A_i, \qquad a_1 \otimes \cdots \otimes a_n := \bigotimes_i a_i.$$

The K-algebra (6) is called the *tensor product* of the K-algebras A_i, for $i \in I$. All its elements have the form

$$\sum_{\alpha = (a_i) \in M} c_\alpha \left(\bigotimes_i a_i \right),$$

but this representation is no longer unique in general. For each $j \in I$ there is a map

$$\sigma_j : A_j \to \bigotimes_i A_i$$

taking $a \in A_j$ to the element $\bigotimes_i a_i$ defined by $a_j = a$ and $a_i = 1_{A_i}$ for $i \neq j$. By definition, $\sigma_j(1_{A_j})$ is the unity element in $\bigotimes_i A_i$, and for all $a, b \in A_j$ we have $\sigma_j(ab) = \sigma_j(a)\sigma_j(b)$. As can easily be seen from (4) and (5), we also have $\sigma_j(a + b) = \sigma_j(a) + \sigma_j(b)$ and $\sigma_j(ca) = c\sigma_j(a)$ for $c \in K$. *Thus every σ_j is a homomorphism of K-algebras.* This whole construction was aimed at showing that (6) yields a *K-algebra*, which, together with the σ_j defined above, enjoys certain functorial properties:

F7. *Let $(\varphi_i)_{i \in I}$ be a family of K-algebra homomorphisms $\varphi_i : A_i \to A$. Then if*

(7)
$$\varphi_i(a)\varphi_j(b) = \varphi_j(b)\varphi_i(a)$$

for all $i \neq j$ and all $a \in A_i, b \in A_j$, there exists a unique homomorphism of K-algebras

$$\varphi : \bigotimes_{i \in I} A_i \to A$$

such that $\varphi \circ \sigma_i = \varphi_i$ for all $i \in I$.

Proof. (a) Suppose φ has already been found. Then

$$\varphi \left(\bigotimes_i a_i \right) = \varphi \left(\prod_i \sigma_i(a_i) \right) = \prod_i \varphi(\sigma_i(a_i)) = \prod_i \varphi_i(a_i),$$

so φ is uniquely determined. Note that the expression $\prod_i \varphi_i(a_i)$ is well-determined because of assumption (7).

(b) It is clear from (a) how φ is to be defined; but we must prove that the definition is consistent. The map $(a_i) \mapsto \prod_i \varphi_i(a_i)$ from M into A is multiplicative and thus can be extended to a homomorphism of K-algebras $\psi : KM \to A$ satisfying

$$\psi\left(\sum_\alpha c_\alpha \alpha\right) = \sum_{\alpha=(a_i)} c_\alpha \left(\prod_i \varphi_i(a_i)\right).$$

It is easy to ascertain that elements of the form (4) and (5) are in the kernel of ψ. Thus $\psi(U) = 0$, so ψ gives rise to a K-algebra homomorphism $\varphi : KM/U \to A$ such that

(8) $$\varphi(\textstyle\bigotimes_i a_i) = \prod_i \varphi_i(a_i).$$

In particular, $\varphi(\sigma_j(a)) = \varphi_j(a)$ for all $a \in A_j$, which proves the assertion. $\square$

Remark. All of this continues to work if K is a commutative ring with unity rather than a field. The only change is that when K is a ring we talk of K-*modules* rather than K-vector fields (though the name K-algebra remains). Now, not all K-modules are lucky enough to be *free*, that is, to contain a set of K-linearly independent elements whose K-linear combinations make up the whole module. If such a set exists, it is called a *basis*. Our KM does have a basis — in fact a canonical basis, the set $(e_\mu)_{\mu \in M}$ (page 60). Thus KM is a free K-module. Other constructions so far in this section also work just as well in this more general setting, as the reader should check, with only the change that U is a K-submodule of KM. But watch out: our vector space intuition does not work so well with modules, and it is possible, for example, for U to be the whole of KM! So the tensor product (6) might be the zero ring, even if all the A_i are nonzero. This does not happen, however, in the following special situation:

F8. *Let K be a commutative ring with unity and suppose $(A_i)_{i \in I}$ are K-algebras such that each A_i has a K-basis M_i with $1_{A_i} \in M_i$. (In particular, the A_i are free K-modules.) Then the family of elements of the form*

(9) $$\textstyle\bigotimes_i b_i \quad \text{with } b_i \in M_i \text{ and } b_i = 1_{A_i} \text{ for almost all } i$$

is a K-basis of the tensor product $\bigotimes_{i \in I} A_i$.

Proof. Clearly the elements (9) span $\bigotimes_{i \in I} A_i$ as a K-module. Let $\alpha = \bigotimes_i a_i$ be a fixed but arbitrary element among those in (9). For each $i \in I$, let $f_i : A_i \to K$ be the *linear functional* that assigns to each $a \in A_i$ the coordinate of a corresponding to the basis element a_i. Then consider the K-linear map

$$h_\alpha : \bigotimes_i A_i \to K$$

such that $h_\alpha(\bigotimes_i x_i) = \prod_i f_i(x_i)$. This map is well-defined. For an arbitrary element $\beta = \bigotimes_i b_i$ of the form (9) we have

$$h_\alpha(\beta) = \prod_i f_i(b_i) = \begin{cases} 0 & \text{if } (b_i) \neq (a_i), \\ 1 & \text{if } (b_i) = (a_i). \end{cases}$$

Now suppose we have a linear dependence

$$\sum_{\beta=(b_i)} c_\beta \left(\bigotimes_i b_i \right) = 0,$$

where the sum is over the elements $\beta = (b_i)_i$ in (9). Applying h_α we see that $c_\alpha = 0$. Thus the elements (9) are linearly independent over K.   □

Here is an immediate consequence of F8:

F9. *We keep the assumptions of* F8. *The multiplicative identity* $1 = \bigotimes_i 1_{A_i}$ *of* $\bigotimes_i A_i$ *is nonzero. For every* j *the map* $\sigma_j : A_j \to \bigotimes_i A_i$ *is injective; hence* A_j *can be regarded as a subalgebra of* $\bigotimes_i A_i$.

Example. Let $\mathscr{X}$ be a nonempty set. For each $X \in \mathscr{X}$, let $A_X := K[X]$ be the polynomial ring in one variable X over K. Then

$$K[\mathscr{X}] := \bigotimes_{X \in \mathscr{X}} A_X$$

is called the *polynomial ring in the variables* $X \in \mathscr{X}$ *over* K. The elements of $K[\mathscr{X}]$ are called *polynomials in the variables* $X \in \mathscr{X}$. We regard $K[X]$ as a subalgebra of $K[\mathscr{X}]$. In view of F8, the family of *monomials*

$$M_\nu(\mathscr{X}) := \prod_X X^{\nu(X)}, \quad \text{with } \nu = (\nu(X))_X \in (\mathbb{N} \cup \{0\})^{(\mathscr{X})},$$

forms a basis of $K[\mathscr{X}]$.

If $\mathscr{X} = \{X_1, \dots, X_n\}$ has n elements, we set $K[X_1, \dots, X_n] := K[\mathscr{X}]$; in this case the basis representation of an $f \in K[\mathscr{X}]$ has the form

$$f = \sum_{\nu=(\nu_1,\dots,\nu_n)} c_\nu X_1^{\nu_1} X_2^{\nu_2} \dots X_n^{\nu_n}.$$

The following functorial property comes directly from the definition of $K[\mathscr{X}]$ together with F7:

F10. *Let* K *be a commutative ring with unity and* $\mathscr{X}$ *a nonempty set. If* A *is a commutative* K-algebra, *any map* $\mathscr{X} \to A$ *can be uniquely extended to a homomorphism of* K-algebras $\varphi : K[\mathscr{X}] \to A$ *(the substitution homomorphism).*

If $\mathscr{X} = \{X_1, \dots, X_n\}$ and $\alpha_i := \varphi(X_i)$, denote by $f(\alpha_1, \dots, \alpha_n)$ the image $\varphi(f)$ of a polynomial $f \in K[X_1, \dots, X_n]$ under φ. In particular, $f(X_1, \dots, X_n) = f$. Note also that $K[X_1, \dots, X_n] = K[X_1, \dots, X_{n-1}][X_n]$.

3. We now turn to an important algebraic application of *Zorn's Lemma.*

Zorn's Lemma. *Let* M *be a (partially) ordered set in which every chain has an upper bound. Then* M *has a maximal element.*

Recall that a set M is called *partially ordered*, or simply *ordered*, if it is endowed with a relation $\leq$ satisfying the following properties: (i) $a \leq a$; (ii) $a \leq b$ and $b \leq c$ imply $a \leq c$; (iii) $a \leq b$ and $b \leq a$ imply $a = b$. Let $(M, \leq)$ be an ordered set. A subset N of M is a *chain* if it is *totally ordered* with the order induced from M; that is, if for $a, b \in N$ we either have $a \leq b$ or $b \leq a$. Again let $N \subseteq M$. An element $a \in M$ is an *upper bound for* N if $x \leq a$ for every $x \in N$. Finally, $m \in M$ is a *maximal element of* M if any $m' \in M$ such that $m \leq m'$ actually equals m.

We take Zorn's Lemma as a well-known fundamental statement of set theory; see for example the Wikipedia entry at http:/en.wikipedia.org/wiki/Zorn's_lemma. As a typical example of its application, consider:

F11. *Let V be a vector space over a field K, and let T be a linearly independent subset of V. Then V has a basis B such that $T \subseteq B$.*

Proof. The set M of all linearly independent subsets of V containing T is ordered by inclusion. It is also nonempty, since $T \in M$. Let $N \neq \varnothing$ be a chain in M. The union U of all $Y \in N$ is then linearly independent as well. Otherwise there would exist a finite subset A of U exhibiting a linear dependence; because A is finite and N is a chain, there would exist $Y \in N$ such that $A \subseteq Y$. But this is impossible, since Y was assumed linearly independent.

By *Zorn's Lemma*, then, there exists a maximal linearly independent subset B of V such that $T \subseteq B$. But such a set must be a basis for V. (Why?)   □

The following result is important in our context:

F12. *Let I be an ideal in a ring R with unity. If $I \neq R$, there exists a maximal ideal of R containing I.*

Proof. Consider the set M of ideals J of R such that $1 \notin J \supseteq I$, and order M by inclusion. M is nonempty since it contains I. If $N \neq \varnothing$ is a chain in M, consider the union $\tilde{J}$ of all $J \in N$. One checks easily that $\tilde{J} \in M$; also $\tilde{J}$ is obviously an upper bound for N. By Zorn's Lemma, then, M has a maximal element J. From the definitions and the fact $J \neq R$ (since $1 \notin J$) we see that J is a *maximal ideal* of R and J contains I.   □

Remark. For noncommutative rings, the exact same proof yields the corresponding statement about left ideals (instead of two-sided ideals).

We are finally ready to pick up our long-awaited proof:

Proof of Theorem 1. Consider the *tensor product* $A = \bigotimes_{i \in I} E_i$ of the extensions E_i of K (regarded as K-algebras), together with the corresponding homomorphisms of K-algebras $\sigma_i : E_i \to A$. Then A is a (commutative) *K-algebra* with $1 \neq 0$ (see F9; but in general A is not a field). By F12, there exists a *maximal ideal* M in A. Therefore A/M is a *field*, by F14 in Chapter 4. Being a K-algebra, A/M can be regarded as an extension E of K. The desired K-homomorphisms $\tau_i : E_i \to E$ are obtained by composing the σ_i with the quotient map $A \to A/M = E$. It is clear that E arises from K by adjunction of the union of the sets $\tau_i E_i$.   □

7

Separable Extensions

1. Definition 1. Let K be a field and C an algebraic closure of K. Two elements $\alpha, \beta \in C$ are called *conjugate over* K if there is an automorphism σ of C/K with $\sigma(\alpha) = \beta$. The elements of C conjugate to $\alpha \in C$ over K are called the *K-conjugates* of α (in C).

F1. *With the notations of Definition 1, the following statements are equivalent:*

(i) *β and α are conjugate over K in C.*

(ii) *β is a root of* $\mathrm{MiPo}_K(\alpha)$.

(iii) *There is an isomorphism* $\tau: K(\alpha)/K \to K(\beta)/K$ *such that* $\tau(\alpha) = \beta$.

(iv) $\mathrm{MiPo}_K(\alpha) = \mathrm{MiPo}_K(\beta)$.

In particular, any $\alpha \in C$ has at most $[\alpha : K] = K(\alpha) : K$ distinct K-conjugates in C.

Proof. (iii) $\Rightarrow$ (i): Extend τ to an automorphism σ of C/K (Chapter 6, Theorem 3 and F3).

(i) $\Rightarrow$ (ii): By assumption there is an automorphism σ of C/K such that $\sigma(\alpha) = \beta$. Set $f = \mathrm{MiPo}_K(\alpha)$. Then $f(\beta) = f(\sigma\alpha) = 0$ (Chapter 6, F1).

(ii) $\Rightarrow$ (iv): Set $f = \mathrm{MiPo}_K(\alpha)$ and suppose $f(\beta) = 0$. Since f is irreducible and normalized, $f = \mathrm{MiPo}_K(\beta)$.

(iv) $\Rightarrow$ (iii): This follows from Chapter 6, F1(b).

The last statement of F1 is clear, since $f = \mathrm{MiPo}_K(\alpha)$ can have at most $[\alpha : K] = \deg f$ distinct zeros in C. □

F2. *If K is a field and C an algebraic closure of K, the following statements are equivalent for a given $\alpha \in C$ with $f = \mathrm{MiPo}_K(\alpha)$:*

(i) *α has exactly n distinct K-conjugates in C.*

(ii) *There are exactly n distinct homomorphisms from $K(\alpha)/K$ into C/K.*

(iii) *f has exactly n distinct roots in C.*

Proof. A homomorphism $\tau: K(\alpha)/K \to C/K$ is determined by the image $\tau(\alpha)$. Thus F2 follows from F1. □

Definition 2. Let E/K be a field extension and suppose $\alpha \in E$ is algebraic over K. The number of distinct roots of $f = \text{MiPo}_K(\alpha)$ in any splitting field of f over K is called the *separable degree* of α over K, and is denoted by

$$[\alpha : K]_s.$$

An element α is said to be *separable over* K if $[\alpha : K]_s = [\alpha : K]$; otherwise it is *inseparable over* K. Thus α is separable over K if and only if its minimal polynomial over K only has simple roots (in its splitting field).

Remarks. (a) Clearly, $[\alpha : K]_s \leq [\alpha : K]$.

(b) $[\alpha : K]_s$ is the number of K-conjugates of α in an algebraic closure C of K.

(c) Any $\alpha \in K$ is separable over K.

So far, so clear. Now we may ask: If α is separable over K, is every $\beta \in K(\alpha)$ also separable over K?

Definition 3. An *algebraic field extension* E/K is called *separable* if every element of E is separable over K; otherwise E/K is *inseparable*.

We also agree on the following conventions: If $E_1/K, E_2/K$ are field extensions, we denote by

$$G(E_1/K, E_2/K)$$

the set of all homomorphisms $E_1/K \to E_2/K$ (see Chapter 6, Definition 1). When $E_1 = E_2 = E$ we use the abbreviation $G(E/K) := G(E/K, E/K)$.

Remark. If E/K is *algebraic*, $G(E/K)$ has a natural group structure (Chapter 6, F3). We thus obtain the *automorphism group* of the algebraic extension E/K.

Theorem 1. *Let E/K be a* **finite** *extension, of degree n, and let C be an algebraic closure of K.*

(I) $G(E/K, C/K)$ *has at most n elements.*

(II) $G(E/K, C/K)$ *has n elements if and only if E/K is separable.*

We postpone for a while the proof of this important theorem; first we bring its content to bear:

Definition 4. Let E/K be an algebraic field extension and C an algebraic closure of K. Then

$$[E : K]_s = |G(E/K, C/K)|$$

is called the *separable degree* of E/K. (This number is independent of the choice of C; see Chapter 6, Theorem 2(II).)

Remarks. Let E/K be *any* field extension, but assume $\alpha \in E$ algebraic over K. Then

(1) $$[K(\alpha) : K]_s = [\alpha : K]_s$$

by F1 (see also Definition 2 and Remark (b) following it). It follows further that

(2) α is separable over K $\iff$ $K(\alpha)/K$ is separable.

For, by definition, α is separable over K if and only if $[\alpha : K]_s = [\alpha : K]$; and because of (1), this is equivalent to $[K(\alpha) : K]_s = [K(\alpha) : K]$. But, by part (II) of Theorem 1, this latter equality holds if and only if $K(\alpha)/K$ is separable. Note that (2) answers in the affirmative the question posed just before Definition 3.

We base the proof of Theorem 1 on the following result:

Lemma. *Let F be an intermediate field of an algebraic extension E/K and let C be an algebraic closure of E. Then there exists a* **bijection**

$$G(E/K, C/K) \longleftrightarrow G(F/K, C/K) \times G(E/F, C/F).$$

Proof. By Chapter 6, Theorem 3, there exist maps

$$G(F/K, C/K) \to G(C/K) \qquad \qquad G(E/F, C/F) \to G(C/F)$$
$$\sigma \mapsto \tilde{\sigma} \qquad\qquad\text{and}\qquad\qquad \tau \mapsto \tilde{\tau}$$

with $\tilde{\sigma}_F = \sigma$ and $\tilde{\tau}_E = \tau$. We claim that the map

$$G(F/K, C/K) \times G(E/F, C/F) \to G(E/K, C/K)$$
$$(\sigma, \tau) \mapsto (\tilde{\sigma}\tilde{\tau})_E$$

is bijective. First we show injectivity: Suppose $(\tilde{\sigma}\tilde{\tau})_E = (\tilde{\sigma}_1\tilde{\tau}_1)_E$. By restriction to F we see first that $\tilde{\sigma}_F = (\tilde{\sigma}_1)_F$, then that $\sigma = \sigma_1$ and therefore $\tilde{\sigma} = \tilde{\sigma}_1$. From the assumption it then follows that $\tilde{\tau}_E = (\tilde{\tau}_1)_E$, which is to say $\tau = \tau_1$.

To prove surjectivity it is enough (by Chapter 6, Theorem 3) to prove that, for every $\psi \in G(C/K)$, there exist σ and τ as above, such that $\psi_E = (\tilde{\sigma}\tilde{\tau})_E$. For a given ψ set $\sigma := \psi_F$ and then $\tau := \tilde{\sigma}^{-1}\psi_E$. Then $\sigma \in G(F/K, C/K)$ and $\tau \in G(E/F, C/F)$, because $\tilde{\sigma}^{-1}\psi_E$ fixes F pointwise. By definition $\tilde{\sigma}^{-1}\psi_E = \tilde{\tau}_E$, so $\psi_E = \tilde{\sigma}\tilde{\tau}_E = (\tilde{\sigma}\tilde{\tau})_E$.   □

Proof of Theorem 1. We work by induction on $n = E : K$. For $n = 1$ the assertion is clear. Suppose $n > 1$; then there exists $\alpha \in E \setminus K$. For $F = K(\alpha)$ we then have $E : F < n$. By the lemma,

(3) $|G(E/K, C/K)| = |G(F/K, C/K)| \cdot |G(E/F, C/F)|.$

The first factor on the right-hand side is at most $F : K$ because of F1 (see also Remark (a) following Definition 2). The second factor is at most $E : F$ by the induction hypothesis. Thus the left-hand side of (3) is worth at most $[F : K][E : F] = [E : K] = n$, proving assertion (I).

For assertion (II), assume E/K is separable. Trivially, F/K is separable; but also E/F is separable, because for each $\beta \in E$ the polynomial $\text{MiPo}_F(\beta)$, being a factor of MiPo$_K(\beta)$, has only simple roots. By the induction hypothesis and (3) we conclude that $\left|G(E/K, C/K)\right| = [F : K][E : F] = [E : K] = n$.

Conversely, assume $\left|G(E/K, C/K)\right| = n$. Taking (3) and (I) into account we see that

$$\left|G(F/K, C/K)\right| = F : K$$

An application of F2 to $F = K(\alpha)$ then shows that α is separable over K. Now recall that this holds for any $\alpha \in E \notin K$. Thus E/K is separable (because each $\alpha \in K$ is separable over K). □

In view of Definition 4 we can reformulate Theorem 1 as follows:

Theorem 1′. *Let E/K be a finite field extension. Then $[E : K]_s$ is finite, and*

(4) $$[E : K]_s \leq [E : K].$$

Equality holds if and only if E/K is separable.

Remark. $[E : K]_s$ actually divides $[E : K]$, as can be proved by induction with a bit more effort. But this result will follow more easily from F17 below.

F3. *If F is an intermediate field of an algebraic extension E/K,*

(5) $$[E : K]_s = [E : F]_s \cdot [F : K]_s.$$

Proof. This is an immediate consequence of the lemma. □

F4. *Given an extension E/K, there is equivalence between:*

  (i) *E/K is finite and separable.*

 (ii) *E is generated over K by finitely many separable algebraic elements $\alpha_1, \ldots, \alpha_m \in E$.*

Proof. (ii) ⇒ (i): E/K is clearly finite (Chapter 2, F7). By induction on m and using Theorem 1′ we get $E : K = [K(\alpha_1, \ldots, \alpha_m) : K(\alpha_1)]_s \cdot [K(\alpha_1) : K]_s$, since α_1 is separable over K. Then F3 implies that $[E : K] = [E : K]_s$, so E/K is separable by Theorem 1′.

The implication (i) ⇒ (ii) is obvious. □

F5. *Consider an algebraic extension E/K. Then*

$$E_s = \{\alpha \in E \mid \alpha \text{ separable over } K\}$$

*is an intermediate field of E/K, called the **separable closure of K in E**.*

Proof. For $\alpha, \beta \in E_s$, consider the subextension $K(\alpha, \beta)/K$ of E/K. By F4 this is separable; therefore $K(\alpha, \beta) \subseteq E_s$, and the assertion follows. □

F6. *Let E/K be an algebraic field extension and let A be a subset of E. If A only contains elements separable over K, the extension $K(A)/K$ is separable.*

Proof. By F5 we have $K(A) \subseteq E_s$. $\square$

F7 (Transitivity of separability). *Let L be an intermediate field of an algebraic field extension E/K. If both E/L and L/K are separable, so is E/K (and conversely).*

Proof. In the case of finite extensions the assertion is clear; see F3 and Theorem 1′. The general case is reduced to the finite case as follows. For $\beta \in E$, let $\alpha_0, \alpha_1, \ldots, \alpha_{n-1}$ be the coefficients of $f = \mathrm{MiPo}_L(\beta)$. Then f is also the minimal polynomial of β over the subfield $F := K(\alpha_0, \alpha_1, \ldots, \alpha_{n-1})$. Since β is separable over L by assumption, f has only simple roots. This implies the separability of $F(\beta)/F$. But in view of F4, F/K too is separable, because by assumption $\alpha_0, \alpha_1, \ldots, \alpha_{n-1}$ are separable over K. $\square$

2. And now at long last:

Definition 5. Let K be a field. A polynomial $f \in K[X]$ of degree $n \geq 1$ is called *separable* if it has n distinct roots in the splitting field of f over K.

Remarks. (a) Let L/K be an extension and let $\alpha \in L$ be algebraic over K. In view of the definition, α is separable over K if and only if the minimal polynomial of α is separable over K (see F2).

(b) Let $f \in K[X]$ be any nonconstant polynomial and let E by a splitting field of f over K. The prime factorization of f in $E[X]$ has the form

$$f(X) = \gamma(X - \alpha_1)^{e_1}(X - \alpha_2)^{e_2} \cdots (X - \alpha_r)^{e_r},$$

with the α_i all distinct. Then f is separable if and only if all the e_i equal 1, that is, f has no multiple roots.

(c) In analysis, the differential calculus is a useful tool for dealing with multiple roots. In algebra we make do with the following formal differential calculus on polynomials: Given a polynomial $f = \sum_{i=0}^{n} a_i X^i$ in $K[X]$, the (*formal*) *derivative* of f is $f' := \sum_{i=1}^{n} i a_i X^{i-1}$. The map $K[X] \rightarrow K[X]$ defined by $f \mapsto f'$ is obviously linear:

$$(af + bg)' = af' + bg',$$

and satisfies the *product rule* with respect to multiplication:

$$(fg)' = f'g + g'f.$$

(Because of linearity this just has to be verified in the case that $f = X^i$ and $g = X^j$ are monomials.)

F8. *With the preceding notation, suppose that $f(\alpha) = 0$ for some $\alpha \in E$. Then α is a multiple root of f if and only if $f'(\alpha) = 0$.*

Proof. By assumption we have $f(X) = (X - \alpha)^e g(X)$ in $E[X]$, with $e \geq 1$ and $g(\alpha) \neq 0$. Differentiation gives

$$f'(X) = e(X - \alpha)^{e-1} g(X) + (X - \alpha)^e g'(X).$$

Substituting α of X shows that $f'(\alpha) = 0 \iff e \geq 2$. $\square$

F9. *A polynomial $f \in K[X]$ is separable if and only if f and f' are relatively prime in $K[X]$.*

Proof. Let E be a splitting field of f over K. By F8, f being separable is equivalent to f and f' being relatively prime in $E[X]$. But for polynomials in $K[X]$, relative primeness in $E[X]$ is the same as in $K[X]$; see §4.7 in the Appendix.   □

F10. *If $f \in K[X]$ is irreducible, then f is separable if and only if $f' \neq 0$.*

Proof. We use F9. If f, f' are relatively prime, $f' \neq 0$. Now suppose instead that f, f' are not relatively prime in $K[X]$. Because f is irreducible, f must divide f'; but since deg $f' <$ deg f, this is only possible if $f' = 0$.   □

F11. *Let K be a field of characteristic 0. Every irreducible polynomial $f \in K[X]$ is separable. Thus every algebraic extension E/K is separable.*

Proof. Write $f = \sum_{i=0}^{n} a_i X^i$, so that $f' = \sum_{i=1}^{n} i a_i X^{i-1}$. Saying that $f' = 0$ is the same as saying that

$$(6) \qquad\qquad i a_i = 0 \quad \text{for all } 1 \leq i \leq n.$$

But if char $K = 0$ this condition is only satisfied if $a_1 = a_2 = \ldots = a_n = 0$, that is, if $f = a_0$ is a constant polynomial.   □

3. We now turn to the case char $K = p > 0$. Condition (6) is then equivalent to

$$a_i = 0 \quad \text{for all } i \text{ such that } i \not\equiv 0 \bmod p.$$

As a consequence:

F12. *If char $K = p > 0$,*

$$(7) \qquad\qquad f' = 0 \iff f \in K[X^p]$$

for $f \in K[X]$. Thus, if f is assumed irreducible, it is separable if and only if does not lie in $K[X^p]$.

Thus, over a field of characteristic $p > 0$ there exist nonconstant polynomials with zero derivative; any polynomial of the form

$$c_0 + c_1 X^p + c_2 X^{2p} + \cdots + X^{kp}$$

(for $k \geq 1$) has this property. (Whether there really are any irreducible polynomials in $K[X^p]$ is a question we have not broached yet.)

F13. *Let E/K be a field extension with char $K = p > 0$. If $\alpha \in E$ is algebraic over K, there exists an integer $m \geq 0$ such that α^{p^m} is separable over K.*

Proof. Set $f = \text{MiPo}_K(\alpha)$. Clearly there is an integer $m \geq 0$ such that $f \in K[X^{p^m}]$ but $f \notin K[X^{p^{m+1}}]$. Thus there is a polynomial $g \in K[X]$ such that

(8) $$f(X) = g(X^{p^m}) \quad \text{but} \quad g \notin K[X^p].$$

We wish to show that g is irreducible. Suppose $g(X) = h_1(X)h_2(X)$ in $K[X]$. Then $g(X^{p^m}) = h_1(X^{p^m})h_2(X^{p^m})$. But f in (8) is irreducible, so $h_1 \in K^\times$ or $h_2 \in K^\times$. Since $g(\alpha^{p^m}) = f(\alpha) = 0$ we then have $g = \text{MiPo}_K(\alpha^{p^m})$. By F12, g is separable. $\square$

F14. *For an arbitrary field K of characteristic $p > 0$, the map*

(9) $$\alpha \mapsto \alpha^p$$

of K into itself is a homomorphism; in particular, $\alpha^p = 1$ in K if and only if $\alpha = 1$.

Proof. Obviously, $(\alpha\beta)^p = \alpha^p\beta^p$ and $1^p = 1$. Also, by (15) in Chapter 5, $(\alpha+\beta)^p = \alpha^p + \beta^p$. For the last assertion see again Chapter 4, F15. $\square$

Remark. In the case of a *finite field* K of characteristic p, the injective map (9) is also surjective. Thus we get an *automorphism* σ_p of K. Consequently, for any power $q = p^n$ of p, the map $\sigma_q : \alpha \to \alpha^q$ is likewise an automorphism of K, since σ_q is the n-th power of σ_p in Aut K.

Definition 6. An algebraic extension E/K is called *purely inseparable* if any $\alpha \in E$ not belonging to K is inseparable over K.

F15. *Let E/K be purely inseparable with $E \neq K$ (which implies that char $K = p > 0$). Given any $\alpha \in E$ one can find values of $m \in \mathbb{N}$ such that $\alpha^{p^m} \in K$; if m is taken as the smallest such integer,*

(10) $$X^{p^m} - \alpha^{p^m}$$

is the minimal polynomial of α over K.

Proof. By F13 there exists $m \geq 0$ such that α^{p^m} is separable over K, and hence lies in K by the inseparability assumption. Let m be minimal with this property, and let $f(X)$ the polynomial (10). Then $f(X) \in K[X]$. But $f(X) = (X - \alpha)^{p^m}$ since char $K = p$, so if g is an irreducible factor of f in $K[X]$, the prime factorization of f in $K[X]$ must have the form

$$f(X) = g(X)^j, \quad \text{with } 1 \leq j \leq p^m.$$

But $p^m = \deg f = j \deg g$, so $\deg g = p^n$ with $n \leq m$. Then

$$g(X) = (X - \alpha)^{p^n} = X^{p^n} - \alpha^{p^n},$$

so $\alpha^{p^n} \in K$ because $g(X) \in K[X]$. But m was assumed minimal, so $n = m$. Thus $f(X) = g(X)$, and f is irreducible. $\square$

F16. *Let E/K be an algebraic extension and C an algebraic closure of E. The following statements are equivalent:*

  (i) *E/K is purely inseparable.*

  (ii) *$[E:K]_s = 1$.*

  (iii) *Every K-homomorphism $\sigma : E \to C$ fixes E pointwise.*

  (iv) *No $\alpha \in E$ is conjugate over K to $\beta \in C$ distinct from α.*

Proof. (i) $\Rightarrow$ (ii): Take $\sigma \in G(E/K, C/K)$ and $\alpha \in E$. By F15 we have $\gamma := \alpha^{p^m} \in K$ for some $m \in \mathbb{N}$. An application of σ yields

$$\sigma(\alpha)^{p^m} = \sigma(\alpha^{p^m}) = \sigma(\gamma) = \gamma = \alpha^{p^m},$$

which by F14 implies $\sigma(\alpha) = \alpha$. Thus $G(E/K, C/K)$ contains a single element, the inclusion $E \to C$.

(ii) $\Rightarrow$ (iii) and (iii) $\Rightarrow$ (iv) are clear from the definitions.

(iv) $\Rightarrow$ (i): Take $\alpha \in E$ and $f = \mathrm{MiPo}_K(\alpha)$. All roots of f in C are conjugate to α over K. By the hypothesis, then, α can only be separable over K if $\deg f = 1$, that is, if $\alpha \in K$.   □

Remark. Let F be an intermediate field of an algebraic extension E/K. By F3 we have $[E:K]_s = [E:F]_s \cdot [F:K]_s$. It follows from this and F16 that *E/K is purely inseparable if and only if E/F and F/K are purely inseparable.*

F17. *Let E/K be a algebraic extension and E_s the separable closure of K in E.*

  (a) *E/E_s is purely inseparable.*

  (b) *If E_s/K is finite, so is $[E:K]_s$; moreover $[E:K]_s = E_s:K$. In particular, if E/K is a finite extension, $[E:K]_s$ divides $E:K$.*

  (c) *If E/E_s is finite, $E:E_s$ is a power of $p = \operatorname{char} K$.*

Proof. (a) Let $\alpha \in E$ be separable over E_s. Then $E_s(\alpha)/E_s$ is separable, and by F7 so is $E_s(\alpha)/K$. It follows that $E_s(\alpha) \subseteq E_s$, that is, $\alpha \in E_s$.

(b) By F3 we have $[E:K]_s = [E:E_s]_s \cdot [E_s:K]_s$. By part (a) and F16, the first factor must equal 1. The second coincides with $E_s:K$, since E_s/K is separable and, by assumption, finite. (By the way, from $[E:K]_s < \infty$ it follows conversely that $E_s:K < \infty$. Think about it for a while. A proof can be given using Theorem 3 in Chapter 8, and is left as an exercise; see also §8.22 in the Appendix.)

(c) We must show that the degree of a finite, purely inseparable extension E/K is a power of p. Take $\alpha \in E \smallsetminus K$. Since $E/K(\alpha)$, too, is purely inseparable (see the preceding Remark), it can be assumed by induction that $E:K(\alpha)$ is a power of p. By F15, $K(\alpha):K$ is also a power of p, and thus so is $E:K = [E:K(\alpha)] \cdot [K(\alpha):K]$.   □

Remark. Keep the notation of F17. Then

(11) $[E:K]_i := E:E_s$

is called the *inseparable degree* of E/K. (See also §7.9 in the Appendix.)

Definition 7. A field K is called *perfect* if every algebraic extension over K is separable.

By F11, every field of characteristic 0 is perfect. Assume char $K = p > 0$. Let

(12) $K^p := \{\alpha^p \mid \alpha \in K\}$

be the set of p-th powers of elements in K. By F14, this is a *subfield* of K.

F18. *If K is perfect and K'/K is an algebraic field extension, K' is also perfect.*

Proof. Let E/K' be any algebraic extension. Since E/K' and K'/K are algebraic, so is E/K. Since K is perfect, E/K is separable. Then E/K' is also separable. □

F19. *A field K with char $K = p > 0$ is perfect if and only if $K^p = K$. In particular, every finite field is perfect.*

Proof. Let K be perfect. For a given $\alpha \in K$, let E be a splitting field of the polynomial $f(X) = X^p - \alpha$ over K. In E this polynomial has a root, that is, there exists a $\beta \in E$ such that $\beta^p = \alpha$. Since the map $x \mapsto x^p$ of E into itself is a field homomorphism and hence injective, f has only this one root β in E (alternatively, this follows from the equality $X^p - \beta^p = (X - \beta)^p$). Thus $f(X) = (X - \beta)^p$.

Now set $g = \text{MiPo}_K(\beta)$. By assumption, g is separable. But g divides f, so it must be of the form $g(X) = X - \beta$. Thus $\beta \in K$.

Conversely, assume $K = K^p$. Suppose there is an inseparable irreducible polynomial $f \in K[X]$. By F12, it must have the form $f(X) = g(X^p)$, with $g \in K[X]$. Since $K = K^p$, we have $g(X) = \sum b_i^p X^i$ with $b_i \in K$. It follows that $f(X) = g(X^p) = \sum b_i^p X^{pi} = \left(\sum b_i X^i\right)^p$; that is, f is not irreducible in $K[X]$, contradicting the assumption.

Finally, if K is finite, $K^p = K$ by the Remark following F14.   □

Remark. As an example of a nonperfect field, consider $K = \mathbb{F}_p(t)$, the field of rational functions in one variable over the field $\mathbb{F}_p$ with p elements. Indeed, the polynomial $f(X) = X^p - t \in K[X]$ is irreducible (Eisenstein's criterion, F10 in Chapter 5), but not separable (F12). Associated with f is the purely inseparable extension $K(t^{1/p})/K$ of degree p (where $t^{1/p}$ is defined as the unique p-root of t in a fixed algebraic closure of K).

8

Galois Extensions

1. Let E be a field. If G is a group of automorphisms of E, the set

$$E^G := \{\alpha \in E \mid \sigma\alpha = \alpha \text{ for all } \sigma \in G\}$$

is called the *fixed field of G*. (It is clear that E^G really is a subfield of E.)

F1. *Let E be a field and G a **group** of automorphisms of E; denote by $K = E^G$ the fixed field of G. Take $\alpha \in E$. If the set*

$$G\alpha := \{\sigma\alpha \mid \sigma \in G\}$$

*is **finite**, α is algebraic over K. If $G\alpha$ contains exactly n distinct elements, say $\alpha_1, \alpha_2, \ldots, \alpha_n$ (one of them being α), the minimal polynomial of α over K is the separable, normalized polynomial*

$$
\text{(1)} \qquad\qquad f(X) = \prod_{i=1}^{n} (X - \alpha_i).
$$

Proof. Each $\tau \in G$ gives rise to a ring isomorphism $\tau : E[X] \to E[X]$, sending $g(X) = \sum b_i X^i$ to $g^\tau(X) = \sum \tau(b_i) X^i$. Thus the polynomial f in (1) satisfies

$$
\text{(2)} \qquad\qquad f^\tau(X) = \prod_{i=1}^{n} (X - \tau(\alpha_i)).
$$

But τ gives rise to a *permutation* of $\alpha_1, \alpha_2, \ldots, \alpha_n$, since for every $\sigma \in G$ we have $\tau(\sigma\alpha) = (\tau\sigma)\alpha \in G\alpha$, and plus τ is injective. Thus (2) implies that $f^\tau(X) = f(X)$ for every $\tau \in G$. By definition, then, all the coefficients of f already lie in the fixed field K of G. But α is a root of $f(X) \in K[X]$, and therefore algebraic over K.

Let $g = \mathrm{MiPo}_K(\alpha)$. Then in any case g divides f in $K[X]$. Since $g(\alpha) = 0$ we have $g(\sigma\alpha) = 0$ for every $\sigma \in G$, that is, $g(\alpha_i) = 0$ for $i = 1, 2, \ldots, n$. Thus g has at least degree n. It follows that $f = g$, since both polynomials are normalized. $\square$

Definition 1. An algebraic field extension E/K is a *Galois extension* if

(3) $$K = E^{G(E/K)}.$$

The group $G(E/K)$ is then called the *Galois group of E/K*.

Remarks. (a) For an arbitrary extension E/K, it is clear that $K \subseteq E^{G(E/K)}$.

(b) *Let G be a group of automorphisms of a field E, with fixed field $K = E^G$. If E/K is algebraic, it is a Galois extension.* This is because $G \subseteq G(E/K)$ by definition, and hence $E^{G(E/K)} \subseteq E^G = K$. Because of (a) equality (3) holds.

Theorem 1. *For an algebraic extension E/K there is equivalence between*:

(i) *E/K is Galois.*

(ii) *E/K is normal and separable.*

Proof. (ii) $\Rightarrow$ (i): Let α be any element of E not lying in K. We must show that there exists $\tau \in G(E/K)$ such that $\tau\alpha \neq \alpha$. Let C be an algebraic closure of E. By assumption, α is separable over K, so $[K(\alpha):K]_s = K(\alpha):K \neq 1$. Thus there exists $\sigma \in G(C/K)$ such that $\sigma\alpha \neq \alpha$ (see Chapter 7). Since E/K is normal, σ restricts to an automorphism $\tau \in G(E/K)$ by Chapter 6, Theorem 4. We thus have $\tau\alpha = \sigma\alpha \neq \alpha$ as required.

(i) $\Rightarrow$ (ii): Take $\alpha \in E$ and set $f = \text{MiPo}_K(\alpha)$. We must show that f is separable and that it splits into linear factors **over E**. But this follows immediately from F1, with $G = G(E/K)$. The necessary assumption that $\{\sigma\alpha \mid \sigma \in G\}$ be finite is satisfied since all the $\sigma\alpha$ are roots of f.   □

F2. *Let $f \in K[X]$ be a separable polynomial with splitting field E over K. Then E/K is a finite Galois extension. (Instead of assuming that f is separable, it is enough to assume that the prime factors of f in $K[X]$ are separable.)*

Proof. E/K is normal, according to F5 in Chapter 6. Let $\alpha_1, \ldots, \alpha_n$ be the roots of f in E. Then $E = K(\alpha_1, \ldots, \alpha_n)$. Each α_i is separable over K (see Definition 2 in Chapter 7). Thus E/K is finite and separable, by F4 in Chapter 7.   □

Theorem 2. *Let E/K be a Galois extension. For every intermediate field F of E/K, the extension E/F is also Galois*:

$$F = E^{G(E/F)}.$$

The map

$$F \mapsto G(E/F)$$

*from the set of intermediate fields of E/K into the set of subgroups of $G(E/K)$ is therefore **injective**.*

Proof. Clearly E/F is algebraic, separable and normal if E/K has each of these properties. Thus E/F is a Galois extension, by Theorem 1.   □

Is F/K a Galois extension in the situation of Theorem 2? In general, no. True, F/K is trivially separable, but it need not be normal, because it is not necessarily the case that $\sigma F \subseteq F$ for each $\sigma \in G(E/K)$.

F3. *If F is an intermediate field of a Galois extension E/K,*

$$(4) \qquad G(E/\sigma F) = \sigma G(E/F)\sigma^{-1} \quad \text{for any } \sigma \in G(E/K).$$

Moreover, the following statements are equivalent:

(i) F/K *is a Galois extension.*

(ii) $\sigma F = F$ *for all $\sigma \in G(E/K)$.*

(iii) $\sigma G(E/F)\sigma^{-1} = G(E/F)$ *for all $\sigma \in G(E/K)$.*

Proof. To prove (4), note that, for an arbitrary $\tau \in G(E/K)$,

$$\tau \in G(E/\sigma F) \iff \tau(\sigma\alpha) = \sigma\alpha \text{ for all } \alpha \in F \iff (\sigma^{-1}\tau\sigma)\alpha = \alpha \text{ for all } \alpha \in F$$
$$\iff \sigma^{-1}\tau\sigma \in G(E/F) \iff \tau \in \sigma G(E/F)\sigma^{-1}.$$

(ii) $\iff$ (iii): In view of (4) and Theorem 2 we have

$$\sigma F = F \iff G(E/F) = G(E/\sigma F) = \sigma G(E/F)\sigma^{-1}.$$

(i) $\iff$ (ii): In view of Theorem 1 we just have to prove that (ii) is equivalent to F/K being normal (F/K is separable in any case). Let C be an algebraic closure of E (and so also of F and K). We use the normality criterion (ii′) of Chapter 6, Theorem 4.

Any $\sigma \in G(E/K)$ extends to some $\tau \in G(C/K)$ (Chapter 6, Theorem 3 and F3). If F/K is normal we get $\tau F = F$, hence $\sigma F = F$.

Conversely, if $\sigma F = F$ for all $\sigma \in G(E/K)$, it is also true that $\tau F = F$ for all $\tau \in G(C/K)$, since any $\tau \in G(C/K)$ restricts to $\sigma \in G(E/K)$ (by the normality of E/K). Therefore F/K is normal. $\qquad\qquad\square$

The proof shows that if E/K and F/K are normal extensions with $F \subset E$, and C is an algebraic closure of E, there is a commutative diagram

(5)

$$G(C/K) \xrightarrow{\ r\ } G(E/K)$$

with well defined homomorphisms r, p, r', each of them surjective.

Definition 2. Let G be a group. A subgroup H of G is called *normal* if

$$\sigma H\sigma^{-1} = H \quad \text{for all } \sigma \in G.$$

Thus the notion of a normal subgroup comes up quite naturally in the study of fields; but it is also a key notion in group theory: see the remarks following the Fundamental Homomorphism Theorem (F3 in Chapter 3).

F4. *We keep the assumptions of F3.*

(a) *F/K is a Galois extension if and only if $G(E/F)$ is a normal subgroup of $G(E/K)$.*

(b) *If F/K is Galois, the natural map $p : G(E/K) \to G(F/K)$ gives rise to an isomorphism of groups*

(6) $$G(E/K)\big/G(E/F) \to G(F/K).$$

Proof. Part (a) is clear from F3 and Definition 2. We prove (b). We know that p is surjective. By definition,

$$\ker p = \{\sigma \in G(E/K) \mid \sigma x = x \text{ for all } x \in F\} = G(E/F).$$

We conclude by invoking the Fundamental Homomorphism Theorem for groups (remarks following F3 in Chapter 3). □

F5. *Let E' be a normal closure of an algebraic extension E/K see (Chapter 6, F6). If E/K is separable, so is E'/K (and therefore E'/K is Galois).*

Proof. Suppose $E = K(B)$. In an algebraic closure C of E', let A be the set of all roots of minimal polynomials of $\alpha \in B$. Then $E' = K(A)$. If E/K is separable, every element of A is separable over K. By F6 in Chapter 7, this implies E'/K separable. □

2. We now wish to study particularly the implications of finiteness.

F6. *If a Galois extension E/K is finite, the Galois group of E/K is also finite, with*

$$|G(E/K)| = E : K.$$

Proof. Let C be an algebraic closure of E. Since E/K is separable, $[E : K] = [E : K]_s = |G(E/K, C/K)|$. Since E/K is normal, we can identify $G(E/K, C/K)$ with $G(E/K, E/K) = G(E/K)$ (Chapter 6, Theorem 4). □

Theorem 3 (Primitive element theorem). *A field extension E/K that is **finite** and **separable** is also simple, that is, $E = K(\alpha)$ for some $\alpha \in E$.*

Proof. In view of Chapter 3, Theorem 5, it suffices to show that E/K has only finitely many intermediate fields. Let E' be a normal closure of E/K. If E'/K has only finitely many intermediate fields, so does E/K. By F5, E'/K is a Galois extension; it is also finite by F6 in Chapter 6. Thus we might as well assume that E/K is Galois to begin with. Then $G(E/K)$ is a finite group (F6) and as such has only finitely many subgroups. By Theorem 2 this means that E/K has only finitely many intermediate fields. □

Theorem 4. *Let E be a field and G a **finite group** of automorphisms of E, with fixed field $K = E^G$. The extension E/K is finite and Galois; moreover,*

$$G = G(E/K),$$

that is, G coincides with the Galois group of E/K.

Proof. In view of F1, E/K is algebraic, separable and normal. Thus it is a Galois extension. Let $d := |G|$ be the order of G. By F1 we know at first only that

$$(7) \qquad K(\alpha) : K \le d \quad \text{for each } \alpha \in E.$$

There is certainly some $\alpha \in E$ for which $K(\alpha) : K$ is maximal. Take any $\beta \in E$. By Theorem 3, there exists $\gamma \in E$ with $K(\alpha, \beta) = K(\gamma)$. From our choice of α we have $K(\gamma) : K \le K(\alpha) : K$; therefore $K(\alpha) = K(\gamma) \ni \beta$. Thus $E = K(\alpha)$. In particular, E/K is finite, and (7) becomes

$$(8) \qquad E : K \le d = |G|.$$

The inclusion $G \subseteq G(E/K)$ is trivial. But $|G(E/K)| = E : K$ by F6, so in view of (8) we must have $G = G(E/K)$. □

For another justification of (8) see §12.6 in the Appendix.

Theorem 5 (Fundamental theorem of Galois theory for finite Galois extensions). *Let E/K be a finite Galois extension. Then the map*

$$(9) \qquad F \mapsto G(E/F)$$

*is a **bijection** between the set of intermediate fields F of E/K and the set of subgroups H of $G := G(E/K)$. Each extension E/F is Galois and satisfies*

$$(10) \qquad E : F = |G(E/F)|.$$

Moreover there is an equivalence

$$(11) \qquad F_1 \subseteq F_2 \iff G(E/F_1) \supseteq G(E/F_2).$$

The inverse of (9) is the map

$$(12) \qquad H \mapsto E^H = \text{the fixed field of } H.$$

F/K is a Galois extension if and only if $G(E/F)$ is a normal subgroup of G. If F/K is Galois, one obtains by restriction a natural isomorphism

$$(13) \qquad G(F/K) \simeq G(E/K)/G(E/F).$$

Proof. Theorem 2 says that, for every intermediate field F of E/K,

$$(14) \qquad E^{G(E/F)} = F.$$

Since E/K was assumed finite, $G = G(E/K)$ is finite by F6, and hence so is any subgroup of G. By Theorem 4, then,

$$(15) \qquad G(E/E^H) = H$$

for any subgroup H of G. Thus, by (14) and (15), the maps (9) and (12) are indeed inverse to each other. The remainder of Theorem 5 is now clear (look again at Theorem 2, F6, and F4). □

3. As can easily be seen from Theorem 3, *if E/K is a finite Galois extension, E is the splitting field of a separable (and irreducible) polynomial $f \in K[X]$ over K.* By F2 the converse also holds: *If $f \in K[X]$ is a separable polynomial and E is a splitting field of f over K, then E/K is a finite Galois extension.* The following definition is therefore apposite and convenient (and is close to the original definition of a "Galois group"; see Évariste Galois, *Œuvres mathématiques*, Paris, 1897):

Definition 3. Take $f \in K[X]$. Assume that f is *separable* (or just that all prime factors of f are). Let E be a splitting field of f over K. The Galois group of the finite Galois extension E/K is also called the *Galois group of f over K*, or the *Galois group of the equation $f(X) = 0$ over K.*

Example. The Galois group of $X^3 - 2 = 0$ over $\mathbb{Q}$ is isomorphic to the symmetric group S_3.

   Proof: $X^3 - 2$ has three distinct roots $\alpha_1, \alpha_2, \alpha_3$ in $\mathbb{C}$. Let G be the Galois group of $X^3 - 2$ over $\mathbb{Q}$. Every $\sigma \in G$ permutes $\alpha_1, \alpha_2, \alpha_3$. Thus one gets a homomorphism

$$G \to S_3.$$

This map is injective, since σ is uniquely determined by its action on $\alpha_1, \alpha_2, \alpha_3$. Therefore G is isomorphic to a subgroup of S_3. Now let $E := \mathbb{Q}(\alpha_1, \alpha_2, \alpha_3) = \mathbb{Q}(\sqrt[3]{2}, \zeta_3)$ be the splitting field of $X^3 - 2$ over $\mathbb{Q}$; then $E:\mathbb{Q} = 6$ (see F10 in Chapter 5 and F11 in Chapter 2). Therefore $G \simeq S_3$.

Remark. Let $f \in K[X]$ be as in Definition 3. One sees just as in the example that, more generally, *If f has degree n, the Galois group of f over K is isomorphic to a subgroup of the symmetric group S_n.* But it can certainly be a *proper* subgroup; for instance, the group of the equation $X^4 - 2 = 0$ over $\mathbb{Q}$ is of order 8 (whereas S_4 has order 24). To prove this one must show that the splitting field $E = \mathbb{Q}(\sqrt[4]{2}, i)$ of $X^4 - 2$ over $\mathbb{Q}$ (Chapter 6, example (a) after F4) has degree $E:\mathbb{Q} = 8$. Setting $F = \mathbb{Q}(\sqrt[4]{2})$, one sees first that $F:\mathbb{Q} = 4$, since $X^4 - 2$ is irreducible by Eisenstein's criterion (Chapter 5, F10). At the same time $E:F = F(i):F \leq 2$; since $F \subseteq \mathbb{R}$ we have $i \notin F$, so $E:F = 2$. From the degree formula we then get $E:\mathbb{Q} = 8$.

F7. *Let G be the Galois group of a **separable** polynomial $f \in K[X]$ over K, and let N be the set of roots of f in a splitting field E of f over K. There is equivalence between:*

   (i) *f is **irreducible**.*

   (ii) *G acts **transitively** on N, that is, for any $\alpha, \beta \in N$ there exists $\sigma \in G$ taking α to β.*

Proof. (i) $\Rightarrow$ (ii): Any two roots $\alpha, \beta \in N$ of the irreducible polynomial f have the same minimal polynomial, and so are conjugate over K (Chapter 7, F1); in other words — letting C be an algebraic closure of E — some $\tau \in G(C/K)$ maps α to β. Since E/K is normal, τ restricts to some $\sigma \in G(E/K) = G$ such that $\sigma\alpha = \beta$.

(ii) $\Rightarrow$ (i): Fix $\alpha \in N$ and take $g = \mathrm{MiPo}_K(\alpha)$. Certainly g divides f. Given $\beta \in N$ there exists by assumption a $\sigma \in G$ with $\sigma\alpha = \beta$. But since $g(\alpha) = 0$ we must have $g(\beta) = g(\sigma\alpha) = 0$. Every root of f is thus also a root of g. Since f is separable, this means f divides g. $\square$

Definition 4. Let $f \in K[X]$ be normalized of degree $n \geq 1$, and let E be a splitting field of f over K. In $E[X]$ we have

$$f(X) = \prod_{i=1}^{n}(X - \alpha_i),$$

where $\alpha_1, \ldots, \alpha_n$ are not necessarily distinct. Now consider the element

$$(16) \qquad \Delta = \Delta(\alpha_1, \ldots, \alpha_n) := \prod_{i<j}(\alpha_j - \alpha_i)$$

of E. Its square

$$(17) \qquad D = D(f) := \Delta^2 = \prod_{i<j}(\alpha_j - \alpha_i)^2 = (-1)^{n(n-1)/2}\prod_{j\neq i}(\alpha_j - \alpha_i)$$

is obviously independent of the order in which we take $\alpha_1, \ldots, \alpha_n$. We call $D(f)$ the *discriminant* of the polynomial $f \in K[X]$; it is nonzero if and only if f is separable.

By the foregoing, $\sigma D = D$ for every $\sigma \in G(E/K)$. Galois theory then gives

$$(18) \qquad\qquad\qquad D(f) \in K,$$

since in the inseparable case we have $D(f) = 0$, which implies (18) trivially.

Furthermore, $\sqrt{D} = \Delta \in E$ by (16), so $K(\sqrt{D}) \subseteq E$. When is $\sqrt{D}$ not actually in K, that is, when is $K(\sqrt{D}) : K = 2$? This is an interesting question; here we will treat only the following special case, whose proof affords us a nice opportunity to practice our Galois-theoretical skills.

F8. *Suppose the Galois group G of an **irreducible** and separable polynomial $f \in K[X]$ over K is **cyclic** of **even order** n. So long as* char $K \neq 2$, *the discriminant D of f is not a square in K, so $\sqrt{D} \notin K$.*

Proof. Saying that G is cyclic is saying that any element of G can be written as a power σ^j of a fixed $\sigma \in G$. Let α be a root of f in the splitting field E of f over K. Since G is *abelian*, all its subgroups are normal, so $K(\alpha)/K$ is *normal* (see F4). Since f is irreducible, this implies that all roots of f already lie in $K(\alpha)$, that is, $E = K(\alpha)$. Thus f has precisely n distinct roots $\alpha_1, \ldots, \alpha_n$ in E, which we can assume to be numbered as follows: $\alpha_1 = \alpha$, $\alpha_2 = \sigma\alpha$, $\alpha_3 = \sigma^2\alpha$, ..., $\alpha_n = \sigma^{n-1}\alpha$. Then $\Delta = \prod_{i<j}(\alpha_j - \alpha_i) = \prod_{1\leq i<j\leq n}(\sigma^{j-1}\alpha - \sigma^{i-1}\alpha) = \prod_{0\leq i<j\leq n-1}(\sigma^j\alpha - \sigma^i\alpha)$.
It follows that

$$\frac{\sigma\Delta}{\Delta} = \frac{\prod_{i=2}^{n}(\alpha - \alpha_i)}{\prod_{j=2}^{n}(\alpha_j - \alpha)} = (-1)^{n-1} = -1,$$

since we've assumed n *even*. Thus the element $\triangle = \sqrt{D}$ of E is not G-invariant, and so does not lie in K.   □

Perhaps it might have been conceptually clearer to point out the relation $\sigma \triangle = \mathrm{sgn}(\sigma) \triangle$, where $\mathrm{sgn}(\sigma)$ is the *sign* or *parity* of σ regarded as a permutation of S_n (see also Problem 15.9 in the Appendix).

Remark. Keeping the preceding notations and the assumptions of F8, *the field $K(\sqrt{D})$ is the unique intermediate field F of E/K such that $F:K = 2$.* You should now, at the end of this chapter, persuade yourself of this fact, as a consequence of the Fundamental theorem of Galois theory, a later result on cyclic groups (F6 in Chapter 9), and of course F8.

Finite Fields, Cyclic Groups and Roots of Unity

1. One infinite family of finite fields has already come to our attention (Section 3.5): namely, for every prime number p, there is a field $\mathbb{F}_p := \mathbb{Z}/p$ with p elements. It was Galois who first discovered that there are other finite fields out there. In this chapter we will put together a list, so to speak, of all finite fields, and discuss some key properties of these fields.

Let K be a given finite field and let k be its prime field (Section 3.5). Since k is finite one cannot have $k \simeq \mathbb{Q}$, so $k \simeq \mathbb{F}_p$, where p, a prime number, is the characteristic of K. We look at K as a vector space over k, necessarily of finite dimension. Setting $d := K:k$ we then get an isomorphism of k-vector spaces $K \simeq k^d$. Thus the number of elements of K is

$$q := |K| = p^d.$$

The multiplicative group $K^\times$ of K has order

$$|K^\times| = q - 1 = p^d - 1.$$

We claim that $\alpha^{q-1} = 1$ for all $\alpha \in K^\times$. This comes from a more general fact:

F1. *If G is a finite group or order n, then $x^n = 1$ for every $x \in G$.*

Proof. We prove the assertion here only in the case that G is abelian (see also page 95). Suppose $G = \{a_1, \ldots, a_n\}$, and consider $g := a_1 a_2 \ldots a_n$. For a given $x \in G$, the map $a_i \mapsto xa_i$ is a permutation of G, so $g = (xa_1)(xa_2) \ldots (xa_n) = x^n a_1 a_2 \ldots a_n = x^n g$. Therefore $x^n = 1$. $\qquad \square$

Thus any element $\alpha \neq 0$ of a q-element field K satisfies the equation $\alpha^{q-1} = 1$, hence also the equation $\alpha^q = \alpha$. The latter is also satisfied by $\alpha = 0$. Therefore every element of K is a root of $X^q - X$, and by looking at degrees we see that

$$(1) \qquad X^q - X = \prod_{\alpha \in K} (X - \alpha).$$

In particular, K is a splitting field of $X^q - X$ over $\mathbb{F}_p$, and as such is uniquely determined up to isomorphism. So all q-element fields are isomorphic.

How about the *existence* of a finite field of p^n elements (for a given prime p and natural number n)? The preceding discussion shows how one must proceed: For $q := p^n$, consider the polynomial

$$(2) \qquad\qquad f(X) = X^q - X$$

over $\mathbb{F}_p$. Let K be a splitting field of f over $\mathbb{F}_p$ (whose existence is guaranteed by Chapter 6, F4). The next two assertions then imply that K has exactly $q = p^n$ elements:

   (i) *K contains only roots of f.*

   (ii) *f has no multiple roots.*

Proof of the assertions. (i) Let $K' = \{\alpha \in K \mid \alpha^q = \alpha\}$ be the set of roots of f (in K). By definition, K' is the fixed field of the automorphism $\sigma_q : \alpha \mapsto \alpha^q$ of K (Chapter 7, Remark after F14). Hence K' is a field. It contains $\mathbb{F}_p$, since $\mathbb{F}_p$ is the prime field of K. Therefore K' is a splitting field of f over $\mathbb{F}_p$, and because $K' \subseteq K$ we have $K' = K$.

(ii) The derivative of the polynomial in (2) is $f'(X) = qX^{q-1} - 1 = -1$. By Chapter 7, F9, f is in fact separable. (Another proof: If $\alpha \neq 0$ is a root of f, set

$$g := \frac{X^{q-1} - 1}{X - \alpha} = \frac{X^{q-1} - \alpha^{q-1}}{X - \alpha} = X^{q-2} + \alpha X^{q-3} + \cdots + \alpha^{q-2};$$

thus $g(\alpha) = (q-1)\alpha^{q-2} \neq 0$.) $\qquad\qquad\square$

To summarize the work so far:

Theorem 1. *Let C be a (fixed) algebraic closure of $\mathbb{F}_p$. For every $n \in \mathbb{N}$, there is in C exactly one finite subfield $\mathbb{F}_{p^n}$ having p^n elements, namely the splitting field of the polynomial $X^{p^n} - X$ over $\mathbb{F}_p$ in C. The elements of $\mathbb{F}_{p^n}$ are precisely all the roots of $X^{p^n} - X$. Every finite field (of characteristic p) is isomorphic to one and only one $\mathbb{F}_{p^n}$.*

Let K be a finite field with $q = p^n$ elements. Then $K \simeq \mathbb{F}_q$. If E is an extension of K such that $E : K = m$, we have $|E| = q^m$. As we saw above, E is then a splitting field of $X^{q^m} - X$ over K. Therefore, up to a K-isomorphism, K has at most one extension of degree m over K (Chapter 6, F4). Conversely, $\mathbb{F}_{q^m}$ is an extension of $\mathbb{F}_q$, for any $m \in \mathbb{N}$. Therefore $K \simeq \mathbb{F}_q$ clearly has an extension E such that $E \simeq \mathbb{F}_{q^m}$. Because $|E| = |K|^{E:K}$ we then have $E : K = m$. Putting it all together:

Theorem 1'. *If K is a finite field and $m \in \mathbb{N}$, there is an extension E of K of degree m, and it is unique up to K-isomorphism.*

What can one say about the structure of the multiplicative group $K^\times$ of a finite field K? We shall see that the answer is as simple as can be: $K^\times$ is *cyclic*. Before we prove this fundamental theorem, however, we will indulge in a little detour on cyclic groups.

2. First let G be any group and α an element of G. There is a well defined group homomorphism

(3) $$\varphi : \mathbb{Z} \to G, \quad \text{determined by } 1 \mapsto \alpha.$$

Let $\langle \alpha \rangle := \{\alpha^m \mid m \in \mathbb{Z}\}$ denote its image. By the Fundamental Homomorphism Theorem, $\langle \alpha \rangle$ is isomorphic to $\mathbb{Z}/\ker \varphi$. But $\ker \varphi$, being a subgroup of $\mathbb{Z}$, is actually an ideal of the ring $\mathbb{Z}$. Since $\mathbb{Z}$ is a principal ideal domain, $\ker \varphi$ has the form

(4) $$\ker \varphi = n\mathbb{Z}, \quad \text{with } n \in \mathbb{N} \cup \{0\} \text{ uniquely determined.}$$

Definition. If $n \neq 0$ in (4), we call n the *order of* α, and write $n = \operatorname{ord} \alpha$. If $n = 0$, we set $\operatorname{ord} \alpha = \infty$. Since $\langle \alpha \rangle \simeq \mathbb{Z}/\ker \varphi$, we see that α has finite order if and only if the subgroup $\langle \alpha \rangle$ of G generated by α is finite.

F2. *Let G be a group and α and element of G. Then*

(5) $$\operatorname{ord} \alpha = |\langle \alpha \rangle|.$$

If α has finite order and $m \in \mathbb{Z}$, we have

(6) $$\alpha^m = 1 \iff \operatorname{ord} \alpha \mid m;$$

in particular, $\operatorname{ord} \alpha$ *is the smallest natural number m such that $\alpha^m = 1$.*

Proof. Suppose $n := \operatorname{ord} \alpha < \infty$. Then $\langle \alpha \rangle \simeq \mathbb{Z}/n\mathbb{Z}$, but of course $|\mathbb{Z}/n\mathbb{Z}| = n$. This proves (5). To prove (6), write

$$\alpha^m = 1 \iff m \in \ker \varphi \iff m \in n\mathbb{Z} \iff n \mid m. \qquad \square$$

Definition. A group G is called *cyclic* if there exists $\gamma \in G$ such that $G = \langle \gamma \rangle$; such a γ is called a *generator of G*.

F3. *A group G is cyclic if and only if $G \simeq \mathbb{Z}/n\mathbb{Z}$ for some $n \in \mathbb{N} \cup \{0\}$.*

Proof. If $G = \langle \gamma \rangle$, the map φ in (3), with $\alpha = \gamma$, is surjective. Thus $G \simeq \mathbb{Z}/\ker \varphi = \mathbb{Z}/n\mathbb{Z}$ for some $n \in \mathbb{N} \cup \{0\}$. The converse is clear, since the residue class of 1 in $\mathbb{Z}/n$ is a generator of the group $\mathbb{Z}/n$ (additively written). $\qquad \square$

If G is finite, any $\alpha \in G$ generates a finite group and therefore has finite order. More precisely:

F4. *If G is a finite group and α is any element of it, the order of α divides that of G.*

Proof. If $n = |G|$ we have $\alpha^n = 1$ for any $\alpha \in G$; see F1. The assertion then follows from (6) in F2. $\qquad \square$

Remark. If α is an element of a finite group G, we get from (5):

$$\operatorname{ord} \alpha = |G| \iff G = \langle \alpha \rangle.$$

F5. *If G is a cyclic group, so is any subgroup H of G.*

Proof. Suppose $G = \langle \gamma \rangle$, so the homomorphism (3) is surjective, where $\alpha = \gamma$. The inverse image $\varphi^{-1}(H)$ of H under φ is a subgroup of $\mathbb{Z}$, necessarily of the form $\varphi^{-1}(H) = k\mathbb{Z}$. Since φ is onto, we get $H = \varphi(k\mathbb{Z}) = \langle \gamma^k \rangle$, that is, γ^k generates H.   □

F6. *Let G be a **finite, cyclic** group of order n. The map*

$$
(7) \qquad\qquad H \mapsto |H|
$$

is a bijection between the set of subgroups of G and the set of natural numbers that divide n.

Proof. Suppose $G = \langle \gamma \rangle$, for a fixed generator γ. If d is a natural number and $d \mid n$, set $d' = n/d$. The order of the subgroup $H(d) := \langle \gamma^{d'} \rangle$ coincides with $\mathrm{ord}(\gamma^{d'})$. Because $(\gamma^{d'})^d = \gamma^n = 1$ we have $\mathrm{ord}(\gamma^{d'}) \mid d$, by (6); thus $|H(d)|$ divides d. On the other hand, $(\gamma^{d'})^{|H(d)|} = 1$ by F1, and hence $n = \mathrm{ord}\,\gamma$ divides $d' \cdot |H(d)|$; since $n = dd'$ we conclude that $|H(d)|$ is divisible by d. Putting it all together we get

$$
(8) \qquad\qquad |H(d)| = d.
$$

We claim that $d \mapsto H(d)$ is the inverse map to (7). Let H be any subgroup of G, having order d. We must show that d divides n and that

$$
(9) \qquad\qquad H = H(d).
$$

By F5, H is of the form $H = \langle \gamma^m \rangle$. Since $(\gamma^m)^n = 1$ it is indeed the case that $d = |H| = \mathrm{ord}(\gamma^m)$ divides n. By F1 we get $(\gamma^m)^d = 1$; again from (6) there follows $n \mid md$, which is to say $d' \mid m$. Hence $\gamma^m \in \langle \gamma^{d'} \rangle = H(d)$, so that $H \subseteq H(d)$. Since both groups have the same order d, (9) follows.   □

3. We are now ready for the result promised earlier:

Theorem 2. *The multiplicative group $K^\times$ of a finite field K is **cyclic**.* (A generator of $K^\times$ is called a *primitive root*.)

In fact we prove something more general:

Theorem 2′. *Let K be any field. Then any **finite** subgroup G of $K^\times$ is cyclic.*

The proof relies on the following characterization of cyclic groups:

Lemma. *A finite group G of order n is cyclic if*

$$
(10) \qquad\qquad \left| \{ x \in G \mid x^d = 1 \} \right| \le d \quad \text{for every } d \mid n.
$$

Proof of Theorem 2′ assuming the Lemma. Set $n := |G|$. For any $d \in \mathbb{N}$ there are in K at most d elements x with $x^d = 1$, since that's the most roots that $X^d - 1 \in K[X]$ might have in K. Thus (10) is satisfied, which means G is cyclic.   □

Proof of the Lemma. For $d \mid n$, set

$$\psi_G(d) := \left|\{x \in G \mid \text{ord } x = d\}\right|.$$

Then

(11)
$$\sum_{d \mid n} \psi_G(d) = n = |G|,$$

since any $\alpha \in G$ has a well defined order d, with $d \mid n$. For $m \in \mathbb{N}$, set

(12) $\varphi(m) :=$ the number of elements of $\mathbb{Z}/m$ that are generators.

Then $\varphi(m) \geq 1$. We will show that

(13)
$$\psi_G(d) \leq \varphi(d) \quad \text{for all } d \mid n.$$

For $\psi_G(d) = 0$ this is clear. Thus suppose $\psi_G(d) \geq 1$; this means there exists $\alpha \in G$ with ord $\alpha = d$. Then $H = \langle \alpha \rangle$ is a subgroup of order d in G, and $x^d = 1$ for any $x \in H$. From (10) it follows, in particular, that any order-d element in G already lies in H. Because $H \simeq \mathbb{Z}/d$, we then get $\psi_G(d) = \psi_H(d) = \varphi(d)$, which yields (13).

We remark that in the case of the group $\mathbb{Z}/n$ we must have

(14)
$$\psi_{\mathbb{Z}/n}(d) = \varphi(d),$$

since all order-d elements of $\mathbb{Z}/n$ lie in one and the same order-d cyclic subgroup of $\mathbb{Z}/n$ (see F6 and F5). By summation we get from (13), in conjunction with (11) and (14),

$$n = \sum_{d \mid n} \psi_G(d) \leq \sum_{d \mid n} \varphi(d) = \sum_{d \mid n} \psi_{\mathbb{Z}/n}(d) = n.$$

But in view of (13) this can only happen if $\psi_G(d) = \varphi(d)$ for all $d \mid n$. In particular, $\psi_G(n) = \varphi(n) \geq 1$, so G possesses an element of order n.   □

Remarks. Let the quotient map $\mathbb{Z} \to \mathbb{Z}/n$ be written $k \mapsto \bar{k}$. Then

(15)
$$\text{ord } \bar{k} = \frac{n}{(k,n)}.$$

Proof: We have

$$\frac{n}{(k,n)} \bar{k} = \frac{n}{(k,n)} \cdot k \cdot \bar{1} = 0,$$

so $j := \text{ord } \bar{k}$ divides the right-hand side of (15). Next, $j\bar{k} = 0$, so $n \mid jk$. But this implies that $n/(k,n)$ divides j.

If G is a group and $\alpha \in G$ has finite order n, we have for every $k \in \mathbb{Z}$

(16)
$$\text{ord}(\alpha^k) = \frac{n}{(k,n)}.$$

To prove this observe that $\langle \alpha \rangle \simeq \mathbb{Z}/n$ and apply (15).

According to (16) we have $\mathrm{ord}(\alpha^k) = n$ if and only $(k, n) = 1$, that is, when k and n are relatively prime. Thus the function φ defined in (12) satisfies

$$(17) \qquad \varphi(n) = \left| \{ k \in \mathbb{N} \mid 1 \leq k \leq n \text{ with } (k, n) = 1 \} \right|.$$

For this reason the function φ is of interest in number theory; it is called *Euler's totient function* or *φ-function*. Now note that

$$(n, k) = 1 \iff \mathrm{ord}\,\bar{k} = n \iff \mathbb{Z}/n = \langle \bar{k} \rangle \iff \bar{1} \in \langle \bar{k} \rangle \iff \exists l : \bar{1} = l\bar{k} = \overline{lk}.$$

Therefore $(n, k) = 1$ *if and only if* $\bar{k}$ *is a unit in the ring* $\mathbb{Z}/n\mathbb{Z}$. In particular, the group of units $(\mathbb{Z}/n\mathbb{Z})^\times$ of $\mathbb{Z}/n\mathbb{Z}$ has exactly $\varphi(n)$ elements:

$$(18) \qquad \varphi(n) = \left| (\mathbb{Z}/n)^\times \right|.$$

We call $(\mathbb{Z}/n)^\times$ the *group of prime residue classes modulo n*. The function φ is *multiplicative* in the following sense:

$$(19) \qquad \varphi(n_1 n_2) = \varphi(n_1)\varphi(n_2) \quad \text{if } (n_1, n_2) = 1.$$

To see this, consider the natural ring homomorphism

$$\mathbb{Z}/n_1 n_2 \to \mathbb{Z}/n_1 \times \mathbb{Z}/n_2.$$

Because n_1 and n_2 are assumed relatively prime this map is injective; since both domain and counterdomain have $n_1 n_2$ elements, it must be an isomorphism. But a ring isomorphism implies an isomorphism of the corresponding groups of units, in this case $(\mathbb{Z}/n_1 n_2)^\times \simeq (\mathbb{Z}/n_1)^\times \times (\mathbb{Z}/n_2)^\times$. Keeping (18) in mind, we get (19).

4. We now talk a bit about the third item in the title of this chapter.

Definition. If K is a field, denote by $W(K)$ the set of all elements of finite order in the group $K^\times$. These elements are called *roots of unity*. For $n \in \mathbb{N}$, set

$$W_n(K) = \{ \zeta \in K \mid \zeta^n = 1 \}.$$

The elements of $W_n(K)$ are the *n-th roots of unity* of K. Clearly $W(K)$ and $W_n(K)$ are subgroups of $K^\times$. An element $\zeta \in W(K)$ is a *primitive n-th root of unity* if $\mathrm{ord}\,\zeta = n$.

F7. *For any n, $W_n(K)$ is a **finite cyclic** group whose order divides n. If K has characteristic $p > 0$, then $W_{np}(K) = W_n(K)$.*

Proof. $W_n(K)$ is *finite* because its elements are the roots of the polynomial $X^n - 1$ in K. Being a finite subgroup of $K^\times$, however, $W_n(K)$ must be *cyclic*; see Theorem 2′ in Section 9.3. The order of a generator divides n, by F2.

Now suppose char $K = p > 0$. Since $\zeta^{np} = (\zeta^n)^p = 1$ we have $\zeta^n = 1$ by F14 in Chapter 7. Thus $W_{np}(K) \subseteq W_n(K)$. The opposite inclusion is obvious. $\qquad \square$

Remark. $W_n(\mathbb{C}) = \{e^{2\pi ik/n} \mid k = 0, 1, 2, \ldots, n-1\}$ has n elements.

F8. *Let C be an algebraically closed field, and take $n \in \mathbb{N}$. In case char $C = p > 0$ assume also that $(n, p) = 1$. Then $W_n(C)$ has order n. Thus C contains a primitive n-th root of unity ζ, and any such is a generator of $W_n(C)$.*

Proof. Consider $f(X) = X^n - 1 \in C[X]$. We must show that f has no multiple roots; the rest then follows from F7. Now $f'(X) = nX^{n-1}$; because char K does not divide n, it follows that $f'(\zeta) \neq 0$ for any $\zeta \in C^\times$. $\quad\square$

Incidentally, if char $C = p > 0$ and n is arbitrary, F7 shows that $W_n(C)$ has order $np^{-w_p(n)}$, where $w_p(n)$ is defined as on page 40.

Definition. Let K be a field and take $n \in \mathbb{N}$. The splitting field E of $X^n - 1$ over K is called the *field of n-th roots of unity over K*. We use the notation $E = K^{(n)} = K(\sqrt[n]{1})$.

F9. *Let $E = K(\sqrt[n]{1})$ be the field of n-th roots of unity over K. Then E/K is a (finite) Galois extension, with **abelian** Galois group G. If n is not divisible by char K, then G is canonically isomorphic to a subgroup of $(\mathbb{Z}/n)^\times$.*

Proof. Suppose n is not divisible by char K. The polynomial $f(X) = X^n - 1$ is then separable over K, as seen in the proof of F8. Hence E is a splitting field of a separable polynomial over K, and E/K is a Galois extension by Chapter 8, F2. (That this is still the case for arbitrary n follows from F7.)

Let $\zeta_n \in E$ be a primitive n-th root of unity, so ord $\zeta_n = n$. Given $\sigma \in G$, the image $\sigma\zeta_n$ also has order n, so

$$(20) \qquad \sigma\zeta_n = \zeta_n^k \quad \text{with } (k, n) = 1,$$

where k can be uniquely determined by the condition $1 \leq k \leq n$. Then any $\zeta = \zeta_n^j$ in $W_n(E)$ satisfies $\sigma\zeta = \zeta^k$, so k in (20) is independent of the choice of ζ_n. In this way we get a well defined map

$$(21) \qquad G \to (\mathbb{Z}/n)^\times, \quad \sigma \mapsto \bar{k}.$$

It is easy to see that this is a group homomorphism. It is also injective, since σ is determined by $\sigma\zeta_n$ — recall that $E = K(\zeta_n)$. Now G, being isomorphic to a subgroup of the abelian group $\mathbb{Z}/n$, is abelian as well. $\quad\square$

As can be seen already from $K = \mathbb{R}$ or $\mathbb{C}$, the map (21) is generally not surjective. (For more interesting examples see F11 in Section 9.5.) But in the case of $K = \mathbb{Q}$ we have:

Theorem 3 (Gauss). *Let $E = \mathbb{Q}(\sqrt[n]{1})$ be the field of n-th roots of unity over $\mathbb{Q}$. The Galois group of $E/\mathbb{Q}$ is canonically isomorphic to the group of prime residue classes $(\mathbb{Z}/n)^\times$. In particular, $\mathbb{Q}(\sqrt[n]{1}) : \mathbb{Q} = \varphi(n)$.*

Proof. Let ζ be a primitive n-th root of unity in E. In view of the homomorphism (21), we must show that for each $k \in \mathbb{N}$ relatively prime to n, there exists $\sigma \in G(E/K)$ with $\sigma\zeta = \zeta^k$. By F7 in Chapter 8, this boils down to showing that ζ^k is also a root of $f := \mathrm{MiPo}_{\mathbb{Q}}(\zeta)$:

(22)
$$f(\zeta^k) = 0.$$

(Group-theoretically, ζ^k with $(k,n) = 1$ is certainly specifiable as the image of an automorphism of the group $\langle\zeta\rangle$; however, to derive the existence of a corresponding field automorphism of $\mathbb{Q}(\zeta)$, we must ensure that ζ^k satisfies the same defining equation as ζ, that is, (22) must hold.) By expressing k as a product of prime numbers, one sees that it suffices to prove the assertion for the case of $k = p$ a prime, where furthermore p does not divide n. In $\mathbb{Q}[X]$ we have the decomposition

(23)
$$X^n - 1 = f(X)g(X).$$

Then Gauss's Lemma (Chapter 5, F7) implies that $f(X), g(X) \in \mathbb{Z}[X]$. Now we assume, contrary to claim, that $f(\zeta^p) \neq 0$. By (23) we have $g(\zeta^p) = 0$. Therefore ζ is a root of the polynomial $g(X^p)$, and so we get, again by using Gauss's Lemma, a decomposition

$$g(X^p) = f(X)h(X) \quad \text{with } h(X) \in \mathbb{Z}[X].$$

Passing to polynomials over $\mathbb{F}_p = \mathbb{Z}/p$ via the natural map $\mathbb{Z}[X] \to (\mathbb{Z}/p)[X]$, we get the decomposition

(24)
$$\bar{g}(X^p) = \bar{f}(X)\bar{h}(X) \quad \text{in } \mathbb{F}_p[X].$$

But for any polynomial $\bar{g}(X) = \sum \alpha_i X^i$ over $\mathbb{F}_p$ we have $\bar{g}(X)^p = \left(\sum \alpha_i X^i\right)^p = \sum \alpha_i^p X^{ip} = \sum \alpha_i X^{pi} = \bar{g}(X^p)$. Thus (24) can be written in the form

$$\bar{g}(X)^p = \bar{f}(X)\bar{h}(X) \quad \text{in } \mathbb{F}_p[X].$$

In an algebraic closure of $\mathbb{F}_p$, therefore, $\bar{f}$ and $\bar{g}$ must have a common root, so $X^n - \bar{1} = \bar{f}(X)\bar{g}(X)$ cannot be separable. This contradicts F8, because $(p,n) = 1$. $\qquad\square$

In the sequel let C be an algebraically closed extension of $\mathbb{Q}$ (for instance $C = \mathbb{C}$, or the field of algebraic numbers in $\mathbb{C}$). Set $W_n = W_n(C)$. Then

(25)
$$X^n - 1 = \prod_{\zeta \in W_n} (X - \zeta).$$

Definition. The polynomial

$$F_n(X) = \prod_{\mathrm{ord}\,\zeta = n} (X - \zeta),$$

where the product is taken over all primitive n-th roots of unity of C, is called the *n-th cyclotomic polynomial*.

F10. *The n-th cyclotomic polynomial F_n has the following properties*:

(a) F_n *is normalized.*

(b) $\deg F_n = \varphi(n)$.

(c) $X^n - 1 = \prod_{d \mid n} F_d(X)$.

(d) $F_n(X) \in \mathbb{Z}[X]$.

Proof. Parts (a) and (b) are immediate. Since any $\zeta \in W_n$ has a well defined order d with $d \mid n$, part (c) follows from (25). Part (d) can be proved by induction on n: Part (c) implies that

$$F_n(X) = (X^n - 1) \Big/ \prod_{\substack{d \mid n \\ d < n}} F_d(X),$$

and in $\mathbb{Z}[X]$ one can always divide with remainder by *normalized* polynomials. Another proof of (d): Any σ in the Galois group G of $\mathbb{Q}(\sqrt[n]{1})/\mathbb{Q}$ satisfies

$$F_n^\sigma(X) = \prod_{\mathrm{ord}\,\zeta=n}(X - \sigma(\zeta)) = F_n(X),$$

so all coefficients of $F_n(X)$ lie in the fixed field of G, and so, by Galois theory, also in $\mathbb{Q}$. But once we know that $F_n \in \mathbb{Q}[X]$, it follows that F_n, being a normalized factor of $X^n - 1$, must lie in $\mathbb{Z}[X]$ (Chapter 5, F7).   □

Theorem 3′ (Gauss). *The n-th cyclotomic polynomial $F_n(X)$ is irreducible in $\mathbb{Q}[X]$.*

Proof. Let ζ be a primitive n-th root of unity. We must show that

$$F_n(X) = \mathrm{MiPo}_{\mathbb{Q}}(\zeta),$$

since we already know that $F_n(X)$ is a normalized polynomial in $\mathbb{Q}[X]$ vanishing at ζ. By Theorem 3, $\mathbb{Q}(\zeta):\mathbb{Q} = \varphi(n) = \deg F_n$, which is all is needed.   □

Remark. Let p be a prime number. Then

(26) $$F_p(X) = \frac{X^p - 1}{X - 1} = 1 + X + \cdots + X^{p-1},$$

and more generally

(27) $$F_{p^m}(X) = \frac{X^{p^m} - 1}{X^{p^{m-1}} - 1} = 1 + X^{p^{m-1}} + \cdots + X^{(p-1)p^{m-1}};$$

in particular, then,

(28) $$\varphi(p^m) = (p-1)p^{m-1} = p^m - p^{m-1}.$$

The irreducibility of (27) had already been proved in Chapter 5, F13 using a different method (yet one based on the same principles in a way). Of course (28) is also easily derived from (17). And one may observe that granting the validity of Theorem 3′ the assertion of Theorem 3 can be derived immediately.

If $n = p_1^{e_1} p_2^{e_2} \ldots p_r^{e_r}$ is the prime factorization of a natural number $n > 1$, it follows from (19) together with (28) that

$$(29) \qquad \varphi(n) = \prod_{i=1}^{r} (p_i - 1) \, p_i^{e_i - 1}.$$

5. We now return once again to finite fields, considering them from the viewpoint of Galois theory. Though simple, the relation here is of greatest significance for deeper arithmetic questions.

Theorem 4. *Let E/K be any extension of finite fields and let q be the number of elements of K. Then E/K is Galois with a cyclic Galois group, and in fact $G(E/K)$ is generated by the automorphism $\sigma_q : \alpha \to \alpha^q$ of E. (We call σ_q the Frobenius automorphism, or simply the Frobenius, of E/K.)*

Proof. Let G be the subgroup of Aut E generated by σ_q. Then K is the fixed field of G in E, since by Section 9.1 K contains exactly those $\alpha \in E$ such that $\alpha^q = \alpha$. By Galois theory (Chapter 8, Theorem 4), E/K is a Galois extension and $G = G(E/K)$ is its Galois group. □

We could also have proved this result without depending of Chapter 8; by and large, Galois theory results for finite fields can be verified directly without much trouble.

F11. *Let K be a finite field with q elements. Given a natural number n relatively prime to q, denote by $q \bmod n$ the residue class of q in $(\mathbb{Z}/n)^{\times}$. Then*

$$K(\sqrt[n]{1}) : K = \mathrm{ord}\,(q \bmod n).$$

Proof. Set $E = K(\sqrt[n]{1})$ and let ζ be a primitive n-th root of unity in E. Theorem 4 shows that $E : K$ equals the order of $\sigma_q \in G(E/K)$. Thus, since $E = K(\zeta)$, the degree $E : K$ is the smallest natural number f such that $\zeta^{q^f} = \zeta$; that is, $q^f \equiv 1 \bmod n$. In other words, $E : K$ is the order of $q \bmod n$ in $(\mathbb{Z}/n)^{\times}$. □

For practice, derive F11 directly from Theorems 1 and 2, without appealing to Theorem 4.

10

Group Actions

According to the Fundamental Theorem of Galois theory, the intermediate fields of a finite Galois extension E/K are in one-to-one correspondence with the subgroups of the Galois group of E/K. This by itself would be reason enough to study groups, and thus it is time for us to turn our attention to some key notions of group theory.

Natural examples of groups usually come up in mathematics as automorphism groups of certain structures. Our investigations in Chapter 8 illustrate this typical trend (among others). Think also of concepts such as the linear group of a vector space, the orthogonal group of a quadratic form, etc.; or even, if you will, the group of bijections of a nonempty set M — where the structure in question is a bare set.

In the sequel, G will denote a *group* and M a *nonempty set*.

Definition 1. We say that G *acts* (or *operates*) *on* M if there exists a group homomorphism

$$T : G \to S(M)$$
$$\sigma \mapsto T(\sigma) = T_\sigma$$

from G into the group of permutations of M. The result of applying $\sigma \in G$ to $x \in M$ is denoted also by

$$\sigma.x = \sigma x = T_\sigma(x).$$

Thus one gets a map

(1)
$$G \times M \to M$$
$$(\sigma, x) \mapsto \sigma x$$

satisfying

$$\text{(i) } 1x = x \quad \text{and} \quad \text{(ii) } (\sigma\tau)x = \sigma(\tau x).$$

Conversely: Given a map (1) satisfying (i) and (ii), the group G acts on M by means of the map $T : G \to S(M)$ defined by $T(\sigma)(x) = \sigma x$. Indeed, (ii) says that $T(\sigma\tau) = T(\sigma) \circ T(\tau)$, and (i) says that $T(1) = \text{id}_M$, so $T(\sigma) \circ T(\sigma^{-1}) = T(1) = \text{id}_M = T(\sigma^{-1}) \circ T(\sigma)$. Thus we really do have $T(\sigma) \in S(M)$, with $T(\sigma)^{-1} = T(\sigma^{-1})$. Either the map $G \to S(M)$ or the map $G \times M \to M$ can be called the corresponding *action* of G on M.

Example 1. Let E/K be a Galois extension with Galois group $G = G(E/K)$. The group G acts on E via $(\sigma, x) \mapsto \sigma(x)$, and it acts on $E^\times$ likewise. If E is a splitting field of a polynomial $f \in K[X]$ having degree n and roots $\alpha_1, \ldots, \alpha_n$, the group G also acts on $M = \{\alpha_1, \ldots, \alpha_n\}$.

There are many other examples of group actions, arising in very diverse ways. It is a group's nature, so to speak, to act on something.

Definition 2. Let G act on M. For $x \in M$, the set

$$Gx = \{\sigma x \mid \sigma \in G\}$$

is called the *orbit of x* (relative to the given group action). Compare Chapter 8, F1.

F1. *Let G act on M. Any two distinct orbits (relative to the given group action) are disjoint, and M is the union of all such orbits.*

Proof. For $x, y \in M$, write $x \sim y$ if there is $\sigma \in G$ such that $\sigma x = y$. You can easily persuade yourself that $\sim$ defines an equivalence relation on M. The equivalence class of $x \in M$ under $\sim$ is none other than the orbit Gx of x. This is enough to prove the assertion. $\qquad\square$

Definition 3. Let G act on M. The action is *transitive* if it has a single orbit; i.e., if, for some (or for that matter, for any) $x \in M$,

$$M = Gx.$$

See also Chapter 8, F7.

F1′. *Let the group G act on a finite set M. Then*

$$|M| = \sum_C |C|,$$

where the sum is over all distinct orbits C.

Remark. The cardinality $|Gx|$ is called the *size* (or *length*) of the orbit of x (under G).

Example 2. Let G be a group. Then G acts on $M := G$ via the action $(\sigma, \tau) \mapsto \sigma\tau$. The corresponding homomorphism $T : G \to S(G)$ associates to every $\sigma \in G$ the *left translation T_σ* corresponding to σ, defined by $T_\sigma(\tau) = \sigma\tau$. Clearly, T is injective. If $|G| = n$ we have $S(G) \simeq S_n$, so we obtain:

F2 (Cayley's Theorem). *Every group G of order n is isomorphic to a subgroup of the symmetric group S_n.*

Example 3. Let H be a subgroup of a group G. Then H acts on $M = G$ by left translations. The orbit of $\sigma \in G$ is

$$H\sigma = \{\rho\sigma \mid \rho \in H\}.$$

This is called the *(right) coset* of σ mod H. The set of such right cosets is denoted by $H\backslash G$, and its cardinality by

$$G : H = |H\backslash G|.$$

If G is finite, so are H and $H\backslash G$. Since $|H\sigma| = |H|$ for all $\sigma \in G$, we see that F1$'$ implies the next result:

F3 (Euler–Lagrange). *For any subgroup H of a finite group G,*

(2) $$|G| = (G : H) \cdot |H|.$$

In particular, the order of H divides that of G.

Remarks. (a) Of course (2) is also valid for infinite groups, if regarded as an equation between cardinals.

(b) We denote the subgroup $\{1\}$ by 1. Then $|H| = H : 1$ for any H, and moreover

(2$'$) $$G : 1 = (G : H)(H : 1).$$

(c) We complete here the proof of Chapter 9, F1, picking up the general (nonabelian) case. Given $\alpha \in G$ we look at $H = \langle \alpha \rangle$. Since $|H| = \operatorname{ord} \alpha$, this number divides $n = |G|$, according to F3. In particular, $\alpha^n = 1$.

(d) For H a given subgroup of G, write

$$G/H$$

for the set of *left* cosets σH, for $\sigma \in G$. *Then G is the disjoint union of all the distinct left cosets* mod H. *There are as many left cosets* mod H *as right cosets* mod H; *thus*

$$|G/H| = G : H.$$

To see this, note that $(\sigma, \tau) \mapsto \tau\sigma^{-1}$ also defines an action of H on $M = G$, which associates to each $\sigma \in H$ the *right translation* by σ^{-1}. The corresponding orbits are precisely the sets τH. The map $\sigma \mapsto \sigma^{-1}$ of G onto itself yields a bijection $H\backslash G \to G/H$, since $(H\sigma)^{-1} = \sigma^{-1}H^{-1} = \sigma^{-1}H$.

We call $G : H$ the *index of H in G*. If G is finite,

(2$''$) $$G : H = |G| : |H|$$

by (2); in particular; $G : H$ then divides $|G|$.

Definition 4. Let G act on M and take $x \in M$. The subgroup

$$G_x = \{\sigma \in G \mid \sigma x = x\}$$

is called the *stabilizer* of x.

F4. *Let G act on M and take $x \in M$. The map*

$$G \to Gx, \quad \sigma \mapsto \sigma x$$

defines a bijection $i : G/G_x \to Gx$. Thus, if Gx is finite, so is $G : G_x$, and

(3) $$|Gx| = G : G_x.$$

If G is finite, so is Gx, and

(4) $$|Gx| = \frac{|G|}{|G_x|};$$

in particular then the size of any orbit divides the order of G.

Proof. Take $H := G_x$. Then $\sigma H = \tau H$ if and only if $\rho := \sigma^{-1}\tau \in H$. It follows that $\tau x = \sigma \rho x = \sigma x$. Thus we get a well defined map $i : G/G_x \to Gx$ with $i(\sigma H) = \sigma x$. Clearly i is surjective. But i is also injective since the equality $\sigma x = \tau x$ implies $\sigma^{-1}\tau x = x$, hence $\sigma^{-1}\tau \in H$ and also $\sigma H = \tau H$. $\square$

F5 (Orbit formula). *Let the group G act on a **finite** set M, and let $x_1, \dots, x_s$ be representatives for each of the s distinct orbits of the action. Then*

$$|M| = \sum_{i=1}^{s} (G : G_{x_i}).$$

Proof. By F1', we have $|M| = \sum_{i=1}^{s} |Gx_i|$. The assertion follows thanks to (3). $\square$

Example 4. Let G be a group. As can easily be checked, G acts on $M = G$ via the map $(\sigma, \tau) \mapsto \sigma\tau\sigma^{-1}$. Let $\sigma \mapsto T_\sigma$ be the corresponding homomorphism from G into $S(G)$, that is, $T_\sigma(\tau) = \sigma\tau\sigma^{-1}$. Now,

$$T_\sigma(\tau_1\tau_2) = \sigma(\tau_1\tau_2)\sigma^{-1} = \sigma\tau_1\sigma^{-1}\sigma\tau_2\sigma^{-1} = T_\sigma(\tau_1)T_\sigma(\tau_2);$$

therefore T_σ is actually an automorphism of G. Thus we obtain a homomorphism

$$T : G \to \operatorname{Aut} G$$

from G into the *automorphism group* of the group G. Elements of $S(G)$ in the image of T are called *inner automorphisms* of G. The kernel of T is the normal subgroup of G given by

$$ZG := \{\sigma \mid \sigma\tau\sigma^{-1} = \tau \text{ for all } \tau \in G\};$$

it is called the *center* of G, and of course can also be characterized as

$$ZG = \{\sigma \mid \sigma\tau = \tau\sigma \text{ for all } \tau \in G\}.$$

The stabilizer G_τ of τ with respect to the action under consideration is the subgroup

(5) $$Z_G(\tau) := \{\sigma \mid \sigma\tau\sigma^{-1} = \tau\},$$

called the *centralizer of τ in G*. The orbit of $\tau \in G$, that is, the set

$$\{\sigma\tau\sigma^{-1} \mid \sigma \in G\},$$

is called the *conjugacy class of τ in G*. Two elements τ_1 and τ_2 in the same conjugacy class are *conjugate in G*; this happens if and only if τ_2 is the image of τ_1 under an inner automorphism: $\tau_2 = \sigma\tau_1\sigma^{-1}$ for some $\sigma \in G$.

We now apply F5 to the action from the preceding example. We obtain:

F6 (Class formula). *If G is a finite (nonabelian) group and $\tau_1, \tau_2, \ldots, \tau_r$ represent each of the conjugacy classes of G that contain more than one element, we have*

$$|G| = |ZG| + \sum_{i=1}^{r} \big(G : Z_G(\tau_i)\big).$$

Proof. The conjugacy class of $\tau \in G$ has a single element if and only if $\tau \in ZG$. $\square$

Definition 5. Let p be a prime. A finite group G is called a *p-group* if $|G|$ is a power of p.

F7. *Every finite p-group $G \neq 1$ has nontrivial center.*

Proof. By F6, p must divide $|ZG|$, so $ZG \neq 1$. $\square$

F8. *If G is a p-group of order p^m, there exists a chain*

$$G = H_0 \supseteq H_1 \supseteq H_2 \supseteq \cdots \supseteq H_m = 1$$

*of **normal subgroups** H_i of G such that*

$$H_{i-1} : H_i = p \quad \text{for all } 1 \leq i \leq m.$$

Proof. Let $m > 0$. By F7, ZG contains an element $\alpha \neq 1$. The cyclic subgroup $\langle\alpha\rangle$ has a subgroup H of order p (see Chapter 9, F6). Since $H \subseteq ZG$, this H is normal in G. Now consider the quotient group $\overline{G} = G/H$. By induction we can assume that there exists a chain $\overline{G} = N_0 \supseteq N_1 \supseteq \cdots \supseteq N_{m-1} = 1$ of normal subgroups N_i of $\overline{G}$ such that $N_{i-1} : N_i = p$. Each map $\varphi_i : G \to \overline{G} \to \overline{G}/N_i$ gives rise to an isomorphism $G/\ker\varphi_i \simeq \overline{G}/N_i$. Now set $H_m = 1$ and $H_i := \ker\varphi_i$ for $0 \leq i \leq m-1$. We get a chain $G = H_0 \supseteq H_1 \supseteq \cdots \supseteq H_{m-1} = H \supseteq H_m = 1$ of normal subgroups of G. For $0 \leq i \leq m-1$ we have, by the definition of H_i, the equalities $G : H_i = \overline{G} : N_i = p^i$, the latter because of Euler–Lagrange, equation (2). Again from (2) we then get $H_{i-1} : H_i = p$. From our choice of H this is also true for $i = m$. Thus we have obtained a chain with the desired properties for G. $\square$

Notation. If H is a subgroup of G we write $H \leq G$. If H is a normal subgroup of G, we indicate this by writing $H \trianglelefteq G$.

Example 5. Let G act on M. Then G acts also on the power set of M, via

$$(\sigma, X) \mapsto \sigma X = \{\sigma x \mid x \in X\}.$$

We have $|\sigma X| = |X|$ for every $X \subseteq M$.

Example 6. In the preceding example, consider in particular the action of G on $M = G$ by inner automorphisms. The orbit of an $X \subseteq G$ is then $\{\sigma X \sigma^{-1} \mid \sigma \in G\}$. We use the notation $X^\sigma := \sigma^{-1} X \sigma$. If now $X = H \leq G$ is a *subgroup* of G, the image $H^\sigma = \sigma^{-1} H \sigma$ is also a subgroup of same cardinality. We say that H^σ and H are *conjugate subgroups* of G. The stabilizer of H,

$$N_G H := \{\sigma \in G \mid \sigma H \sigma^{-1} = H\},$$

is called the *normalizer of H in G*. By definition, $H \trianglelefteq N_G H$. Moreover,

(6) $$H \trianglelefteq G \iff N_G H = G.$$

By F4, $G : N_G H$ is the number of subgroups of G conjugate to H.

Example 7. Take $H \leq G$. Then the action of G on G/H via $(\sigma, \tau H) \mapsto \sigma \tau H$ is *transitive*. The stabilizer of τH is the subgroup $\tau H \tau^{-1}$ of G, since for $\sigma \in G$ we have $\sigma(\tau H) = \tau H \iff \sigma \in \tau H \tau^{-1}$.

Example 8. In the situation of the preceding example let $U \leq G$ be another subgroup besides H. Then U acts on G/H via $(\sigma, \tau H) \mapsto \sigma \tau H$. The orbit of τH is the set

(7) $$\{\sigma \tau H \mid \sigma \in U\}$$

of left cosets mod H; the stabilizer of τH is

(8) $$\tau H \tau^{-1} \cap U,$$

by Example 7. The union of cosets of (7) is the set

$$U \tau H.$$

This is called the *double coset of τ relative to U and H*. Clearly,

(a) *G is the disjoint union of all the distinct double cosets relative to U and H.*

If we now assume that G is finite and let m be the size of the orbit (7) of τH, we get $|U \tau H| = m \cdot |H|$. By virtue of F4, then, $m = U : (\tau H \tau^{-1} \cap U)$; see (8). Thus the cardinality of the double coset $U \tau H$ is

(9) $$|U \tau H| = \frac{|U| \, |H|}{|\tau H \tau^{-1} \cap U|}$$

Moreover

$$|\tau H \tau^{-1} \cap U| = |H \cap \tau^{-1} U \tau| = |H \cap U^\tau|,$$

as can easily be seen by applying the inner automorphism corresponding to τ. Thus:

(b) *Let G be finite and let T be a complete set of representatives of the distinct double cosets of G relative to U and H. Then*

(10)
$$|G| = \sum_{\tau \in T} \frac{|U||H|}{|\tau H \tau^{-1} \cap U|} = \sum_{\tau \in T} \frac{|U||H|}{|H \cap U^{\tau}|}.$$

For the rest of this chapter G will denote a *finite group* and p a *prime number*.

Definition 6. (a) A subgroup $H \leq G$ with $|H| = p^m$ is called a *p-subgroup* of G.

(b) Suppose $|G| = p^n a$ with $(a, p) = 1$, so that p^n is the highest power of p dividing $|G|$. A subgroup $H \leq G$ such that $|H| = p^n$ is called a *Sylow p-subgroup of G*. We denote by

$$\mathrm{Syl}_p G$$

the set of such subgroups. For an arbitrary subgroup H of G, F3 gives

(11)
$$H \in \mathrm{Syl}_p G \iff H \text{ is a } p\text{-group and } G:H \not\equiv 0 \bmod p.$$

Example 9. Consider the group $G = \mathrm{GL}(n, p) = \mathrm{GL}(n, \mathbb{F}_p)$ of invertible $n \times n$ matrices with coefficients in the field $\mathbb{F}_p$. It is easy to see that G has order

(12)
$$|G| = (p^n - 1)(p^n - p) \dots (p^n - p^{n-1}),$$

since there are $p^n - 1$ possibilities for what the matrix does to the first element of the canonical basis of $\mathbb{F}_p^n$, then $p^n - p$ for the second, and so on. The highest power of p that fits in $|G|$ is thus

(13)
$$p^{1+2+\dots+(n-1)} = p^{n(n-1)/2}.$$

The subgroup

$$P := \left\{ \begin{pmatrix} 1 & & & * \\ & 1 & & \\ & & \ddots & \\ 0 & & & 1 \end{pmatrix} \right\} \leq \mathrm{GL}(n, p)$$

of all upper triangular matrices with 1 in the diagonal obviously has order (13). Thus P is a Sylow p-subgroup of G.

Lemma. *Suppose $H \leq G$ and let P be a Sylow p-subgroup of G. Then there exists $\tau \in G$ such that*

$$H \cap P^{\tau}$$

is a Sylow p-subgroup of H.

Proof. We look at the double coset decomposition of G relative to H and P. By (10) we have

(14)
$$|G| = \sum_{\tau \in T} \frac{|H||P|}{|H \cap P^{\tau}|}.$$

Dividing this equation by the highest possible power of p, namely $p^n = |P|$, we see that for at least one τ we must have

$$\frac{|H|}{|H \cap P^\tau|} = H : H \cap P^\tau \not\equiv 0 \bmod p.$$

On the other hand, $H \cap P^\tau$ is a p-group, since it is a subgroup of P^τ. Thus, by (11), $H \cap P^\tau$ is a Sylow p-subgroup of H. □

Theorem 1 (Sylow's Theorems).

First. *G contains a Sylow p-subgroup. Every p-subgroup of G is contained in some Sylow p-subgroup of G.*

Second. *Any two Sylow p-subgroups of G are conjugate.*

Third. *Let n_p be the number of Sylow p-subgroups of G. Then*
  (a) *n_p divides $G : P$ for $P \in \mathrm{Syl}_p G$, and*
  (b) *$n_p \equiv 1 \bmod p$.*

Remarks. The number n_p actually satisfies
  (a′) $n_p = G : N_G P$, and also
  (b′) $n_p \equiv 1 \bmod p^d$ for every d such that

(15) $P : P \cap P' \equiv 0 \bmod p^d$ for every $P' \in \mathrm{Syl}_p G$ distinct from P.

Now (b) will follow from (b′), since (15) is obviously satisfied for $d = 1$.

As a consequence of Sylow's second theorem, *a **normal** Sylow p-subgroup of G is the **only** Sylow p-subgroup of G*. The converse is clear: *If G only has one Sylow p-subgroup, this group is normal in G*, since any conjugate of a Sylow p-subgroup of G is also a Sylow p-subgroup.

Proof of Sylow's Theorems. (1) Let $n = |G|$. By F2, G is isomorphic to a subgroup of S_n. By associating to each $\pi \in S_n$ the permutation matrix $P_\pi \in \mathrm{GL}(n, \mathbb{F}_p)$ that accounts for the effect of π on the canonical basis $e_1, \ldots, e_n$ (that is, $P_\pi e_i = e_{\pi(i)}$), we get an injective homomorphism of S_n into $\mathrm{GL}(n, \mathbb{F}_p)$. Thus G is isomorphic to a subgroup H of $\mathrm{GL}(n, \mathbb{F}_p)$. But for $\mathrm{GL}(n, \mathbb{F}_p)$ we have produced a Sylow p-subgroup in Example 9. By the Lemma on the previous page, this means that H, too, has a Sylow p-subgroup, and thus so does G.

(2) Take $P \in \mathrm{Syl}_p G$ and let $H \leq G$ be any p-subgroup of G. By the Lemma, there exists $\tau \in G$ such that $H \cap P^\tau \in \mathrm{Syl}_p H$. But H is also a p-group, so $H \cap P^\tau = H$, and therefore $H \subseteq P^\tau \in \mathrm{Syl}_p G$. This completes the proof of Sylow's first theorem. Now, if $H \in \mathrm{Syl}_p G$, it follows that $H = P^\tau$, since both groups have same order. This proves Sylow's second theorem.

(3) Take $P \in \mathrm{Syl}_p G$. Sylow's second theorem yields $n_p = |\{P^\tau \mid \tau \in G\}| = G : N_G P$ (see Example 6). This proves (a′), hence also (a).

(4) There remains to prove (b'). To do this consider equation (14) with $H = N_G P$. Then

$$P \tau N_G P, \quad \text{for } \tau \in T,$$

runs over all the distinct double cosets of G relative to P and $N_G P$. We may as well assume that $1 \in T$. For $\tau \neq 1$ in T, then, we have $\tau \notin N_G P$, so $P^\tau \neq P$. Being a p-group, $P^\tau \cap N_G P$ is contained in a Sylow p-subgroup of $N_G P$, thanks to Sylow's first theorem. But then, being normal in $N_G P$, the group P must be the unique Sylow p-subgroup of $N_G P$, so that $P^\tau \cap N_G P \subseteq P$. It follows that $P^\tau \cap N_G P \subseteq P^\tau \cap P$, and hence

$$P^\tau \cap N_G P = P^\tau \cap P.$$

Dividing (14) by $|H| = |N_G P|$ we get

$$G : N_G P = 1 + \sum_{\tau \in T \setminus \{1\}} (P : P^\tau \cap P).$$

Using (a') it now follows that $n_p \equiv 1 \mod p^d$ for any d that obeys (15). □

Remarks. Together with F8, Sylow's first theorem immediately implies that *for any prime power p^k dividing the order of G there exists a subgroup H of G such that $|H| = p^k$*. For $k = 1$, in particular, we get a theorem that goes back to Cauchy: *For any prime p dividing the order of G, there is an element of order p in G.*

The Norwegian mathematician Ludwig Sylow (1832–1918) recognized the significance of Cauchy's Theorem and fleshed it out into the three statements that bear his name, which have played, ever since their publication (in 1872 in the *Mathematische Annalen*), a fundamental role in group theory.

If a finite group G is *abelian* (or just *nilpotent*—see §10.19 in the Appendix), it is actually the case that there is a subgroup of *any* order dividing the order of G. But not so for general groups; already the alternating group A_4 has no subgroup of order 6 (see §15.11 in the Appendix).

11

Applications of Galois Theory to Cyclotomic Fields

1. We start by considering, in the light of Galois theory, the general problem of constructibility with ruler and compass, which has served as our lodestar, so to speak, since Chapter 1.

Theorem 1. *Let $M \subseteq \mathbb{C}$ contain* 0 *and* 1, *and set*

$$K := \mathbb{Q}(M \cup \overline{M}).$$

Given $z \in \mathbb{C}$, there is equivalence between:

(i) $z \in \triangle M$ *(that is, z is constructible with ruler and compass);*

(ii) *z is algebraic over K, and the degree over K of the normal closure E/K of $K(z)/K$ is a power of* 2.

Proof. The implication (ii) $\Rightarrow$ (i) is especially interesting, and we prove it first. By assumption the extension E/K is Galois, and its Galois group $G = G(E/K)$ is a 2-group. Thus, by F8 in Chapter 10, there exists a chain

$$(1) \qquad\qquad G = H_0 \supseteq H_1 \supseteq \ldots \supseteq H_n = 1$$

of subgroups of G (normal, in fact) such that $H_{i-1} : H_i = 2$. By Galois theory (Chapter 8, Theorem 5) this chain of *subgroups H_i* of G has a corresponding chain

$$(1') \qquad\qquad K = K_0 \subseteq K_1 \subseteq \cdots \subseteq K_n = E$$

of *intermediate fields K_i* of E/K, with $K_i : K_{i-1} = H_{i-1} : H_i = 2$. By Theorem 1′ in Chapter 1, this implies that $z \in \triangle M$.

The implication (i) $\Rightarrow$ (ii) just serves to round out the picture; its proof, though elementary, is quite instructive. Taking $z \in \triangle M$, we know from Theorem 1 in Chapter 1 that z lies in a certain extension K_m of K that can be obtained from K by successively adjoining square roots:

$$z \in K_m = K(w_1, w_2, \ldots, w_m),$$

where $w_i^2 \in K(w_1, \ldots, w_{i-1})$ and $w_i \notin K(w_1, \ldots, w_{i-1})$ for $1 \leq i \leq m$.

Let E_m/K be the normal closure of K_m/K. Then E_m/K is a finite Galois extension. Clearly, to prove our assertion it is enough to show that $E_m : K$ is a power of 2. This we do by induction on m. For $m = 1$ we have $K_1 = K(w_1)$ with $w_1^2 \in K$, $w_1 \notin K$; in this case K_1/K is normal, so $E_1 = K_1$ and hence $E_1 : K = 2$. Now suppose $m > 1$ and set $K_{m-1} := K(w_1, \ldots, w_{m-1})$. By the induction hypothesis, the normal closure E_{m-1}/K of K_{m-1}/K has degree $E_{m-1} : K$ equal to a power of 2. We have $K_m = K_{m-1}(w_m)$.

Now let $\alpha_1 = w_m, \alpha_2, \ldots, \alpha_s$ be the distinct conjugates of w_m over K (in $\mathbb{C}$). Then $\alpha_i^2 \in E_{m-1}$, since α_i^2 is conjugate over K to the $w_m^2 \in K_{m-1}$. Now of course $E_m = E_{m-1}(\alpha_1, \ldots, \alpha_s)$, so $E_m : E_{m-1}$ must be a power of 2. But then $E_m : K = (E_m : E_{m-1})(E_{m-1} : K)$ too is a power of 2. $\qquad \square$

2. We now apply Theorem 1 to the problem of dividing the circle into n parts. As we saw in Chapter 1, the question is to decide, for a given $n \in \mathbb{N}$, whether or not the complex number

$$\zeta = e^{2\pi i/n}$$

lies in the field $\triangle \mathbb{Q}$. The extension $\mathbb{Q}(\zeta)/\mathbb{Q}$ is normal; by Theorem 1, then, $\zeta \in \triangle \mathbb{Q}$ if and only if $\mathbb{Q}(\zeta) : \mathbb{Q}$ is a power of 2. By Gauss's Theorem (Chapter 9, Theorem 3) we have $\mathbb{Q}(\zeta) : \mathbb{Q} = \varphi(n)$, so we must figure out for what values of n the natural number $\varphi(n)$ is a power of 2. Let

$$(2) \qquad\qquad n = 2^e p_1^{e_1} \ldots p_r^{e_r}$$

be the prime factorization of n, where e is nonnegative, the primes $p_1, \ldots, p_r$ are odd and pairwise distinct, and each e_i is at least 1. Applying φ to (2) and taking multiplicativity into account (see (19) in Chapter 9), we get

$$\varphi(n) = \varphi(2^e) \varphi(p_1^{e_1}) \ldots \varphi(p_r^{e_r}).$$

Since $\varphi(p^m) = (p-1)p^{m-1}$ for $m \geq 1$, we see that $\varphi(n)$ is a power of 2 if and only if all the e_i's are 1 and $p_i - 1$ is a power of 2 for every i. This leads to:

Theorem 2 (Gauss). *A regular n-gon is constructible with ruler and compass if and only if*

$$n = 2^e p_1 p_2 \ldots p_r,$$

where $e \geq 0$ is arbitrary and $p_1, \ldots, p_r$ are distinct primes of the form

$$p_i = 1 + 2^{2^{k_i}}.$$

The only thing in this result that has not yet been proved is:

Lemma. *For $m \in \mathbb{N}$, the integer $1 + 2^m$ cannot be prime unless m is a power of 2.*

Proof. Suppose $m = m_1 m_2$, with $m_2 > 1$ odd. Then

$$1 + 2^m = 1 - (-2^{m_1})^{m_2} = (1 + 2^{m_1})(1 - 2^{m_1} + 2^{m_1 2} - \cdots + 2^{m_1(m_2-1)})$$

is composite. $\qquad \square$

Remarks. For $k \in \mathbb{N} \cup \{0\}$, the number

$$(3) \qquad\qquad F_k = 1 + 2^{2^k}$$

is called the k-th *Fermat number*. It is easy to show that

$$(4) \qquad\qquad F_0 = 3, \; F_1 = 5, \; F_2 = 17, \; F_3 = 257, \; F_4 = 65537$$

are prime. Fermat (1601–1665), the great reviver of number theory in the modern era (see Winfried Scharlau and Hans Opolka, *From Fermat to Minkowski*, Springer, 1985), occupied himself, in a context totally different from circle division, with the numbers that now bear his name; see Pierre de Fermat, *Œuvres* I, 127ff. He stated that they are all primes (*Œuvres* II, p. 309). But already the next Fermat number,

$$F_5 = 1 + 2^{32} = 641 \cdot 6700417,$$

is composite, as Euler (1707–1783) found out and showed more or less as follows: Because $641 = 5 \cdot 2^7 + 1 = 5^4 + 2^4$, there is a congruence

$$2^{32} = 2^4 \cdot (2^7)^4 \equiv -5^4 \cdot (2^7)^4 = -(5 \cdot 2^7)^4 \equiv -1 \bmod 641.$$

Fermat's mistake is noteworthy because he himself had found counterexamples of the caliber of Euler's in the case of the *Mersenne numbers* $2^p - 1$; but he seems to have relied on Frenicle, who apparently agreed with the claim (Fermat, *Œuvres* II, p. 208).

Today we know that F_k is composite at least for $5 \le k \le 32$, and we don't know whether there are any prime Fermat numbers after those in (4). (Note that F_{17} already has 39457 decimal digits, and F_{32} has over a billion.)

It is easy to see from the definition (3) that $F_{k+1} - 2 = (F_k - 2) F_k$; by induction,

$$(5) \qquad\qquad F_m - 2 = \prod_{0 \le k < m} F_k.$$

Thus $n = F_5 - 2 = 2^{32} - 1 = F_0 F_1 F_2 F_3 F_4$ is the largest known *odd* number of sides that a constructible regular polygon can have; its decimal representation is 4294967295.

Incidentally, (5) implies that any two distinct Fermat numbers are relatively prime. Thus there appear infinitely many primes in the factorization of Fermat numbers.

3. Given a *prime number* $p \neq 2$, we now look at the field $\mathbb{Q}(\zeta_p)$ of p-th roots of unity over $\mathbb{Q}$ (where ζ_p denotes a p-th root of unity $\neq 1$ in $\mathbb{C}$, for example $\zeta_p = e^{2\pi i/p}$) and ask *what square roots of nonzero rational numbers lie in this field*. For $d_1, d_2 \in \mathbb{Q}^\times$, the extensions $\mathbb{Q}(\sqrt{d_1})$ and $\mathbb{Q}(\sqrt{d_2})$ coincide if and only if $d_1 = x^2 d_2$, for $x \in \mathbb{Q}^\times$. Thus we should investigate *quadratic subfields* of $E = \mathbb{Q}(\zeta_p)$, that is, subfields F of E such that $F : \mathbb{Q} = 2$. Now, $E/\mathbb{Q}$ is a Galois extension, with Galois group

$G = G(E/\mathbb{Q})$ isomorphic to $(\mathbb{Z}/p)^{\times}$; see Chapter 9, Theorem 3. The group $(\mathbb{Z}/p)^{\times}$ is *cyclic* of order $p-1$, by Theorem 2 in Chapter 9. Thus for any factor t of $p-1$ the group G has a unique subgroup of order t (Chapter 9, F6); then by Galois theory (Theorem 5 in Chapter 8) E has exactly one subfield F such that $E:F = t$. Setting $t = (p-1)/2$, we conclude that $E = \mathbb{Q}(\zeta_p)$ *has a unique subfield F with* $F:\mathbb{Q} = 2$. In other words: *There exists a unique square-free integer $d \neq 1$ such that* $\sqrt{d} \in \mathbb{Q}(\zeta_p)$.

What then is this number d, as a function of p?

Notation. For a prime number $p \neq 2$, set

(6)
$$p^* = \begin{cases} p & \text{for } p \equiv 1 \bmod 4, \\ -p & \text{for } p \equiv 3 \bmod 4. \end{cases}$$

If desired, this can also be written as

(6')
$$p^* = (-1)^{(p-1)/2} p.$$

Theorem 3. *For every prime $p \neq 2$, the extension $\mathbb{Q}(\sqrt{p^*})$ is the unique quadratic subfield of $\mathbb{Q}(\zeta_p)$, and thus p^* is the unique square-free integer (apart from 1) such that $\sqrt{p^*} \in \mathbb{Q}(\zeta_p)$.*

Proof. As seen earlier, $\mathbb{Q}(\zeta_p)$ has a unique subfield F such that $F:\mathbb{Q} = 2$, and $F = \mathbb{Q}(\sqrt{d})$ for a unique square-free $d = d(p) \in \mathbb{Z}$. Instead of finding d "naively" (but see the remark in the proof of F1 below), we draw on F8 of Chapter 8, which is perfect for the situation. According to that result, $F = \mathbb{Q}(\sqrt{D})$, where $D = D(f)$ is the discriminant of the p-th cyclotomic polynomial

(7)
$$f(X) = 1 + X + X^2 + \cdots + X^{p-1}.$$

Then $d(p)$ is simply the square-free part of $D(f)$. So all we have to do is compute the discriminant $D(f)$; the result is

(8)
$$D(f) = (-1)^{(p-1)/2} p^{p-2},$$

according to the following lemma. Thus, as asserted,

$$d(p) = (-1)^{(p-1)/2} p = p^*. \qquad \square$$

Lemma. *For $p \neq 2$ prime, the discriminant of the polynomial $f = F_p$ is given by (8).*

Proof. For simplicity we write ζ for ζ_p. Substituting $X = 1$ into the polynomial $f(X) = \prod_{k=1}^{p-1}(X - \zeta^k)$ and taking (7) into account we get

(9)
$$\prod_{k=1}^{p-1}(1 - \zeta^k) = p.$$

By (17) in Section 8.3, we have

(10)
$$\varepsilon D(f) = \prod_{j \neq i} (\zeta^i - \zeta^j) = \prod_{i=1}^{p-1} f'(\zeta^i),$$

where $\varepsilon = (-1)^{n(n-1)/2} = (-1)^{(p-1)/2}$. Since $X^p - 1 = f(X)(X-1)$, this gives

$$p\zeta^{p-1} = f'(\zeta)(\zeta - 1).$$

Now taking the product over all conjugates, and applying (9) and (10), one obtains what is needed:

$$p^{p-1} = \varepsilon D(f) p. \qquad \square$$

We now wish to study more closely the action of the Galois group G on the quadratic subfield F of $\mathbb{Q}(\zeta)$ (where $\zeta = \zeta_p$ as above). By Theorem 3 in Chapter 9, there is first of all a natural isomorphism

(11)
$$(\mathbb{Z}/p)^\times \to G = G(\mathbb{Q}(\zeta)/\mathbb{Q}),$$

sending each element a mod p of $(\mathbb{Z}/p)^\times$ to the automorphism σ_a characterized by

(12)
$$\sigma_a(\zeta) = \zeta^a.$$

For each $\sigma \in G$, then, we have

(13)
$$\sigma(\sqrt{p^*}) = \chi(\sigma)\sqrt{p^*},$$

with a well defined $\chi(\sigma) \in \{1, -1\}$. The map $\chi : G \to \{1, -1\}$ is of course a group homomorphism; it is called the *sign character* (of p).

Definition. For $a \in \mathbb{Z}$ such that $(a, p) = 1$, set

(14)
$$\left(\frac{a}{p}\right) := \chi(\sigma_a).$$

This is called the *Legendre symbol* of a (relative to p). The element $\left(\frac{a}{p}\right)$ of $\{1, -1\}$ is thus fixed by the equation

(15)
$$\sigma_a(\sqrt{p^*}) = \left(\frac{a}{p}\right)\sqrt{p^*}.$$

As already observed,

(16)
$$\left(\frac{ab}{p}\right) = \left(\frac{a}{p}\right)\left(\frac{b}{p}\right).$$

By definition, $H := \ker \chi$ is the subgroup of G associated by Galois theory to the subfield $F = \mathbb{Q}(\sqrt{p^*})$. But since G is cyclic and of order $p-1$, it has only one subgroup of index 2, namely the group

(17)
$$H = \{\tau^2 \mid \tau \in G\}$$

of all squares in G (Chapter 9, F6). This, in view of the isomorphism (11), implies that

(18) $$\left(\frac{a}{p}\right) = 1 \iff a \bmod p \text{ is a square in } (\mathbb{Z}/p)^{\times}.$$

We see, then, that the Legendre symbol (14) has an elementary number-theoretic description:

(19) $$\left(\frac{a}{p}\right) = \begin{cases} 1 & \text{if } X^2 \equiv a \bmod p \text{ has a solution in } \mathbb{Z}, \\ -1 & \text{if } X^2 \equiv a \bmod p \text{ has no solution in } \mathbb{Z}. \end{cases}$$

(The choice of 1 and -1 instead of 0 and 1 to label the branches of this dichotomy is made so that (16) works.) Since H in (17) is the unique subgroup of order $(p-1)/2$ in G is, we obtain *Euler's criterion*:

(20) $$\left(\frac{a}{p}\right) = 1 \iff a^{(p-1)/2} \equiv 1 \bmod p.$$

Substituting $a = -1$ (and considering that then, since $p \neq 2$, congruence implies equality), we obtain

(21) $$\left(\frac{-1}{p}\right) = (-1)^{(p-1)/2} = \begin{cases} 1 & \text{for } p \equiv 1 \bmod 4, \\ -1 & \text{for } p \equiv 3 \bmod 4. \end{cases}$$

We know that $1, \zeta, \ldots, \zeta^{p-2}$ form a basis of the $\mathbb{Q}$-vector spaces $\mathbb{Q}(\zeta)$, since the minimal polynomial F_p of ζ over $\mathbb{Q}$ has degree $p - 1$. Multiplication by ζ transforms this basis into the basis

(22) $$\zeta, \zeta^2, \ldots, \zeta^{p-1},$$

which is ideal for Galois-theoretic considerations, consisting as it does of all the $\mathbb{Q}$-conjugates of the one element ζ. Now $\sqrt{p^*}$, like any other element of $\mathbb{Q}(\zeta)$, has a unique representation in the form

(23) $$\sqrt{p^*} = \sum_{\sigma \in G} a_{\sigma} \sigma(\zeta) \quad \text{with } a_{\sigma} \in \mathbb{Q}.$$

Applying some $\tau \in G$ we get $\chi(\tau)\sqrt{p^*} = \sum_{\sigma} a_{\sigma} \tau\sigma(\zeta)$; this yields a relation $\chi(\tau)a_1 = a_{\tau^{-1}}$ between coefficients. Thus

$$a_{\sigma} = a_1 \chi(\sigma) \quad \text{for any } \sigma \in G.$$

Then (23) becomes

(24) $$\sqrt{p^*} = a_1 \cdot \sum_{\sigma \in G} \chi(\sigma)\sigma(\zeta).$$

More can be said:

F1. *With the preceding notation,*

$$(25) \qquad \sqrt{p^*} = \pm \sum_{\sigma \in G} \chi(\sigma)\sigma(\zeta).$$

Thus, in terms of the Legendre symbol (see (14) and (12)), *we have*

$$(25') \qquad \sum_{a=1}^{p-1} \left(\frac{a}{p}\right)\zeta^a = \pm\sqrt{p^*}.$$

Proof. We must show that the rational factor a_1 in (24) satisfies $a_1^2 = 1$. There seems to be no tidy conceptual argument at hand to let us get around the dirty work. So we make a virtue of necessity and determine the square of

$$(26) \qquad \alpha := \sum_{\sigma} \chi(\sigma)\sigma(\zeta)$$

by a direct computation — whereby we incidentally establish again that the quadratic irrationality $\sqrt{p^*}$ does belong to $\mathbb{Q}(\zeta_p)$. Observing that $\chi(\sigma) = \chi(\sigma^{-1})$, we have

$$\alpha^2 = \sum_{\sigma,\tau} \chi(\sigma)\chi(\tau)\sigma(\zeta)\tau(\zeta) = \sum_{\sigma,\tau} \chi(\sigma^{-1}\tau)\sigma(\zeta)\tau(\zeta),$$

Let $\rho = \sigma^{-1}\tau$, so $\tau = \sigma\rho$. Since a sum over all (σ,τ) is also a sum over all (σ,ρ), we get

$$\alpha^2 = \sum_{\rho,\sigma} \chi(\rho)\sigma(\zeta)\sigma\rho(\zeta) = \sum_{\rho} \chi(\rho)\sum_{\sigma} \sigma(\zeta\rho(\zeta)).$$

We can evaluate the rightmost sum by observing that $\zeta\rho(\zeta)$ is a p-th root of unity, primitive unless ρ is the particular automorphism σ_{-1} defined by (12). For a p-th root of unity η,

$$\sum_{\sigma} \sigma\eta = \eta + \eta^2 + \cdots + \eta^{p-1} = \begin{cases} p-1 & \text{if } \eta = 1, \\ -1 & \text{if } \eta \neq 1. \end{cases}$$

Applying this to $\zeta\rho(\zeta)$, we get

$$\alpha^2 = \chi(\sigma_{-1})(p-1) - \sum_{\rho \neq \sigma_{-1}} \chi(\rho)$$

$$= \chi(\sigma_{-1})(p-1) + \chi(\sigma_{-1}) - \sum_{\rho} \chi(\rho) = p\chi(\sigma_{-1}),$$

where $\sum_{\rho} \chi(\rho)$ vanishes because $\ker \chi$ has index 2 in G. From (14) and (21) we know that

$$\chi(\sigma_{-1}) = \left(\frac{-1}{p}\right) = (-1)^{(p-1)/2},$$

so the desired equality $\alpha^2 = p^*$ follows. $\qquad\square$

Remark. In the preceding discussion $\sqrt{p^*}$ denoted, as usual for algebraic purposes, an *arbitrary* (but fixed) solution of the equation $X^2 - p^* = 0$ in $\mathbb{C}$ — which of the two is immaterial. In this light, the occurrence of the $\pm$ in (25) and (25') is not surprising. From the analytic viewpoint, however, it is relevant to ask which sign is the right one, if one understands by $\sqrt{p^*}$ the positive square root $\sqrt{p}$ in the case $p^* = p$ and the complex number $i\sqrt{p}$ in the case $p^* = -p$. For a given p, the factor $a_1 = a_1(\zeta) \in \{1, -1\}$ that appears in (24) depends at most on the choice of ζ; that it does so depend is easy to see because when one replaces ζ by another primitive p-th root ζ^k the sign behaves as follows:

$$(27) \qquad a_1(\zeta^k) = \left(\frac{k}{p}\right) a_1(\zeta).$$

From the point of view of analysis there does exist a *canonical* choice for ζ, namely $\zeta = e^{2\pi i/p}$, and the question then is which sign is appropriate for this choice. Gauss agonized over this problem for years, as he himself confessed, until he finally found the answer in 1805: the sign is always $+1$, independently of p. Readers interested in knowing the proof can find a matrix-theoretic one by Isaac Schur in his *Werke*, vol. II, and a function-theoretic one by Carl Ludwig Siegel in Chandrasekharan's *Introduction to Analytic Number Theory*.

In pursuit of explicit applications of Galois theory, we have wandered unawares into the realm of number theory. We might as well go one step further and prove the famous *quadratic reciprocity law* of Gauss. Without having to fuss about the exact sign determination of the *Gaussian sums* (26), we already have in hand a key to the problem, in the form of the relation $\alpha^2 = p^*$.

Let q be any prime distinct from the original prime p. It will prove convenient to work in the ring $R = \mathbb{Z}[\zeta]$ (still with $\zeta = \zeta_p$). For an arbitrary element of R, say

$$\alpha = \sum_\sigma a_\sigma \sigma(\zeta) \quad \text{with } a_\sigma \in \mathbb{Z},$$

we have

$$\sigma_q(\alpha) = \sum_\sigma a_\sigma \sigma(\zeta)^q \equiv \left(\sum_\sigma a_\sigma \sigma(\zeta)\right)^q \mod q,$$

and hence

$$(28) \qquad \sigma_q(\alpha) \equiv \alpha^q \mod qR.$$

Applying this to $\alpha = \sqrt{p^*}$ given by (25), one gets, using (15),

$$(29) \qquad \left(\frac{q}{p}\right)\sqrt{p^*} \equiv \left(\sqrt{p^*}\right)^q \mod qR.$$

After multiplying by $\sqrt{p^*}$, and assuming $q \neq 2$, we get

$$\left(\frac{q}{p}\right)p^* \equiv (p^*)^{(q+1)/2} \mod qR.$$

Since p^* and q are relatively prime, p^* is invertible modulo q in $\mathbb{Z} \subseteq R$, so

(30) $$\left(\frac{q}{p}\right) \equiv (p^*)^{(q-1)/2} \bmod qR.$$

But $q\mathbb{Z}[\zeta] \cap \mathbb{Z} = q\mathbb{Z}$ (as can be seen easily by comparing coefficients relative to the basis $1, \zeta, \ldots, \zeta^{p-2}$); thus (30) says simply that

$$\left(\frac{q}{p}\right) \equiv (p^*)^{(q-1)/2} \bmod q\mathbb{Z}.$$

Using *Euler's criterion* (20) we then obtain

$$\left(\frac{q}{p}\right) \equiv \left(\frac{p^*}{q}\right) \bmod q\mathbb{Z}.$$

Since $q \neq 2$ this finally yields

$$\left(\frac{q}{p}\right) = \left(\frac{p^*}{q}\right),$$

since the only values that can occur are 1 and -1. To summarize:

Theorem 4 (Quadratic reciprocity law). *If p and q are distinct odd primes,*

(31) $$\left(\frac{q}{p}\right) = \left(\frac{p^*}{q}\right),$$

which, taking into account (6′), (16) and (21), also means that

(32) $$\left(\frac{q}{p}\right) = (-1)^{\frac{1}{2}(p-1)\frac{1}{2}(q-1)}\left(\frac{p}{q}\right).$$

In other words:

(a) *If p or q is congruent to 1 mod 4, then $\left(\frac{q}{p}\right) = \left(\frac{p}{q}\right)$, that is, q is a quadratic residue mod p if and only if p is a quadratic residue mod q.*

(b) *If both p and q are congruent to 3 mod 4, then $\left(\frac{q}{p}\right) = -\left(\frac{p}{q}\right)$, that is, q is a quadratic residue mod p if and only if p is not a quadratic residue mod q.*

Equation (32) is completed by two "supplementary laws" for quadratic residues: For any odd prime p,

(33) $$\left(\frac{-1}{p}\right) = (-1)^{(p-1)/2}$$

and

(34) $$\left(\frac{2}{p}\right) = (-1)^{(p^2-1)/8} = \begin{cases} 1 & \text{for } p \equiv 1 \text{ or } -1 \bmod 8, \\ -1 & \text{for } p \equiv 3 \text{ or } -3 \bmod 8. \end{cases}$$

Proof. All that remains to prove is (34). To do this we work in the field of *eighth roots of unity* over $\mathbb{Q}$. Choose

(35) $$\zeta = e^{\pi i/4},$$

so

(36) $$\sqrt{2} = \zeta + \zeta^{-1},$$

and with this one gets a formula analogous to (25). Now, if ε_p is the sign such that

(37) $$\sigma_p(\sqrt{2}) = \varepsilon_p \sqrt{2},$$

then the relation parallel to (29) is

$$\varepsilon_p \sqrt{2} \equiv \sqrt{2}^p \mod p,$$

and one uses Euler's criterion exactly as above to obtain

(38) $$\varepsilon_p = \left(\frac{2}{p}\right).$$

But since $\sigma_p(\zeta) = \zeta^p$ we have

$$\varepsilon_p = 1 \iff \sigma_p(\sqrt{2}) = \sqrt{2} \iff \zeta^p + \zeta^{-p} = \zeta + \zeta^{-1} \iff p \equiv 1 \text{ or } -1 \mod 8,$$

since for the remaining p's we have $\zeta^p + \zeta^{-p} = -(\zeta + \zeta^{-1})$. Equation (34) follows.
$\square$

Example. Is 221 a quadratic residue modulo the prime 383? To apply Theorem 4, we must first factor 221 into primes. Then

$$\left(\frac{221}{383}\right) = \left(\frac{13}{383}\right)\left(\frac{17}{383}\right) = \left(\frac{383}{13}\right)\left(\frac{383}{17}\right) = \left(\frac{6}{13}\right)\left(\frac{9}{17}\right) = \left(\frac{6}{13}\right)$$
$$= \left(\frac{2}{13}\right)\left(\frac{3}{13}\right) = -\left(\frac{3}{13}\right) = -\left(\frac{13}{3}\right) = -\left(\frac{1}{3}\right) = -1,$$

so the congruence $X^2 \equiv 221 \mod 383$ has no solution in $\mathbb{Z}$.

Remarks. (a) It is clear that the algorithm illustrated by the preceding example will always succeed in determining whether a *fully factorized* number is a quadratic residue. It is possible to avoid the heavy computational burden of a preliminary prime factorization (apart from factoring out 2's). To do this, one extends the Legendre symbol into the *Jacobi symbol*, by setting

(39) $$\left(\frac{a}{b}\right) := \prod_p \left(\frac{a}{p}\right)^{w_p(b)}$$

whenever a, b are relatively prime integers such that $a \neq 0$ and b is odd and positive. Using Theorem 4 one can easily prove (exercise) the relations

(40) $$\left(\frac{b}{a}\right) = (-1)^{\frac{1}{2}(a-1)\frac{1}{2}(b-1)}\left(\frac{a}{b}\right),$$

where a is also assumed odd and positive, and

(41) $$\left(\frac{-1}{b}\right) = (-1)^{(b-1)/2}, \quad \left(\frac{2}{b}\right) = (-1)^{(b^2-1)/8}.$$

Since clearly (39) depends only on a modulo b, the Jacobi symbol achieves what is desired. For example, although $1363 = 29 \cdot 47$ is not a prime, we can write, letting b be the Fermat prime 65537:

$$\left(\frac{1363}{65537}\right) = \left(\frac{113}{1363}\right) = \left(\frac{7}{113}\right) = \left(\frac{1}{7}\right) = 1.$$

Thus 1363 is a quadratic residue mod 65537.

(b) The deeper meaning of the quadratic reciprocity law, which far transcends its nice algorithmic handiness, is beyond our scope. Nonetheless, there is an important fact that we can touch upon, and which can be read off from the properties (40) and (41) of the Jacobi symbol, namely: *The value of $\left(\frac{a}{b}\right)$ depends on $b \in \mathbb{N}$ only modulo $4a$ (and if $a \equiv 1$ mod 4, then $\left(\frac{a}{b}\right)$ depends on b only modulo a).* For a given integer a, then, we have: If p and p' are odd primes not dividing a, then

$$\left(\frac{a}{p}\right) = \left(\frac{a}{p'}\right), \quad \text{if } p \equiv p' \text{ mod } 4a.$$

Thus the quadratic residue behavior of a modulo p is the same for all primes p that belong to the same residue class mod $4a$. This implies (though we cannot go into it here — see F. Lorenz, *Algebraische Zahlentheorie*, BI-Verlag, 1993) that the decomposition behavior of primes p in the quadratic number field $\mathbb{Q}(\sqrt{a})$ depends only on p modulo $4a$. (In the case $a = -1$, compare with §4.18 in the Appendix.)

(c) A word on the history of the quadratic reciprocity law. In 1875 Kronecker (see vol. II of his *Werke*) called attention to the fact that the law was first stated by Euler — as far back as 1744 in its essentials, and in much more developed form again in 1783. Thus Euler preceded Legendre (1785) and Gauss (1801), who seem to have overlooked his remarks. However, the indisputable merit of finding a real proof for what was until then a heuristic observation belongs to Gauss; see his *Disquisitiones Arithmeticae*, where he in fact makes some historical remarks as well (articles 151, 296, 297 and Addenda at the end).

(d) The quadratic reciprocity law is a mathematical gem in itself. It has also proved to be a landmark along a road which, in the realm of *algebraic number theory*, has led toward fuller awareness of class regularity for abelian number field extensions, an awareness that has found deep expression in *Artin's reciprocity law*. (Again, see my *Algebraische Zahlentheorie*.) Furthermore today a trail is being blazed past the abelian case (see the survey article by J. Neukirch in *Ein Jahrhundert Mathematik, 1890–1990: Festschrift zum Jubiläum der DMV*, Deutsche Mathematiker-Vereinigung and Vieweg, 1990).

12

Further Steps into Galois Theory

1. What happens to the Galois group of a polynomial when the ground field is extended? Let $f \in K[X]$ be separable and let G be the Galois group of f over K. If K' is any extension of K, it is easy to see that the Galois group of f *over K'* can be regarded as a subgroup of G. More generally:

Theorem 1 (Translation theorem). *Let E/K be a Galois extension and K'/K any field extension. Assume, without loss of generality, that E and K' are subfields of a field C, and let $EK' = K'(E)$ be their composite in C.*

(a) *EK'/K' is a Galois extension.*

(b) *The map $G(EK'/K') \to G(E/K)$ defined by restriction gives rise to an isomorphism*

$$G(EK'/K') \simeq G(E/E \cap K').$$

The Galois group $G(EK'/K')$ of the extension EK'/K' can thus be identified with a subgroup H of $G(E/K)$, namely the subgroup corresponding to the intermediate field $E \cap K'$ of E/K.

Proof. Clearly EK'/K' is algebraic and separable. Because E/K is normal, E is the splitting field of some set $M \subseteq K[X]$ of polynomials over K. Thus $EK' = K'(E)$ is the splitting field of $M \subseteq K'[X]$ over K', meaning that EK'/K' is normal. Thus EK'/K' is Galois. Obviously, the restriction

$$r : G(EK'/K') \to G(E/K)$$

(1)

$$\sigma \mapsto \sigma_E$$

is a homomorphism. If $\sigma \in G(EK'/K')$ acts trivially on E, it acts trivially on $EK' = K'(E)$, since σ acts trivially on K' by definition. Thus r is injective.

There remains to show that the image of r equals $G(E/E \cap K')$. Let H be this image. Since every $\sigma \in G(EK'/K')$ fixes K' pointwise, $K' \cap E$ is contained in the fixed field E^H of H in E. Conversely, anything in E^H is left fixed by all elements of $G(EK'/K')$, and therefore lies in K'. Therefore

(2)

$$E^H = K' \cap E.$$

If E/K is assumed *finite*, this equality implies the desired conclusion that $H = G(E/K' \cap E)$, by Theorem 5 in Chapter 8. For the case of an infinite Galois extension E/K, readers might try to give a justification on their own. It will soon become apparent how nice it would be to have some appropriate generalization of the *Fundamental theorem of Galois theory* (Chapter 8) applicable to *infinite Galois extensions*. We will get to that in Section 12.4. Given the results there, the conclusion that $H = G(E/K' \cap E)$ easily follows (see Remarks on page 130). $\qquad \square$

The state of affairs described by Theorem 1 can be conveniently visualized through a diagram:

(3)

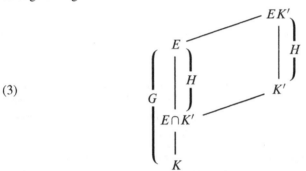

F1. *In the same situation as Theorem 1, assume further that E/K is finite. Then*

(4) $$EK' : K' \text{ divides } E : K.$$

Proof. By Theorem 1, the degree $EK' : K'$ coincides with the order of the subgroup $G(E/E \cap K')$ of $G(E/K)$, and so divides $E : K$. $\qquad \square$

Remark. Relation (4) need not hold when E/K is not Galois. For instance, consider

$$K = \mathbb{Q}, \quad E = \mathbb{Q}(\sqrt[3]{2}), \quad K' = \mathbb{Q}(\zeta_3 \sqrt[3]{2}),$$

where ζ_3 denotes a primitive third root of unity. It is easy to see that $EK' = \mathbb{Q}(\sqrt[3]{2}, \zeta_3)$, $E : K = K' : K = 3$, and $EK' : E = 2$, so $EK' : K' = 2$.

F2. *Let E_1/K and E_2/K be Galois extensions. Assume, without loss of generality, that E_1 and E_2 are subfields of an algebraically closed field C, and let $E_1 E_2$ be their composite in C. Then:*

(a) *$E_1 E_2/K$ is a Galois extension.*

(b) *The group homomorphism*

$$h : G(E_1 E_2/K) \to G(E_1/K) \times G(E_2/K)$$
$$\sigma \mapsto (\sigma_{E_1}, \sigma_{E_2})$$

is injective. If $E_1 \cap E_2 = K$, the map h is an isomorphism.

Proof. (a) Clearly $E_1 E_2/K$ is separable; see Chapter 7, F7.

Let $\sigma : E_1 E_2 \to C$ be a K-homomorphism. For $i = 1, 2$, we have $\sigma(E_i) \subseteq E_i$, since E_i/K is normal; there follows $\sigma(E_1 E_2) \subseteq E_1 E_2$, so $E_1 E_2/K$ is normal.

(b) If $\sigma \in G(E_1 E_2/K)$ acts as the identity on both E_1 and E_2, it does the same on $E_1 E_2$. Therefore h is injective. Now assume $E_1 \cap E_2 = K$. By Theorem 1, there exists for any $\sigma_1 \in G(E_1/K)$ some $\rho \in G(E_1 E_2/E_2)$ such that $\sigma_1 = \rho_{E_1}$. Likewise, for any $\sigma_2 \in G(E_2/K)$ there exists some $\tau \in G(E_1 E_2/E_1)$ such that $\sigma_2 = \tau_{E_2}$. Set $\sigma = \rho\tau$. Then $\sigma_{E_1} = \rho_{E_1}\tau_{E_1} = \rho_{E_1} = \sigma_1$, so $\sigma_{E_1} = \sigma_1$; similarly, $\sigma_{E_2} = \sigma_2$. This shows that h is surjective. $\qquad\square$

Remark. Let things be as in F2, and set $L := E_1 \cap E_2$. Clearly L/K is normal; let $\Delta = G(L/K)$ be its Galois group. If $p_i : G(E_i/K) \to G(L/K)$ is the canonical map, for $i = 1, 2$, define

$$G(E_1/K) \times_\Delta G(E_2/K) := \{(\sigma_1, \sigma_2) \in G(E_1/K) \times G(E_2/K) \mid p_1\sigma_1 = p_2\sigma_2\}.$$

One easily persuades oneself that *the image of the homomorphism h defined above is precisely the subgroup $G(E_1/K) \times_\Delta G(E_2/K)$ of $G(E_1/K) \times G(E_2/K)$, and thus yields an isomorphism*

$$(5) \qquad G(E_1 E_2/K) \simeq G(E_1/K) \times_\Delta G(E_2/K).$$

Indeed: For $h(\sigma) = (\sigma_{E_1}, \sigma_{E_2})$ we have $(\sigma_{E_1})_L = \sigma_L = (\sigma_{E_2})_L$, so $h(\sigma)$ lies in $G(E_1/K) \times_\Delta G(E_2/K)$. Conversely, if (σ_1, σ_2) is an element of this product, we find first an $\sigma \in G(E_1 E_2/K)$ such that $\sigma_L = (\sigma_1)_L = (\sigma_2)_L$. Then, for $i = 1, 2$, we have $\sigma_{E_i}^{-1}\sigma_i \in G(E_i/L)$; but by F2, part (b), $G(E_1 E_2/L) \to G(E_1/L) \times G(E_2/L)$ is surjective, so there exists $\tau \in G(E_1 E_2/L)$ such that $\tau_{E_i} = \sigma_{E_i}^{-1}\sigma_i$. Now the element $\tilde{\sigma} := \sigma\tau$ satisfies $\tilde{\sigma}_{E_i} = \sigma_i$ for $i = 1, 2$.

2. We now discuss a result of R. Dedekind that is of fundamental importance to field theory.

Theorem 2 (Linear independence of field homomorphisms). *Let E/K be a finite separable field extension of degree n, and let C be an algebraically closed extension of K. Denote by $\sigma_1, \sigma_2, \ldots, \sigma_n$ the distinct K-homomorphisms of E in C. Then $\sigma_1, \sigma_2, \ldots, \sigma_n$ are linearly independent over C; that is, for any $c_1, c_2, \ldots, c_n$ in C, the condition*

$$(6) \qquad \sum c_i\sigma_i(\beta) = 0 \quad \text{for all } \beta \in E$$

implies $c_1 = c_2 = \cdots = c_n = 0$.

The following demonstration is somewhat redundant because this theorem is contained in a result shortly to be stated (Theorem 2'), for which we give a different and simple proof. But the method of this first proof is of great intrinsic interest.

Proof of Theorem 2. Let $\beta_1, \ldots, \beta_n$ be a basis of E/K. Clearly, (6) is equivalent to

$$(7) \qquad \sum_{i=1}^{n} c_i\sigma_i(\beta_j) = 0 \quad \text{for } 1 \leq j \leq n.$$

The assertion of Theorem 2 is thus equivalent to the nonvanishing of the determinant of the $n \times n$ matrix $(\sigma_i(\beta_j))_{i,j}$. Set

$$(8) \qquad \Delta(\beta_1, \ldots, \beta_n) = \det(\sigma_i(\beta_j))_{i,j}.$$

So what we have to show is that for some (and therefore for *every*) basis $\beta_1, \ldots, \beta_n$ of E/K we have

$$(9) \qquad \Delta(\beta_1, \ldots, \beta_n) \neq 0.$$

Since E/K was assumed separable, the primitive element theorem (Chapter 8) yields an $\alpha \in E$ such that $E = K(\alpha)$, and we can consider the particular basis of E/K given by

$$(10) \qquad \beta_1 = 1, \quad \beta_2 = \alpha, \quad \beta_3 = \alpha^2, \quad \ldots, \quad \beta_n = \alpha^{n-1}.$$

Then $\sigma_i(\beta_j) = \sigma_i(\alpha^{j-1}) = \sigma_i(\alpha)^{j-1}$. Thus $\Delta(1, \alpha, \ldots, \alpha^{n-1}) = \det(\sigma_i(\alpha)^{j-1})_{i,j}$ is a Vandermonde determinant (see LA I, p. 155), and as such it evaluates to

$$(11) \qquad \Delta(1, \alpha, \ldots, \alpha^{n-1}) = \prod_{i<j} \left(\sigma_j(\alpha) - \sigma_i(\alpha)\right).$$

But since $\sigma_1, \ldots, \sigma_n$ are all distinct and α generates E over K, the $\sigma_1(\alpha), \ldots, \sigma_n(\alpha)$ are all distinct. Thus the particular basis (10) does satisfy (9). $\qquad \square$

As mentioned, Theorem 2 also follows from the next result:

Theorem 2' (Artin). *Let M be a monoid and F a field. If $\sigma_1, \ldots, \sigma_n$ are pairwise distinct homomorphisms of M into the multiplicative group $F^\times$ of F, then $\sigma_1, \ldots, \sigma_n$ are linearly independent over F.*

Proof. The proof rests on a simple fact from linear algebra:

Let B be a set of endomorphisms of an F-vector space V. Let $v_1, \ldots, v_n \in V$ be simultaneous eigenvectors of all the $\beta \in B$; that is, suppose that for each $\beta \in B$ and each $1 \leq i \leq n$ there is a unique $\lambda_i(\beta) \in F$ such that

$$\beta(v_i) = \lambda_i(\beta) v_i.$$

If the functions $\lambda_1, \ldots, \lambda_n : B \to F$ are all distinct, the vectors $v_1, \ldots, v_n$ are linearly independent.

To see this, consider a nontrivial linear dependence involving as few v_i's as possible; write it (after renumbering if necessary) as $v_1 = \sum_{i=2}^{k} c_i v_i$, with $k > 1$ and $c_i \neq 0$ for each i. Then $0 = \beta v_1 - \lambda_1(\beta) v_1 = \sum_{i=2}^{k} c_i(\lambda_i(\beta) - \lambda_1(\beta)) v_i$, leading to a shorter linear dependence if we take β such that $\lambda_2(\beta) \neq \lambda_1(\beta)$. Contradiction.

Now, to see why Theorem 2' follows from the linear algebra statement, consider the F-vector space $V = F^M$ of all maps from M into F. For every $\beta \in M$ we take an endomorphism of V — also denoted by β — as follows: For any $\sigma : M \to F$

in V, the image $\beta(\sigma) : M \to F$ is the map $\gamma \mapsto \sigma(\beta\gamma)$. Then, for each σ_i in the statement of the theorem, we have $\beta(\sigma_i)(\gamma) = \sigma_i(\beta\gamma) = \sigma_i(\beta)\sigma_i(\gamma)$, so

$$\beta(\sigma_i) = \sigma_i(\beta)\sigma_i.$$

Therefore each σ_i is an eigenvector of every β, with eigenvalue $\sigma_i(\beta)$ — here we used the fact that $\sigma_i \neq 0$. Since the maps $\sigma_1, \ldots, \sigma_n$ were assumed to be all distinct, the linear algebra statement applies, with σ_i playing the role of both λ_i and v_i. This proves that the σ_i are linearly independent.   □

Remark. Let E/K be a Galois extension of degree n, with Galois group G. By extension of the field of constants, we can make the tensor product $E \otimes_K E$ into a module over the group algebra EG in a natural way:

(12) $$c\sigma(x \otimes y) = \sigma x \otimes cy.$$

(For the definition of the group algebra see Section 6.2.) Now let $\sigma_1, \ldots, \sigma_n$ be the n distinct elements of G. Consider the homomorphism of EG-modules $f :$ $E \otimes_K E \to EG$ such that

$$f(x \otimes y) = \sum_{i=1}^{n} (\sigma_i^{-1}(x)\,y)\sigma_i.$$

We claim that f is an isomorphism, so we get a *canonical EG-module isomorphism*

(13) $$E \otimes_K E \simeq EG.$$

To prove this it is enough to show that f is injective, since the E-vector spaces $E \otimes_K E$ and EG both have dimension n. Let $\beta_1, \ldots, \beta_n$ be a basis of E/K. Every element z of $E \otimes_K E$ can be expressed (uniquely) in the form

$$z = \sum_{j=1}^{n} \beta_j \otimes c_j \quad \text{with } c_j \in E.$$

Now, if $f(z) = 0$, we have

$$0 = \sum_{j=1}^{n} \left(\sum_{i=1}^{n} \sigma_i^{-1}(\beta_j)c_j\sigma_i \right) = \sum_{i=1}^{n} \left(\sum_{j=1}^{n} \sigma_i^{-1}(\beta_j)c_j \right)\sigma_i.$$

There follows

$$\sum_{j=1}^{n} \sigma_i^{-1}(\beta_j)c_j = 0 \quad \text{for all } i.$$

But in the proof of Theorem 2 we saw that the matrix $\left(\sigma_i^{-1}(\beta_j) \right)_{i,j}$ cannot be markrightThe existence of a normal basissingular; consequently all the c_j vanish, proving the claim.

It is worth remarking that the isomorphism statement (13) can be viewed as a deep reason why Galois theory works. In this connection, E. Artin has shown that Theorem 2′ can serve as the starting point for a logical treatment of Galois theory; see E. Artin, *Galois theory*, Notre Dame mathematical lectures, 1942. See also the hint to §12.6 in the Appendix.

3. We now come to a beautiful and momentous theorem of Galois theory, which actually amounts to a strengthening of the isomorphism statement (13), as will be explained more precisely later.

Theorem 3 (Existence of normal bases). *Let E/K be a finite Galois extension with Galois group G. There exists an element α in E such that the family*

$$(14) \qquad (\sigma(\alpha))_{\sigma \in G}$$

*is a basis of E/K. Such a family is called a **normal basis** of E/K.*

Remarks. (i) Any $\tau \in G$ permutes the elements of a normal basis and is uniquely determined by this permutation.

(ii) If $(\sigma(\alpha))_\sigma$ is a normal basis of E/K, a minute's thought shows that $E = K(\alpha)$; in other words, α is a *primitive element* of E/K.

(iii) Let p be prime and let ζ_p denote a primitive p-th root of unity (in $\mathbb{C}$). Then $\alpha := \zeta_p$ gives rise to a normal basis of $\mathbb{Q}(\zeta_p)/\mathbb{Q}$. Indeed, as we know, the $\mathbb{Q}$-conjugates of ζ_p are

$$(15) \qquad \zeta_p, \zeta_p^2, \ldots, \zeta_p^{p-1};$$

but since $1, \zeta_p, \zeta_p^2, \ldots, \zeta_p^{p-2}$ form a basis of $\mathbb{Q}(\zeta_p)/\mathbb{Q}$, so do the elements in (15). The primeness of p is essential; the corresponding statement for, say, the field $\mathbb{Q}(i)$ of fourth roots of unity would be false, since $i, -i$ are linearly dependent over $\mathbb{Q}$.

Proof of Theorem 3. We first take an arbitrary element α of E and assume there is a relation

$$\sum_{\sigma \in G} a_\sigma \sigma(\alpha) = 0, \quad \text{with } a_\sigma \in K.$$

For any $\tau \in G$ we can apply τ^{-1} to the sum, obtaining

$$\sum_{\sigma \in G} a_\sigma \tau^{-1} \sigma(\alpha) = 0.$$

From this we see that to force all the a_σ to vanish, it suffices to ensure that

$$(16) \qquad \det\left(\tau^{-1}\sigma(\alpha)\right)_{\tau,\sigma \in G} \neq 0,$$

so our task is to prove that there exists $\alpha \in E$ with this property. By the *primitive element theorem* there exists $\beta \in E$ such that $E = K(\beta)$. Then

$$f(X) = \prod_{\sigma \in G} (X - \sigma\beta)$$

is the minimal polynomial of β over K; see Chapter 8, F1. For each $\sigma \in G$, consider
the polynomial

$$g^\sigma(X) = \frac{f(X)}{X - \sigma\beta} \in E[X];$$

then

(17)
$$g^\sigma(\beta) = 0 \quad \text{for} \quad \sigma \neq 1, \quad \text{but}$$
$$g^\sigma(\beta) \neq 0 \quad \text{for} \quad \sigma = 1.$$

Now let $d(X)$ be the determinant of the matrix

$$\left(g^{\tau^{-1}\sigma}(X)\right)_{\tau,\sigma} = \left(\frac{f(X)}{X - \tau^{-1}\sigma\beta}\right)_{\tau,\sigma} \in M_n(E[X]).$$

When we plug $X = \beta$ into this polynomial matrix we get a diagonal matrix whose
diagonal entries are equal and nonzero — all of this by (17). Taking the determinant
we get $d(\beta) \neq 0$, so $d(X)$ cannot be the zero polynomial:

(18) $d(X) \neq 0.$

Now assume that the field K has *infinitely many* elements; then, by (18), there exists
γ *in* K such that $d(\gamma) \neq 0$. So, for such a γ, the matrix

$$\left(\frac{f(\gamma)}{\gamma - \tau^{-1}\sigma\beta}\right)_{\tau,\sigma} = \left(\tau^{-1}\sigma\left(\frac{f(\gamma)}{\gamma - \beta}\right)\right)_{\tau,\sigma}$$

has nonzero determinant; thus the element

$$\alpha := \frac{f(\gamma)}{\gamma - \beta}$$

satisfies the desired condition (16). The existence of a normal basis is proved in the
case where the ground field K is infinite.[1]

In the case of *finite fields* we must resort to a different argument. In this case
Galois groups are necessarily *cyclic* (see Theorem 4 in Section 9.5). The finiteness
of K does not come in other than via this fact; for this reason we may as well
assume simply that

$$G = G(E/K) \text{ is cyclic.}$$

So let σ be a generating element of G and let $n = E : K$ be the order of σ. We
regard σ as an *endomorphism of the K-vector space* E, and show that the minimal
polynomial of this endomorphism is $X^n - 1$. Since $\sigma^n = 1$, clearly σ is a root
of $X^n - 1$. On the other hand, σ cannot be a root of a nonzero polynomial of
smaller degree (over K), because this would amount to a linear dependence relation
among the automorphisms $1, \sigma, \sigma^2, \ldots, \sigma^{n-1}$, contradicting Theorem 2'. Now, the

---

[1] It should be clear how the proof needs to be modified in order to obtain an α that gives rise
to a normal basis of E/F for every intermediate field F simultaneously.

dimension of the K-vector space E is n; therefore the *minimal polynomial* and the *characteristic polynomial* of σ coincide. But then there exists a *cyclic vector* for σ (see LA II, p. 168), which is to say some $\beta \in E$ such that

$$(19) \qquad \beta, \sigma\beta, \sigma^2\beta, \ldots, \sigma^{n-1}\beta$$

form a basis of E/K. By definition, the elements (19) form a normal basis of E/K.

$\square$

The existence theorem for normal bases can also be expressed as follows:

Theorem 3′. *Let E/K be a finite Galois field extension with Galois group G. The group algebra KG and E are isomorphic as KG-modules*:

$$(20) \qquad E \simeq KG \quad \text{as } KG\text{-modules.}$$

Proof. Every element x of KG has a unique representation

$$x = \sum_{\sigma \in G} a_\sigma \sigma.$$

E has a canonical KG-module structure given by the map

$$\left(\sum_\sigma a_\sigma \sigma, \alpha \right) \mapsto \sum_\sigma a_\sigma \sigma(\alpha).$$

Like any ring, KG is a module over itself via the map $(x, y) \mapsto xy$. Now, for any $\alpha \in E$, the map

$$(21) \qquad \sum_\sigma a_\sigma \sigma \mapsto \sum_\sigma a_\sigma \sigma\alpha$$

is obviously a KG-module homomorphism from KG into E. Conversely, given a KG-module homomorphism $\varphi : KG \to E$, set $\alpha := \varphi(1)$; then φ must have the form (21). But clearly (21) is an isomorphism if and only if the images $\sigma\alpha$ of the $\sigma \in G$ make up a basis of E/K. Thus we have shown the equivalence between Theorems 3 and 3′.

$\square$

Remarks. (1) Let E/K be an extension of *finite fields*. Since we know certain things about finite fields (for instance, that the multiplicative group $E^\times$ of E is cyclic), it does not seem totally unreasonable to ask whether one can exhibit more concretely a normal basis for E/K. It is true that the author has never come across a solution, but the problem is hereby posed anyway (see also §12.3 in the Appendix).

(2) The *existence theorem for normal bases* strengthens the earlier statement that

$$(22) \qquad E \otimes_K E \simeq EG$$

as EG-modules (see (13) in the remark following Theorem 2'). This is clear, because the existence of a normal basis for E/K implies, as we have seen, the isomorphism of KG-modules

$$(23) \qquad\qquad E \simeq KG;$$

by tensoring one then gets $E \otimes_K E \simeq KG \otimes_K E \simeq EG$, which is (22). Note, however, that in contrast with (22), the isomorphism (23) is not canonical.

(3) The existence proof we gave for normal bases is less than fully satisfying, in that it requires separate treatment for finite and infinite ground fields. It would be much nicer anyway to be able to derive the existence of the isomorphism (23) from the canonical isomorphism (22). This turns out to be possible, as proved by M. Deuring. More precisely:

Let M and M' be modules over a K-algebra A, both finite-dimensional over K. If for a finite field extension E/K there is an isomorphism $M \otimes_K E \simeq M' \otimes_K E$ of $A \otimes_K E$-modules, then there is an isomorphism $M \simeq M'$ of A-modules.

Sketch of proof. Every A-module M with $M : K < \infty$ is of course the direct sum of *directly indecomposable* submodules. By a fundamental theorem of Krull, Remak and Schmidt (which we will prove in Volume II, Chapter 28), such a decomposition is unique up to isomorphism and reordering of the indecomposable summands. Now, if we assume that $M \otimes_K E$ and $M' \otimes_K E$ are isomorphic as $A \otimes_K E$-modules, they are also isomorphic as A-modules. But *as A-modules* they also clearly satisfy (setting $n = E : K$)

$$M \otimes_K E \simeq M^n, \quad M' \otimes_K E \simeq M'^n,$$

and so also
$$M^n \simeq M'^n.$$

Applying the Krull–Remak–Schmidt Theorem we get $M \simeq M'$. □

Definition. Consider a finite Galois extension E/K with Galois group G. Let H be a subgroup of G and let F be the corresponding intermediate field of E/K. For $x \in E$, set

$$(24) \qquad\qquad \mathrm{Tr}_H(x) = \sum_{\sigma \in H} \sigma x.$$

Clearly $\mathrm{Tr}_H(x)$ is invariant under all the $\tau \in H$, and so lies in the fixed field F of H. Thus we get a function $\mathrm{Tr}_H : E \to F$, which we denote also by $\mathrm{Tr}_{E/F}$ and which we call the *trace* with respect to H (or to E/F).

F3. *Let E/K be a finite Galois extension with Galois group G. For a given subgroup H of G with fixed field F, let $\sigma_1 H, \ldots, \sigma_m H$ be the distinct left cosets of G with respect to H (so $m = G : H$ and an element of G lies in $\sigma_i H$ if and only if it coincides* *with σ_i on F). If for some $\alpha \in E$ the conjugates $\sigma\alpha$, for $\sigma \in G$, form a normal basis of E/K, then the elements*

$$(25) \qquad \sigma_1^{-1} \operatorname{Tr}_{E/\sigma_1 F}(\alpha), \quad \sigma_2^{-1} \operatorname{Tr}_{E/\sigma_2 F}(\alpha), \quad \ldots, \quad \sigma_m^{-1} \operatorname{Tr}_{E/\sigma_m F}(\alpha)$$

form a basis of F/K. If H is a normal subgroup of G (equivalently, if F/K is normal), the basis given by (25) is a normal basis of F/K. In any case we have $F = K(\operatorname{Tr}_{E/F}(\alpha))$.

Proof. A given $x \in E$ has a unique representation

$$x = \sum_{\sigma \in G} a_\sigma \sigma\alpha \quad \text{with } a_\sigma \in K,$$

and it lies in F if and only if it is invariant under every $\tau \in H$. But

$$\tau x = \sum_{\sigma \in G} a_\sigma \tau\sigma\alpha = \sum_{\rho \in G} a_{\tau^{-1}\rho}\rho\alpha = \sum_{\sigma \in G} a_{\tau^{-1}\sigma}\sigma\alpha,$$

so $\tau x = x$ for every $\tau \in H$ if and only if $a_{\tau^{-1}\sigma} = a_\sigma$ for every $\tau \in H$ and every $\sigma \in G$, that is, if and only if a is constant on every right coset $H\sigma$. The elements $\sigma_1^{-1}, \ldots, \sigma_m^{-1}$ each represent a different right coset of G modulo H, so x lies in F if and only if x is of the form

$$x = \sum_{i=1}^{m} a_i \left(\sum_{\tau \in H} \tau\sigma_i^{-1}\alpha \right), \quad \text{with } a_i \in K.$$

Notice here that $\tau\sigma_i^{-1} = \sigma_i^{-1}(\sigma_i\tau\sigma_i^{-1})$ and that the subgroup $\sigma_i H\sigma_i^{-1}$ is associated with the fixed field $\sigma_i F$, so we finally get, as a necessary and sufficient condition for x to lie in F, that x be of the form

$$x = \sum_{i=1}^{m} a_i \sigma_i^{-1} \operatorname{Tr}_{E/\sigma_i F}(\alpha), \quad \text{with } a_i \in K.$$

Such a representation is unique by construction, so we have in fact shown that the elements listed in (25) form a basis of F/K — note that they are indeed in F, since $\sigma_i^{-1}\operatorname{Tr}_{E/\sigma_i F}(\alpha) \in \sigma_i^{-1}(\sigma_i F) = F$.

If H is a normal subgroup we have $\sigma_i F = F$, so (25) lists precisely the conjugates of $\operatorname{Tr}_{E/F}(\alpha)$.

As for the last assertion of F3, one easily sees that in any case $\sigma_i \operatorname{Tr}_{E/F}(\alpha) \neq \sigma_j \operatorname{Tr}_{E/F}(\alpha)$ for $i \neq j$; thus $\operatorname{Tr}_{E/F}(\alpha)$ has at least $m = F:K$ distinct conjugates in E, and the statement follows. $\qquad \square$

4. Let K be a field and C a fixed algebraic closure of K. Denote by C_s the *separable closure* of K in C. Then C_s/K is a Galois extension, and any Galois extension E/K with ground field K can be regarded as intermediate to C_s/K.

Thus the Galois extension C_s/K might prove to be an especially worthy object of study. However, C_s/K is generally not a finite extension. The question, then, is to *find an appropriate generalization of the fundamental theorem of Galois theory from finite to arbitrary Galois extensions.* That the theorem as given in Chapter 8, Theorem 5 does not apply to infinite Galois extensions is shown by the following example. What happens is that the one-to-one correspondence between intermediate fields and subgroups of the Galois group breaks down — we know from Chapter 8, Theorem 2 that there is an injective map from the former to the latter always, but in the infinite case different subgroups of the Galois group may have the same fixed field.

Example. Let $K = \mathbb{F}_p$ be the prime field of characteristic $p > 0$ and set $\mathbb{F}_p\infty := C = C_s$. Denote by $\varphi \in G(C/\mathbb{F}_p)$ the corresponding *Frobenius automorphism*, defined by $\varphi x = x^p$. The *fixed field* of the subgroup $H = \langle \varphi \rangle$ of $G(C/\mathbb{F}_p)$ generated by φ consists of all elements $x \in C$ such that $x^p - x = 0$; therefore

(26)   fixed field of $\langle \varphi \rangle = \mathbb{F}_p$.

Nevertheless, we will show that the group corresponding to the intermediate field $\mathbb{F}_p$ does not coincide with H:

(27) $G(\mathbb{F}_p\infty/\mathbb{F}_p) \neq \langle \varphi \rangle$.

To do this we take any prime q and consider the subfield

(28) $$F = \bigcup_m \mathbb{F}_{p^{q^m}}$$

of $\mathbb{F}_p\infty$ consisting of all $x \in \mathbb{F}_p\infty$ such that $x^{p^{q^m}} = x$ for some $m \in \mathbb{N}$. For every $x \in F$, the degree $\mathbb{F}_p(x):\mathbb{F}_p$ is a q-power. Therefore $F \neq \mathbb{F}_p\infty$, and consequently there exists

$$\tau \in G(\mathbb{F}_p\infty/F) \quad \text{such that } \tau \neq 1.$$

Now, if (27) were not satisfied, τ would be a power of φ, say $\tau = \varphi^n$, where we may as well assume $n \in \mathbb{N}$ (otherwise replace τ by τ^{-1}). Then F would be contained in the fixed field $\mathbb{F}_{p^n}$ of $\varphi^n = \tau$. But this is impossible, since $\mathbb{F}_{p^n}/\mathbb{F}_p$ has degree n, whereas the field F in (28) has infinite degree over $\mathbb{F}_p$.   □

In the sequel, assume given *an arbitrary Galois extension* E/K, with Galois group $G = G(E/K)$. If F is an intermediate field of E/K that is *Galois over K*, and if $\sigma \in G$, we denote by

$$\sigma^F$$

the automorphism of F arising from σ. (We switch away from the more natural notation σ_F used up to now for reasons of convenience, which will soon become obvious.)

Now, we know that any element of E lies in a subfield L of E such that L/K is *finite* and *Galois*. Thus, if we let L run over all intermediate fields of E/K with this property, the natural homomorphism

(29)
$$h : G \to \prod_L G(L/K)$$
$$\sigma \mapsto (\sigma^L)_L$$

is *injective*. In the sequel we will continue to use L as a running index to designate all intermediate fields L of E/K such that L/K is finite and Galois.

How can the *image* of h in (29) be characterized? For $L \subseteq L'$ there is a canonical map

$$f_{L/L'} : G(L'/K) \to G(L/K)$$
$$\tau \mapsto \tau^L$$

and obviously if $L \subseteq L' \subseteq L''$ we have

$$f_{L/L''} = f_{L/L'} \circ f_{L'/L''}.$$

Moreover for $\sigma \in G$ the components of $h(\sigma)$ satisfy

$$f_{L/L'}(\sigma^{L'}) = \sigma^L \quad \text{if } L \subseteq L'.$$

This leads us to the following notion:

Definition. Let I be a (partially) *ordered* set of indexes; assume further that I is *directed*, which means that for any $i, i' \in I$ there exists $j \in I$ with $i \le j$ and $i' \le j$. Assume given a family $(G_i)_{i \in I}$ of sets (groups, rings, topological spaces, etc.) together with maps (homomorphisms)

$$f_{ij} : G_j \to G_i$$

for each pair (i, j) of indices in I such that $i \le j$. This setup is called a *projective system* if in addition we have

$$f_{ik} = f_{ij} \circ f_{jk} \quad \text{whenever } i \le j \le k.$$

The *projective limit* of such a projective system is defined as the following subset of the cartesian product of the G_i:

(30)
$$\varprojlim_{i \in I} G_i := \left\{ (\sigma_i)_i \in \prod_{i \in I} G_i \;\middle|\; f_{ij}(\sigma_j) = \sigma_i \text{ for } i \le j \right\}.$$

When the G_i are *groups* or *rings*, the projective limit is obviously a *subgroup* or *subring* of $\prod G_i$; if the G_i are *topological spaces*, the projective limit is a subspace of the topological space $\prod G_i$, and because all the f_{ij} are assumed continuous, it is in fact *closed* in $\prod G_i$ if all the G_i are *Hausdorff*.

Now let's get back to the situation that we had set up starting from an arbitrary Galois extension E/K. Thanks to the notion just introduced, we can state:

F4. *If E/K is an arbitrary Galois extension, the map* (29) *yields an isomorphism*

$$(31) \qquad G(E/K) \longrightarrow \varprojlim_{L} G(L/K),$$

where L runs over the set of all intermediate fields of E/K that are finite and Galois over K.

Proof. Everything is clear except the surjectivity of (31). So let

$$(32) \qquad (\sigma_L)_L \in \varprojlim_{L} G(L/K)$$

be given; we must show that there exists $\sigma \in G(E/K)$ such that

$$(33) \qquad \sigma^L = \sigma_L \quad \text{for every } L.$$

Since, as mentioned, E is the union of the L, an element $\sigma \in G(E/K)$ is fully determined by the conditions (33). Conversely, the existence of such a σ will be obvious if we can prove that, for any L_1, L_2 in our index set, the maps σ_{L_1} and σ_{L_2} coincide on $L_0 := L_1 \cap L_2$; in other words, that

$$\sigma_{L_1}^{L_0} = \sigma_{L_2}^{L_0}.$$

Let $L := L_1 L_2$ be the composite of L_1 and L_2 in E. Because of (32) we have

$$\sigma_L^{L_1} = f_{L_1/L}(\sigma_L) = \sigma_{L_1}, \quad \sigma_L^{L_2} = f_{L_2/L}(\sigma_L) = \sigma_{L_2}.$$

This indeed implies that

$$\sigma_{L_1}^{L_0} = (\sigma_L^{L_1})^{L_0} = \sigma_L^{L_0} = (\sigma_L^{L_2})^{L_0} = \sigma_{L_2}^{L_0}. \qquad \square$$

Definition. We talk of a *topological group* G when G has, besides a group structure, also a topology such that the map $(x, y) \mapsto xy^{-1}$ of $G \times G$ in G is *continuous*.

Remark. If $(G_i)_{i \in I}$ is a family of topological groups G_i, the cartesian product

$$\prod_{i \in I} G_i$$

is of course also a topological group. If $(G_i)_{i \in I}$ is a projective system, the projective limit

$$\varprojlim_{i \in I} G_i$$

is a topological group, since it is a subgroup of $\prod G_i$. As remarked earlier, the projective limit is in fact *closed* in G_i if the G_i are Hausdorff.

Definition. Let E/K be any Galois extension. We endow the Galois group $G(E/K)$ of E/K with a natural topological group structure as follows: Give each *finite* group $G(L/K)$ in (31) the *discrete topology* and then simply transfer to $G(E/K)$ the topological group structure of

$$(34) \qquad \qquad \tilde{G} = \varprojlim_{L} G(L/K),$$

via the isomorphism (31). The resulting topology on $G(E/K)$ is called the *Krull topology*.

F5. *Let E/K be a Galois field extension. The Galois group $G = G(E/K)$ of E/K becomes a **compact** topological group with the Krull topology. The family*

$$(35) \qquad \qquad (G(E/L))_L,$$

where L runs over all intermediate fields of E/K that are finite and Galois over K, is a fundamental system of open neighborhoods of 1 in G; that means, first, that each $G(E/L)$ is open, and second, that any neighborhood of 1 in G contains some $G(E/L)$.

Proof. By the well-known theorem of Tichonov (or Tychonoff), the cartesian product of compact topological spaces is compact. The spaces $G(L/K)$ are trivially compact, being finite and discrete; thus the projective limit (34), being a closed subset of the cartesian product of the $G(L/K)$, is also compact.

Let I be the set indexing the intermediate fields L. If S runs over the finite subsets of I, the sets

$$U_S := \prod_{L \in S} \{1\} \times \prod_{L \notin S} G(L/K)$$

form, by the definition of the product topology, a fundamental system of open neighborhoods of 1 in the cartesian product of the $G(L/K)$. Thus their intersections with

$$\tilde{G} = \varprojlim G(L/K)$$

also form a fundamental system of open neighborhoods of 1 in the topological group $\tilde{G}$. For given S, let L be the composite of all the $L' \in S$. It follows easily from the definition of the projective limit that

$$U_S \cap \tilde{G} = U_{\{L\}} \cap \tilde{G}.$$

But the inverse image of $U_{\{L\}} \cap \tilde{G}$ under the map in (31) is no other than the subgroup $G(E/L)$ of $G = G(E/K)$. This proves F5. $\qquad \square$

We can now extend to *infinite Galois extensions* the fundamental theorem of Galois theory, stated in Chapter 8, Theorem 5 for the finite case:

Theorem 4. *Let E/K be any Galois extension. The map*

(36) $$F \mapsto G(E/F)$$

*is a bijection between the set of intermediate fields of E/K and the set of **closed** subgroups of $G = G(E/K)$. This bijection maps intermediate fields of finite degree over K to open subgroups of G, and vice versa.*

Proof. (i) Any open subgroup H of a topological group G is closed in G, because $G \smallsetminus H$ is a union of cosets $gH \neq H$, and every gH is open, being the image of an open set H under the homeomorphism $x \mapsto gx$ of G.

(ii) Let F/K be a *finite* intermediate extension of E/K and denote by L the normal closure of F/K in E. Then L/K is also finite. By F5, $G(E/L)$ is open in G. Since $G(E/L)$ is contained in $G(E/F)$, the latter group is also *open* (being the union of cosets $gG(E/L)$, all of which are open).

(iii) As we know, for every intermediate field F of E/K the extension E/F is Galois, so F is the fixed field of $G(E/F)$ in E:

(37) $$F = E^{G(E/F)};$$

see Chapter 8, Definition 1 and Theorem 2. From (37) we immediately deduce that the map (36) is injective.

(iv) We claim that $G(E/F)$ is always *closed* in G. For let σ be an element of G and take $\sigma \notin G(E/F)$. There exists an intermediate field F_0 of F/K for which F_0/K is finite and on which σ is nontrivial — meaning that $\sigma \notin G(E/F_0)$. Therefore

$$\sigma G(E/F_0) \cap G(E/F) = \varnothing.$$

This justifies the claim, since the finiteness of F_0/K implies that $\sigma G(E/F_0)$ is an open neighborhood of σ in G, by (ii).

(v) Let H be an *open* subgroup of G. By F5, there exists a Galois subextension L/K of E/K such that
$$G(E/L) \subseteq H.$$
Let $F = E^H$ be the fixed field of H. From (37) we then get
$$F = E^H \subseteq E^{G(E/L)} = L,$$

so F/K is *finite* because L/K is.

(vi) To complete the proof of the theorem all we need to do is show (and this is the nub) that every *closed* subgroup H of G satisfies

(38) $$H = G(E/E^H).$$

First let H be any subgroup of G. Trivially, $H \subseteq G(E/E^H)$. We show that

(39) $$G(E/E^H) = H^-,$$

where H^- is the closure of H in G. Let $F = E^H$ be the fixed field of H in E, and let σ be any element of $G(E/F)$. To prove that $\sigma \in H^-$ it is sufficient to check that H intersects every fundamental neighborhood $\sigma G(E/L)$ of σ in $G = G(E/K)$:

$$(40) \qquad \sigma G(E/L) \cap H \neq \varnothing.$$

To see this, let H_0 be the image of H under the canonical map $G(E/F) \to G(LF/F)$. Then F is also the fixed field of H_0 in LF. But the Galois extension LF/F is *finite*, so the fundamental theorem for finite Galois extensions (Chapter 8, Theorem 5) yields

$$H_0 = G(LF/F).$$

Thus for the given $\sigma \in G(E/F)$ there is $\tau \in H$ such that $\sigma^L = \tau^L$. Setting $\rho = \sigma^{-1}\tau \in G(E/L)$ we conclude that $\sigma\rho = \tau$ lies in the intersection (40). $\qquad\square$

Remarks. Let F/K and F'/K' be Galois extensions with $K \subseteq K'$ and $F \subseteq F'$. The natural homomorphism

$$(41) \qquad r : G(F'/K') \to G(F/K)$$

is continuous, because if $G(F/L)$ is a basic open neighborhood of 1 in $G(F/K)$, then $G(F'/LK')$ is an open neighborhood of 1 in $G(F'/K')$, and $G(F'/LK') \subseteq r^{-1}(G(F/L))$.

The fixed field of $r(G(F'/K'))$ in F is obviously $F \cap K'$. Since $r(G(F'/K'))$ is compact (being a continuous image of the compact $G(F'/K')$) and hence also closed in $G(F/K)$, it follows from Galois theory (Theorem 4) that we have an equality

$$(42) \qquad r(G(F'/K')) = G(F/F \cap K').$$

Thus the map r in (41) gives rise to a *surjective* homomorphism

$$(43) \qquad G(F'/K') \to G(F/F \cap K')$$

of topological groups; its kernel is $G(F'/FK')$. We claim that the map (43) is *open*. To justify this we must show that the earlier map r in (41) is open if $F \cap K' = K$. So take $G(F'/L')$, a basic open neighborhood of 1 in $G(F'/K')$; we must show that the image $r(G(F'/L')) = G(F/F \cap L')$ is open in $G(F/K)$. By assumption, $G(F'/L')$ is a normal subgroup of finite index in $G(F'/K')$; this carries over to their homomorphic images, so that $G(F/F \cap L')$ is a normal subgroup of finite index in $G(F/K)$. Thus $F \cap L'/K$ is Galois with finite Galois group $G(F \cap L'/K) \simeq G(F/K)/G(F/F \cap L')$, and therefore $r(G(F'/L')) = G(F/F \cap L')$ really is an open neighborhood of 1 in $G(F/K)$.

In particular, if F is an intermediate field of a Galois extension E/K and F/K is also Galois, the restriction homomorphism $G(E/K) \to G(F/K)$ is *continuous* and *open*, and so gives rise to a canonical isomorphism

$$G(E/K)/G(E/F) \simeq G(F/K).$$

of *topological groups*, where the quotient group $G(E/K)/G(E/F)$ is given the *quotient topology* (that is, the finest topology for which the quotient map $G(E/K) \to G(E/K)/G(E/F)$ is still *continuous*).

Incidentally, one can easily check that, if F is an intermediate field of a Galois extension E/K, the Krull topology of $G(E/K)$ induces on the subgroup $G(E/F)$ the Krull topology of the Galois group $G(E/F)$.

Another thing that can be checked easily: If H is an open subgroup of the Galois group $G = G(E/K)$, then H has finite index in G, and in fact $G:H = F:K$, where F is the fixed field of H in E.

13

Norm and Trace

1. We mentioned in Chapter 1 that, by regarding an extension E of a field K as a vector space, we gain the ability to use the powerful tools of *linear algebra*. It is true that after Chapter 3 this viewpoint receded to the background, and only in Chapter 12 did we start making frequent use of it again. But in this chapter we will examine in our context some simple but effective concepts from linear algebra.

Notation. Departing from our practice so far, we will use the letter K in this chapter to denote not necessarily a field, but *any commutative ring with unity*.

If A is a K-algebra and M is an A-module, we consider for each $\alpha \in A$ the K-endomorphism
$$\alpha_M : x \mapsto \alpha x$$
of M. We also write $\alpha_{M/K}$ instead of α_M.

Definition. In the situation above, assume in addition that M is a finitely generated free K-module; this means M has a finite K-basis. The characteristic polynomial $P(\alpha_{M/K})$ of $\alpha_{M/K} \in \operatorname{End}_K(M)$ is called the *characteristic polynomial* of α with respect to the A-module M. It is an element of $K[X]$, and we denote it by
$$P_{M/K}(\alpha) = P_{M/K}(\alpha; X).$$
Likewise, we call
$$\operatorname{Tr}_{M/K}(\alpha) := \operatorname{Trace} \alpha_{M/K}$$
the *trace* of α, and
$$N_{M/K}(\alpha) := \det \alpha_{M/K}$$
the *norm* of α, always with respect to the A-module M.

The following properties of the norm and trace are obvious:

(1) $$\operatorname{Tr}_{M/K}(\alpha + \beta) = \operatorname{Tr}_{M/K}(\alpha) + \operatorname{Tr}_{M/K}(\beta),$$

(2) $$\operatorname{Tr}_{M/K}(a\alpha) = a \operatorname{Tr}_{M/K}(\alpha) \quad \text{for } a \in K,$$

(3) $$\operatorname{Tr}_{M/K}(\alpha\beta) = \operatorname{Tr}_{M/K}(\beta\alpha).$$

It follows from (1) and (2) that $\text{Tr}_{M/K} : A \to K$ is a K-linear form on A, and by using (3) as well we see that the map $(\alpha, \beta) \to \text{Tr}_{M/K}(\alpha\beta)$ is a *symmetric bilinear form* on A.

Next we have

$$(4) \qquad N_{M/K}(\alpha\beta) = N_{M/K}(\alpha)N_{M/K}(\beta).$$

If $C(\alpha) = (c_{ij}(\alpha))_{i,j}$ is the matrix expressing $\alpha_{M/K}$ with respect to a K-basis $e_1, \ldots, e_n$ of M, we have

$$(5) \qquad \text{Tr}_{M/K}(\alpha) = \text{Trace } C(\alpha) = \sum_i c_{ii}(\alpha),$$

$$(6) \qquad N_{M/K}(\alpha) = \det C(\alpha),$$

$$(7) \qquad P_{M/K}(\alpha; X) = \det(X I_n - C(\alpha)),$$

where I_n is the $n \times n$ identity matrix. If $P_{M/K}(\alpha)$ has the form $X^n + a_{n-1}X^{n-1} + \cdots + a_0$, we have

$$(8) \qquad \text{Tr}_{M/K}(\alpha) = -a_{n-1},$$

$$(9) \qquad N_{M/K}(\alpha) = (-1)^n a_0.$$

F1. *In the preceding situation let M_1 be a submodule of the A-module M and set $M_2 := M/M_1$. If M_1 and M_2 are both finitely generated free K-modules, M is also one, and for each $\alpha \in A$ we have*

$$\text{Tr}_{M/K}(\alpha) = \text{Tr}_{M_1/K}(\alpha) + \text{Tr}_{M_2/K}(\alpha),$$
$$N_{M/K}(\alpha) = N_{M_1/K}(\alpha) \cdot N_{M_2/K}(\alpha),$$
$$P_{M/K}(\alpha; X) = P_{M_1/K}(\alpha; X) \cdot P_{M_2/K}(\alpha; X).$$

In particular this is true when $M \simeq M_1 \oplus M_2$.

Proof. Let $e_1, \ldots, e_m$ be a K-basis of M_1. Choosing representatives $f_1, \ldots, f_n$ for the elements of a basis of $M/M_1 = M_2$ obviously yields a basis of a submodule of M complementary to the K-module M_1. Thus $e_1, \ldots, e_m, f_1, \ldots, f_n$ form a K-basis of M. With respect to this basis the matrix of some α_M has the form

$$\begin{pmatrix} C_1(\alpha) & * \\ 0 & C_2(\alpha) \end{pmatrix},$$

where $C_1(\alpha)$ is the matrix of α_{M_1} relative to $e_1, \ldots, e_m$ and $C_2(\alpha)$ is the matrix of α_{M_2} relative to $\bar{f}_1, \ldots, \bar{f}_n$. The rest follows.   □

Remark. Let the situation be as in Definition 1, and let L be a commutative K-algebra. The L-module $M_L = M \otimes_K L$ can be seen in a natural way as a module over the algebra $A_L = A \otimes_K L$. For all $\alpha \in A$ we then have

$$(10) \qquad \text{Tr}_{M_L/L}(\alpha \otimes 1) = \text{Tr}_{M/K}(\alpha), \quad N_{M_L/L}(\alpha \otimes 1) = N_{M/K}(\alpha),$$

$$(11) \qquad P_{M_L/L}(\alpha \otimes 1; X) = P_{M/K}(\alpha; X).$$

For if $e_1, \ldots, e_n$ form a K-basis of M, then $e_1 \otimes 1, \ldots, e_n \otimes 1$ form an L-basis of $M_L = M \otimes_K L$, with respect to which α_{M_L} has exactly the same matrix that α_M has with respect to $e_1, \ldots, e_n$.

In a nutshell, (10) and (11) say that the trace, the norm and the characteristic polynomial are invariant under *extension of the ground ring*. Carefully distinguish this situation from the next one:

Let A and A' be K-algebras, M an A-module and M' an A'-module. As K-modules, let M and M' be free and finitely generated of dimension n and n', respectively. It is easy to check that the $A \otimes_K A'$-module $M \otimes_K M'$ satisfies

(12) $$\mathrm{Tr}_{M \otimes M'/K}(\alpha \otimes \alpha') = \mathrm{Tr}_{M/K}(\alpha) \cdot \mathrm{Tr}_{M'/K}(\alpha'),$$

(13) $$N_{M \otimes M'/K}(\alpha \otimes \alpha') = N_{M/K}(\alpha)^{n'} \cdot N_{M'/K}(\alpha')^n.$$

Definition. Let A be a K-algebra and suppose that A is free and finitely generated as a K-module. The characteristic polynomial of an element $\alpha \in A$ with respect to the A-module A is called the *(regular) characteristic polynomial* of α, and is denoted by $P_{A/K}(\alpha) = P_{A/K}(\alpha; X)$. We define similarly the *(regular) norm* and *trace* of α, and denote them by

$$N_{A/K}(\alpha), \qquad \mathrm{Tr}_{A/K}(\alpha).$$

Remark. Let A be as in Definition 2. Since $N_{A/K}, \mathrm{Tr}_{A/K}, P_{A/K}$ are special cases of Definition 1, they satisfy properties (1)–(13) above. In the role of F1 we have at least the following fact: If $A = A_1 \times A_2 \times \cdots \times A_n$ is a *direct product* of finite-dimensional K-algebras A_i, then (by F1) we have, for every $\alpha = (\alpha_1, \alpha_2, \ldots, \alpha_n) \in A$,

(14) $$\mathrm{Tr}_{A/K}(\alpha) = \sum_i \mathrm{Tr}_{A_i/K}(\alpha_i), \quad N_{A/K}(\alpha) = \prod_i N_{A_i/K}(\alpha_i),$$

(15) $$P_{A/K}(\alpha; X) = \prod_i P_{A_i}(\alpha_i; X).$$

F2. *Let B be a subalgebra of a K-algebra A. Suppose that both B as a K-module and A as a B-left module are free and finitely generated. Then the same is true of the K-module A, and*

(16) $$P_{A/K}(\beta; X) = P_{B/K}(\beta; X)^m \quad \text{for } \beta \in B,$$

where m is the number of elements of a B-basis of A. Similarly,

(17) $$\mathrm{Tr}_{A/K}(\beta) = m \, \mathrm{Tr}_{B/K}(\beta) \quad \text{and} \quad N_{A/K}(\beta) = N_{B/K}(\beta)^m \quad \text{for } \beta \in B.$$

All three equalities follow directly from the isomorphism $A \simeq B^m$ of B-modules together with F1. But note that they are only valid for elements β *in B*. In trying to generalize for arbitrary elements in A one runs into obstacles; but see F5 and also §13.1 in the Appendix.

F3. *Let A be a K-algebra as in Definition 2. An element $\alpha \in A$ is invertible in A if and only if $N_{A/K}(\alpha)$ is invertible in K.*

Proof. If $\alpha\alpha^{-1} = 1$ then $N(\alpha)N(\alpha^{-1}) = 1$, so $N(\alpha)$ is invertible. Conversely, if $N(\alpha) = \det \alpha_{A/K}$ is invertible, $\alpha_{A/K}$ is invertible in $\mathrm{End}_K A$; in particular there exists $x \in A$ such that $\alpha x = 1$. But then $N(\alpha)N(x) = 1$, so $N(x)$ is invertible, and repeating the argument with x instead of α we obtain an $y \in A$ such that $xy = 1$. Left-multiplying by α yields $y = \alpha$, so α is invertible (has a two-sided inverse x). $\square$

2. We are especially interested in the norm and trace as applied to *finite field extensions* E/K (that is, the case where $A = E$ is a field). Thus, let E/K be a finite field extension and α an element of E. Consider first the intermediate field $K(\alpha)$ of E/K. If

(18) $$f(X) = X^n + a_{n-1}X^{n-1} + \cdots + a_1 X + a_0$$

is the *minimal polynomial* of α over K, comparing degrees shows that α coincides with the *characteristic polynomial* of α as an element of the K-algebra $K(\alpha)$:

(19) $$f(X) = \mathrm{MiPo}_K(\alpha) = P_{K(\alpha)/K}(\alpha).$$

Thus, in view of (8) and (9), we also have

(20) $$\mathrm{Tr}_{K(\alpha)/K}(\alpha) = -a_{n-1},$$

(21) $$N_{K(\alpha)/K}(\alpha) = (-1)^n a_0.$$

Using F2, then, we have for the extension E/K the equality

(22) $$P_{E/K}(\alpha; X) = f(X)^m, \quad \text{where } m = E : K(\alpha),$$

and also

(23) $$\mathrm{Tr}_{E/K}(\alpha) = -m a_{n-1}, \quad N_{E/K}(\alpha) = (-1)^{nm} a_0^m.$$

We now obtain for the trace $\mathrm{Tr}_{E/K}(\alpha)$ and the norm $N_{E/K}(\alpha)$ of an element α of a finite field extension E/K the following characterization (which reveals, in particular, that our notation is in agreement with the Tr already introduced in F3 of Chapter 12):

F4. *Let E/K be a finite field extension. Let C be an algebraic closure of K and let $G = G(E/K, C/K)$ be the set of all homomorphisms of E/K in C/K. If E/K is* **separable** *we have, for any $\alpha \in E$,*

(24) $$\mathrm{Tr}_{E/K}(\alpha) = \sum_{\sigma \in G} \sigma\alpha$$

and

(25) $$N_{E/K}(\alpha) = \prod_{\sigma \in G} \sigma\alpha.$$

More generally, without the separability assumption we have, for any $\alpha \in E$,

(26) $$\mathrm{Tr}_{E/K}(\alpha) = [E : K]_i \sum_{\sigma \in G} \sigma\alpha$$

and

(27)
$$N_{E/K}(\alpha) = \left(\prod_{\sigma \in G} \sigma\alpha \right)^{[E:K]_i},$$

where $[E:K]_i$ is the inseparable degree of E/K.

Proof. We may as well assume that $E \subseteq C$.

(i) Assume first that E/K is *separable*. For a given $\alpha \in E$, set $F = K(\alpha)$. Let $n = K(\alpha):K$ and denote by $\rho_1, \ldots, \rho_n$ the n distinct K-homomorphisms of $K(\alpha)$ into C. Then $\rho_1\alpha, \ldots, \rho_n\alpha$ are the distinct roots of $f := \mathrm{MiPo}_K(\alpha)$. This, together with (20) and (21), shows that

$$\mathrm{Tr}_{K(\alpha)/K}(\alpha) = -a_{n-1} = \sum_{i=1}^{n} \rho_i\alpha \quad \text{and} \quad N_{K(\alpha)/K}(\alpha) = (-1)^n a_0 = \prod_{i=1}^{n} \rho_i\alpha.$$

We know from Chapter 7 that each $\rho_i \in G(F/K, C/K)$ has exactly $m := E:F$ distinct extensions $\sigma \in G(E/K, C/K)$. Thus we have

$$\sum_{\sigma \in G} \sigma\alpha = m \sum_{i=1}^{n} \rho_i\alpha = m S_{K(\alpha)/K}(\alpha) = \mathrm{Tr}_{E/K}(\alpha),$$

which is (24). The norm is dealt with similarly.

(ii) In the general case, let E_s be the *separable closure* of K in E. By definition,

$$[E:K]_i = [E:E_s].$$

If char $K = 0$ we have $E = E_s$, so $[E:K]_i = 1$ and there is nothing to prove. So we assume char $K = p > 0$; then the degree $[E:K]_i$ is a p-power p^e, by F17 in Chapter 7. For every $\alpha \in E$, the element α^{p^e} lies in E_s, and we have

$$N_{E/K}(\alpha)^{p^e} = N_{E/K}(\alpha^{p^e}) = N_{E_s/K}(\alpha^{p^e})^{p^e},$$

so

$$N_{E/K}(\alpha) = N_{E_s/K}(\alpha^{p^e}) = \prod_{\sigma} \sigma(\alpha^{p^e}),$$

by part (i), where σ runs through the set $G(E_s/K, C/K)$. But the restriction map

$$G(E/K, C/K) \to G(E_s/K, C/K)$$

is bijective, so we finally get

$$N_{E/K}(\alpha) = \prod_{\sigma \in G} \sigma(\alpha)^{p^e},$$

which proves (27). There remains to prove (26). In the case $\alpha \in E_s$, we have $\mathrm{Tr}_{E/K}(\alpha) = [E:E_s]\,\mathrm{Tr}_{E_s/K}(\alpha)$ and the assertion is clear. So take $\alpha \notin E_s$. The minimal polynomial $f(X)$ of α over K then has the form $f(X) = g(X^p)$, for some $g \in K[X]$. Using (20) we see from this that $\mathrm{Tr}_{K(\alpha)/K}(\alpha) = 0$, which also means that $\mathrm{Tr}_{E/K}(\alpha) = 0$. But since char $K = p$ and $[E:K]_i = p^e > 1$, nothing further is needed. $\qquad\square$

F5 (Nesting formulas for trace and norm). *Let E/K be a finite field extension and L an intermediate field of E/K. Then*

$$(28) \qquad\qquad \mathrm{Tr}_{E/K} = \mathrm{Tr}_{L/K} \circ \mathrm{Tr}_{E/L},$$

$$(29) \qquad\qquad N_{E/K} = N_{L/K} \circ N_{E/L}.$$

Proof. Let C be an algebraic closure of K and $R \subseteq G(C/K)$ a set of representatives for $M := G(L/K, C/K)$, meaning that for each $\mu \in M$ there is a unique $\rho \in R$ such that $\rho_L = \mu$. Any $\sigma \in S := G(E/K, C/K)$ then has a unique representation

$$\sigma = \rho\tau, \quad \text{with } \rho \in R \text{ and } \tau \in T := G(E/L, C/L)$$

(note that $G(C/K)$ is a *group*, by F3 in Chapter 6). For every $\alpha \in E$ we then get

$$\prod_{\sigma \in S} \sigma\alpha = \prod_{\substack{\rho \in R \\ \tau \in T}} \rho\tau\alpha = \prod_{\rho \in R} \rho\Big(\prod_{\tau \in T} \tau\alpha\Big) = \prod_{\mu \in M} \mu\Big(\prod_{\tau \in T} \tau\alpha\Big).$$

This, together with (27) and the equality $[E:K]_i = [E:L]_i [L:K]_i$, leads to

$$(30) \qquad\qquad N_{E/K}(\alpha) = N_{L/K}(N_{E/L}(\alpha))$$

as desired. The corresponding formula for the trace is derived similarly.   □

For nesting formulas in the context of linear algebra see §13.1 in the Appendix.

3. After the generalities of the last two sections, we are now ready for some specific properties of the trace and norm as applied to field extensions.

F6. *For every finite **separable** field extension E/K the trace map $\mathrm{Tr}_{E/K} : E \to K$ is surjective.*

Proof. Since $\mathrm{Tr}_{E/K}$ is a linear form on the K-vector space E, we just need to show it's nonzero. But if we had

$$\mathrm{Tr}_{E/K}(\alpha) = \sum_{\sigma} \sigma\alpha = 0 \quad \text{for all } \alpha \in E$$

(where the first equality comes from F4), we would be in contradiction with the linear independence of the elements σ of $G(E/K, C/K)$, guaranteed by Theorem 2 of Chapter 12.   □

In contrast, if E/K is *inseparable*, $\mathrm{Tr}_{E/K}$ is the *zero map*, by (26).

F7. *For every finite **separable** field extension E/K, the map*

$$E \times E \to K$$
$$(\alpha, \beta) \mapsto \mathrm{Tr}_{E/K}(\alpha\beta)$$

*is a **nondegenerate symmetric bilinear form** on the K-vector space E.*

Proof. Take $\alpha \neq 0$. If $\mathrm{Tr}_{E/K}(\alpha\beta) = 0$ for every $\beta \in E$, then $\mathrm{Tr}_{E/K}(\gamma) = 0$ for all $\gamma \in E$, contradicting F6. □

F8. *Let E/K be a finite **Galois** field extension and $G = G(E/K)$ the Galois group of E/K. The kernel of $\mathrm{Tr}_{E/K} : E \to K$ is made up of all finite sums of elements of the form*

$$(31) \qquad \tau\alpha - \alpha, \quad \text{where } \alpha \in E \text{ and } \tau \in G.$$

If G is cyclic and generated by σ, the following statements are equivalent for a given $\gamma \in E$:

(i) $\mathrm{Tr}_{E/K}(\gamma) = 0$.

(ii) *There exists $\alpha \in E$ such that $\gamma = \sigma\alpha - \alpha$.*

Proof. By Theorem 3 in Chapter 12 there exists $\beta \in E$ such that $(\tau\beta)_{\tau \in G}$ is a K-basis for E. Now let

$$(32) \qquad \gamma = \sum_{\tau \in G} a_\tau \tau\beta, \quad \text{with } a_\tau \in K,$$

be an arbitrary element of E. An application of $\mathrm{Tr} := \mathrm{Tr}_{E/K}$ yields

$$(33) \qquad \mathrm{Tr}\,\gamma = \sum_\tau a_\tau \mathrm{Tr}(\tau\beta).$$

But for any $\alpha \in E$,

$$(34) \qquad \mathrm{Tr}(\tau\alpha) = \sum_{\sigma \in G} \sigma\tau\alpha = \sum_{\rho \in G} \rho\alpha = \mathrm{Tr}\,\alpha,$$

so the element in (32) has trace

$$(35) \qquad \mathrm{Tr}\,\gamma = (\mathrm{Tr}\,\beta)\left(\sum_{\tau \in G} a_\tau\right),$$

by (33). Since $(\tau\beta)_{\tau \in G}$ is a basis for E/K, we have

$$\mathrm{Tr}\,\beta = \sum_{\tau \in G} \tau\beta \neq 0.$$

By (35) we then see that $\mathrm{Tr}\,\gamma = 0$ is equivalent to $\sum_{\tau \in G} a_\tau = 0$. We also have, in full generality,

$$\sum_\tau a_\tau \tau\beta - \left(\sum_\tau a_\tau\right)\beta = \sum_\tau a_\tau(\tau\beta - \beta) = \sum_\tau \big(\tau(a_\tau\beta) - a_\tau\beta\big).$$

Taking it all together we see that $\mathrm{Tr}\,\gamma = 0$ if and only if γ is a sum of elements of the form (31).

Now let G be *cyclic* with σ as a generator. Clearly the set $M := \{\sigma\alpha - \alpha \mid \alpha \in E\}$ is a subgroup of the additive group of E. What remains to show is: For every $\tau = \sigma^k$ in G and every $\alpha \in E$, each difference $\tau\alpha - \alpha$ lies in M. This follows from

$$\sigma^k\alpha - \alpha = \sigma^k\alpha - \sigma^{k-1}\alpha + \sigma^{k-1}\alpha - \cdots + \sigma\alpha - \alpha$$
$$= \sigma(\sigma^{k-1}\alpha + \sigma^{k-2}\alpha + \cdots + \alpha) - (\sigma^{k-1}\alpha + \sigma^{k-2}\alpha + \cdots + \alpha). \quad \square$$

It is of great significance that there is a direct multiplicative analog for F8 in the *cyclic case* (whereas surprisingly this fails to be the case in general; but see §13.4 in the Appendix):

F9 (Hilbert's Theorem 90). *Let E/K be a finite Galois extension with* **cyclic** *Galois group $G = \langle\sigma\rangle$. For a given $\gamma \in E^\times$, there is equivalence between:*

(i) $N_{E/K}(\gamma) = 1$.

(ii) *There exists $\alpha \in E^\times$ such that $\gamma = \dfrac{\alpha}{\sigma(\alpha)}$.*

Proof. That (ii) implies (i) is obvious. We prove the converse. First take an arbitrary, but fixed, $\beta \in E$. Denote by n the order of G. Our first candidate for the desired element α will be

$$\alpha = \sum_{i=0}^{n-1} \lambda_i \sigma^i(\beta) \quad \text{with } \lambda_i \in E,$$

where we choose $\lambda_0 = 1$ without loss of generality. Then

$$\sigma\alpha = \sum_{i=0}^{n-1} \sigma(\lambda_i)\sigma^{i+1}(\beta) = \sum_{i=1}^{n} \sigma(\lambda_{i-1})\sigma^i(\beta).$$

Thus the equation

(36) $\alpha = \gamma\sigma(\alpha)$

will certainly be satisfied if the following conditions hold:

(37) $\lambda_i = \gamma\sigma(\lambda_{i-1})$ for $1 \le i \le n-1$,

(38) $1 = \gamma\sigma(\lambda_{n-1})$.

The conditions (37) are equivalent to

(39) $\lambda_i = \gamma\sigma(\gamma)\sigma^2(\gamma)\ldots\sigma^{i-1}(\gamma)$ for $1 \le i \le n-1$.

This determines all the λ_i (given that we took $\lambda_0 = 1$), independently of β. In particular, we have $\lambda_{n-1} = \gamma\sigma(\gamma)\ldots\sigma^{n-2}(\gamma)$ and consequently

$$\sigma(\lambda_{n-1}) = \sigma(\gamma)\sigma^2(\gamma)\ldots\sigma^{n-1}(\gamma) = \gamma^{-1}N_{E/K}(\gamma).$$

Thus, if $N_{E/K}(\gamma) = 1$, as assumed, (38) is likewise a consequence of (39). We therefore see that the element

(40) $\qquad \alpha = \beta + \gamma\sigma(\beta) + \gamma\sigma(\gamma)\sigma^2(\beta) + \cdots + \left(\gamma\sigma(\gamma)\sigma^{n-2}(\gamma)\right)\sigma^{n-1}(\beta)$

satisfies (36), for every $\beta \in E$. We now have to arrange for α in (40) to be nonzero; this is possible with the right choice of β because $1, \sigma, \ldots, \sigma^{n-1}$ are linearly independent over E. In conclusion, then, we have found $\alpha \in E^\times$ such that

(41) $$\gamma = \frac{\alpha}{\sigma(\alpha)}. \qquad \square$$

F10. If K is a **finite** field and E/K is a finite extension, the norm map $N_{E/K}$ is surjective.

Proof. The norm map $N = N_{E/K}$ gives rise to a *homomorphism*

(42) $\qquad\qquad\qquad\qquad N : E^\times \to K^\times$

of multiplicative groups always; we must show that this map is surjective if K and E are finite. We know that E/K is Galois with cyclic Galois group $G(E/K) = \langle \sigma \rangle$. Now consider, besides (42), the homomorphism

(43) $\qquad\qquad\qquad\qquad \delta : E^\times \to E^\times$

defined by $\delta(\alpha) = \alpha/\sigma\alpha$. By applying the fundamental homomorphism theorem for groups to (42) and (43) we get, in particular, the cardinality relations

$$|\operatorname{im} N| \cdot |\ker N| = |E^\times| = |\operatorname{im} \delta| \cdot |\ker \delta|.$$

We have $\ker \delta = \{\alpha \in E^\times \mid \sigma\alpha = \alpha\}$, so by Galois theory,

$$\ker \delta = K^\times.$$

By Hilbert's Theorem 90 (see F9) we get

$$\operatorname{im} \delta = \ker N.$$

There follows $|\operatorname{im} N| = |K^\times|$ and hence the desired conclusion $K^\times = \operatorname{im} N$. $\square$

The fact stated in F10 is of considerable importance in algebra and in number theory. Here we derived F10 as a nice application of Hilbert's Theorem 90, but it is worth remarking that another proof can be given: We know that the multiplicative group $E^\times$ of E is generated by some element ζ. Letting q denote the number of elements of K, we must show that the element $N_{E/K}(\zeta)$ of $K^\times$ has order $q-1$. This we do by noting that the Galois group $G(E/K)$ is generated by the automorphism $\sigma_q : \alpha \mapsto \alpha^q$. Setting $n = E:K$, we obtain

(44) $\qquad N_{E/K}(\zeta) = \zeta\zeta^q\zeta^{q^2}\ldots\zeta^{q^{n-1}} = \zeta^{1+q+q^2+\cdots+q^{n-1}}.$

Since ζ has order $q^n - 1 = (q-1)(1 + q + q^2 + \cdots + q^{n-1})$, the element $N_{E/K}(\zeta)$ in (44) does indeed have order $q - 1$.

14

Binomial Equations

1. Let K be a field. We consider polynomials of the form

(1) $$f(X) = X^n - \gamma \in K[X],$$

with $n \in \mathbb{N}$. The roots of such a polynomial (in a splitting field E of f over K), that is, the solutions of the *binomial equation* $X^n - \gamma = 0$, are called *n-th roots of γ.* (A binomial equation is also sometimes called a "pure equation".)

In this chapter we will always assume $\gamma \neq 0$. The derivative of $f(X)$ is $f'(X) = nX^{n-1}$. Thus $f(X)$ is *separable* if and only if char K is not a divisor of n.

F1. *Assume that the field K contains a primitive n-th root of unity (in particular, char K is not a divisor of n). Let $\gamma \in K^{\times}$. Then the Galois group G of the binomial equation $X^n - \gamma = 0$ over K is **cyclic**, and its order divides n.*

Moreover, if α is an n-th root of γ in a splitting field E of $X^n - \gamma$ over K, the order of G is the smallest natural number d such that $\alpha^d \in K$. The polynomial $X^d - \alpha^d$ is then the minimal polynomial of α over K. The extension E/K is generated by α.

Proof. Let the notation be as above. In particular, let $\alpha \in E$ be a fixed root of $X^n - \gamma$, so $\alpha^n = \gamma$. Clearly, the elements $\zeta\alpha$, where ζ runs over all n-th roots of unity in K, are roots of $X^n - \gamma$ in E, and there is no other. Thus $E = K(\alpha)$.

Now take $\sigma \in G$. Since α is a root of $X^n - \gamma$, so is $\sigma\alpha$; thus $\sigma\alpha = \zeta\alpha$ for a unique root of unity $\zeta = \zeta(\sigma)$. This defines an *injective* map

(2) $$\sigma \mapsto \frac{\sigma\alpha}{\alpha} = \zeta(\sigma)$$

from G to the group of n-th roots of unity. By assumption, all n-th roots of unity lie in K; this implies that (2) is a *homomorphism*, because by applying $\rho \in G$ to $\sigma\alpha = \zeta(\sigma)\alpha$ we get $\rho\sigma\alpha = \zeta(\sigma)\rho\alpha = \zeta(\sigma)\zeta(\rho)\alpha$, so $\zeta(\rho\sigma) = \zeta(\rho)\zeta(\sigma)$. In summary, the Galois group G is isomorphic to a subgroup of the group of n-th roots of unity. The latter is cyclic of order n, so G is cyclic and its order divides n. Let d be the order of G and σ a generator of G. Then $\zeta = \sigma\alpha/\alpha$ is a *primitive d-th root of unity*. Since

$$\sigma(\alpha^d) = \sigma(\alpha)^d = \zeta^d\alpha^d = \alpha^d,$$

the Galois correspondence shows that α^d is an element of K. Conversely, if we assume that $\alpha^{d'} \in K$, then

$$\alpha^{d'} = \sigma(\alpha^{d'}) = \sigma(\alpha)^{d'} = \zeta^{d'}\alpha^{d'},$$

so $\zeta^{d'} = 1$. Thus d' is divisible by d.

There remains to show only that $\mathrm{MiPo}_K(\alpha) = X^d - \alpha^d$. On the one hand, α is a root of $X^d - \alpha^d$; on the other, $\deg \mathrm{MiPo}_K(\alpha) = K(\alpha):K = E:K = |G| = d$. $\square$

Remarks. Still in the situation of F1, let $\sqrt[n]{\gamma}$ denote an arbitrary n-th root of γ in E. Also let

$$(3) \qquad\qquad d = K(\sqrt[n]{\gamma}):K.$$

By F1 we have $n = rd$ for some $r \in \mathbb{N}$. The element $\beta := (\sqrt[n]{\gamma})^d$ lies in K, and we have $\gamma = \beta^r$ and

$$(4) \qquad\qquad K(\sqrt[n]{\gamma}) = K(\sqrt[d]{\beta}).$$

Also it follows immediately from F1 that the order d of the Galois group of $K(\sqrt[n]{\gamma})/K$ coincides with the order of the element of $K^\times/K^{\times n}$ determined by γ.

Theorem 1. *Let E/K be a finite Galois extension with cyclic Galois group (we call E/K a **cyclic extension**). Let n be the order of the Galois group G of E/K. If K contains a primitive n-th root of unity, E is obtained from K by adjoining an n-th root of an element of K; in other words, $E = K(\alpha)$, where $\alpha \in E$ is a root of a polynomial $X^n - \gamma \in K[X]$.*

Proof. Let K contain a primitive n-th root of unity ζ, and let σ be a generator of G. We seek an element $\alpha \in E^\times$ such that

$$(5) \qquad\qquad \frac{\sigma\alpha}{\alpha} = \zeta.$$

Such an α exists if and only if ζ satisfies $N_{E/K}(\zeta) = 1$ (see Chapter 13, F9, alias Hilbert's Theorem 90). This condition is satisfied since $\zeta \in K$ (so $N_{E/K}(\zeta) = \zeta^n = 1$), so there is indeed α in $E^\times$ such that $\sigma\alpha = \zeta\alpha$. From this we get

$$\sigma(\alpha^n) = \sigma(\alpha)^n = \zeta^n\alpha^n = \alpha^n.$$

Since σ generates all of G, the power α^n must already lie in K. Setting $\gamma := \alpha^n$, we see that α is a root of the polynomial $X^n - \gamma \in K[X]$. Also from (5) we get

$$\sigma^i\alpha = \zeta^i\alpha;$$

thus α has n distinct conjugates $\alpha, \zeta\alpha, \zeta^2\alpha, \ldots, \zeta^{n-1}\alpha$ over K. There follows $K(\alpha):K = n = E:K$, so $E = K(\alpha)$. This concludes the proof. $\square$

Remark. In our proof of Hilbert's Theorem 90 on page 140, we found an α with the desired property by first setting

$$(6) \qquad \alpha = \sum_{i=0}^{n-1} \zeta^{-i} \sigma^i(\beta),$$

with $\beta \in E$. The expression on the right is called the *Lagrange resolvent* and is sometimes denoted by (ζ^{-1}, β). For appropriate $\beta \in E$, as we have seen,

$$(7) \qquad (\zeta^{-1}, \beta) \neq 0,$$

so $\alpha = (\zeta^{-1}, \beta)$ is a primitive element of E/K whose n-th power is in K. Incidentally: if $\beta \in E$ satisfies (7), it is a primitive element of E/K; but the converse is not true in general.

F2. *Let q be a prime number and γ an element of $K^\times$. Either γ is a q-th power in K or the polynomial $X^q - \gamma$ is irreducible in $K[X]$.*

Proof. Suppose that γ is not a q-power in K. Let E be a splitting field of $X^q - \gamma$ over K and α a root of $X^q - \gamma$ in E.

If char $K = q$ we have $X^q - \gamma = X^q - \alpha^q = (X - \alpha)^q$, so $E = K(\alpha)$ and E/K is purely inseparable (see for instance F16 in Chapter 7); then $X^q - \gamma$ is the minimal polynomial of α over K (Chapter F15).

So assume instead that char $K \neq q$. Then $X^q - \gamma$ is separable, hence E contains a primitive q-th root of unity ζ. Suppose, for a contradiction, that $X^q - \gamma$ is reducible over K; then it is *a fortiori* reducible over $K(\zeta)$, and by F1 it must split into linear factors over $K(\zeta)$, since q is prime. Thus

$$K(\alpha) \subseteq K(\zeta).$$

The extension $K(\zeta)/K$ is Galois and its Galois group G is *cyclic* (being isomorphic to a subgroup of the cyclic group $(\mathbb{Z}/q\mathbb{Z})^\times$; see F9 and Theorem 2 in Chapter 9). Let σ be a generator of G; then

$$(8) \qquad \sigma\alpha = \eta\alpha,$$

where η is a q-th root of unity, and in fact a primitive one ($\eta = 1$ is excluded because it would imply that G fixes α, hence $\alpha \in K$, whereas we've assumed that γ is not a q-th power in K). Since $\eta \notin K$, we cannot deduce from (8) things like $\sigma^i\alpha = \eta^i\alpha$, for instance. Instead we take the ratio $\sigma(\eta)/\eta$, which is also a primitive q-th root of unity, so that

$$\eta = \left(\frac{\sigma(\eta)}{\eta}\right)^k \quad \text{for some } 0 \le k \le q-1.$$

Together with (8) this leads to

$$\sigma\left(\frac{\alpha}{\eta^k}\right) = \frac{\eta\alpha}{\sigma(\eta)^k} = \frac{\eta\alpha}{\eta^k\eta} = \frac{\alpha}{\eta^k}.$$

As before this implies that α/η^k lies in K. There follows $(\alpha/\eta^k)^q = \alpha^q = \gamma$, so γ is a q-th power in K after all and we get the desired contradiction. $\qquad \square$

Remark. Still in the situation of F2, let K' denote the field of q-th roots of unity over K. The proof we just gave actually shows something stronger than F2: *Either γ is q-th power in K, or $X^q - \gamma$ is irreducible even over K'.* And we explicitly restate the following consequence of F1: *If $K' = K$, either $X^q - \gamma$ is irreducible in $K[X]$ or $X^q - \gamma$ splits into linear factors over K.*

The statement of F2 enters crucially into the proof of the following general fact:

Theorem 2. *For $\gamma \in K^\times$, the polynomial $X^n - \gamma$ is irreducible over K if and only if the following conditions are met:*

(a) *There is no prime factor q of n such that γ is a q-th power in K.*

(b) *If n is divisible by 4, there is no $\lambda \in K$ such that $\gamma = -4\lambda^4$.*

Proof. Sufficiency. We work by induction on n. For $n = 1$ there is nothing to prove. Take $n > 1$ and assume (a) and (b).

Let q be a prime factor of n and set $m = n/q$; also let α be a root of $X^n - \gamma$. By induction we can assume that $X^m - \gamma$ is irreducible, so $K(\alpha^q) : K = m$. Consider the polynomial $X^q - \alpha^q$ over $K(\alpha^q)$. If this is irreducible, we have

$$K(\alpha) : K = [K(\alpha) : K(\alpha^q)] \cdot [K(\alpha^q) : K] = q \cdot m = n,$$

and therefore $X^n - \gamma$ is irreducible over K, as desired.

If, instead, $X^q - \alpha^q$ is reducible over $K' = K(\alpha^q)$, we know from F2 that α^q is a q-th power in K'. We will analyze various subcases, in each one getting either a contradiction or the irreducibility of $X^n - \gamma$. Write

(9) $$\alpha^q = \beta^q, \quad \text{with } \beta \in K'.$$

Let $N = N_{K'/K}$ be the norm map for the extension K'/K. Applying N to (9) we get

(10) $$N(\beta)^q = N(\beta^q) = N(\alpha^q) = -\gamma(-1)^m,$$

since $X^m - \gamma$ is the minimal polynomial of α^q over K.

If m is odd, (10) expresses γ as a q-th power in K, contradicting (a).

If m is even but q is odd, (10) says that $(-N(\beta))^q = \gamma$, so again γ is a q-th power in K.

Free as we are to choose q, we have covered all cases except

$$n = 2^s, \quad \text{with } s \geq 2.$$

In this latter case (10) says that

(11) $$-\gamma = \delta^2, \quad \text{with } \delta \in K.$$

Suppose $-1 = i^2$ in an extension of K; then $i \notin K$, otherwise γ would be a square in K, by (11). Over $K(i)$ we have the decomposition

$$X^{2^s} - \gamma = \left(X^{2^{s-1}} + i\delta\right)\left(X^{2^{s-1}} - i\delta\right).$$

If the first factor on the right is irreducible over $K(i)$, so is the second, since the two are conjugate; then the uniqueness of prime factorizations in $K[X]$ and in $K(i)[X]$ (together with the fact that $i \notin K$) yields the irreducibility of $X^{2^s} - \gamma$.

Finally, suppose instead that

$$X^{2^{s-1}} + i\delta$$

is reducible over $K(i)$. By the induction hypothesis, either $i\delta$ is a square in $K(i)$ or $i\delta$ equals $-4\tilde{\lambda}^4$ with $\tilde{\lambda} \in K(i)$; in the latter case $i\delta$ is a square in $K(i)$ all the same. Thus there exist $\lambda, \mu \in K$ such that

$$i\delta = (\lambda + i\mu)^2 = \lambda^2 + 2\lambda\mu i - \mu^2.$$

There follows $\lambda^2 = \mu^2$ and $2\lambda\mu = \delta$. By squaring we get $\delta^2 = 4\lambda^4$ and then, taking (11) into account, $\gamma = -4\lambda^4$, contradicting (b).

Necessity. Let $n = qm$ with q prime and $m \geq 1$, and suppose $\gamma = \beta^q$ in K. Then $X^q - \gamma$ is not irreducible over K, and so neither is $(X^m)^q - \gamma$. That shows the necessity of (a).

Let $n = 4m$ with $m \geq 1$, and suppose $\gamma = -4\lambda^4$ with $\lambda \in K$. Then

$$X^4 - \gamma = X^4 + 4\lambda^4 = (X^2 + 2\lambda^2)^2 - (2\lambda X)^2$$
$$= (X^2 - 2\lambda X + 2\lambda^2)(X^2 + 2\lambda X + 2\lambda^2).$$

Thus $X^4 - \gamma$ is not irreducible over K, and so neither is $(X^m)^4 - \gamma$. That takes care of (b). $\qquad\square$

In F1 and Theorem 1 we had to assume that the ground field K contains a primitive n-th root of unity. When $p = \operatorname{char} K > 0$ and $p \mid n$ this condition cannot be met; but we can, at least in the case $p = n$, prove the following substitutes for F1 and Theorem 1:

Theorem 3 (Artin–Schreier). *Let K be a field of characteristic $p > 0$.*

(I) *If E/K is a cyclic extension of degree p, then E arises from K by adjoining a root of a polynomial of the form*

$$X^p - X - \gamma, \quad \text{with } \gamma \in K.$$

(II) *Conversely, given a polynomial of the form*

$$f(X) = X^p - X - \gamma \in K[X],$$

with splitting field E over K, either $E = K$ or f is irreducible over K. In the latter case E/K is a cyclic extension of degree p.

Proof. We prove (II) first. Let $\alpha \in E$ be a root of $f(X)$. For any integer $j \geq 0$, regarded as an element of the prime field of K, we have

$$f(\alpha + j) = (\alpha + j)^p - (\alpha + j) - \gamma = \alpha^p + j^p - \alpha - j - \gamma$$
$$= \alpha^p - \alpha + j^p - j - \gamma = f(\alpha) + j^p - j = 0,$$

where we have used that $j^p - j = 0$. Thus f has in E the p distinct roots

(12) $\alpha, \ \alpha + 1, \ \alpha + 2, \ \ldots, \ \alpha + (p-1);$

in particular, f is a separable polynomial, so the extension E/K is Galois. Moreover $E = K(\alpha)$. Thus, if one root of f lies in K, so do all others. Now assume that f has no root in K. Under any element σ of the Galois group G of E/K, the image of α is a root of f as well, that is,

$$\sigma \alpha = \alpha + j,$$

for a well defined j of the prime field of K. The map

$$\sigma \mapsto \sigma \alpha - \alpha$$

is clearly an injective group homomorphism from the Galois group G into the additive group $\mathbb{Z}/p\mathbb{Z}$ of the prime field of K. The latter is cyclic of prime order p. Since we assumed that $E \neq K$, we get $G \simeq \mathbb{Z}/p\mathbb{Z}$. As claimed, then, E/K is cyclic of degree p. Finally, f is irreducible because $E = K(\alpha)$.

Conversely, to prove (I), let E/K be a cyclic extension of degree p, with Galois group G generated by σ. Since char $K = p$, we have

$$S_{E/K}(1) = [E:K] \cdot 1 = p \cdot 1 = 0.$$

By F8 in Chapter 13, there exists $\alpha \in E$ such that $1 = \sigma \alpha - \alpha$, or yet

(13) $\sigma \alpha = \alpha + 1.$

In particular, we have $\sigma \alpha \neq \alpha$; hence α is not contained in K. This already implies that $E = K(a)$, since E/K has degree p. From (13) we further obtain

$$\sigma(\alpha^p - \alpha) = \sigma(\alpha)^p - \sigma(\alpha) = (\alpha + 1)^p - (\alpha + 1)$$
$$= \alpha^p + 1^p - \alpha - 1 = \alpha^p - \alpha.$$

Hence the element $\gamma := \alpha^p - \alpha$ is fixed by σ and so must lie in K. Therefore α is a root of the polynomial $X^p - X - \gamma$ in $K[X]$.   □

Remarks. (1) Let p be a fixed prime. If C is a field of characteristic p, consider the map

(14) $\wp : C \to C \quad$ defined by $\wp(x) = x^p - x.$

Theorem 3 shows that the solutions of the equation

$$\wp(\alpha) = \gamma$$

in fields of characteristic p play a role analogous to that of p-th roots in fields of characteristic $\neq p$. (See also §14.5 in the Appendix.)

(2) Is it also possible to say something about cyclic extensions E/K whose degree is a *power* of $p = \operatorname{char} K > 0$? By "naive reasoning" only little; see §14.4 in the Appendix. But full information is provided by a subtle theory of E. Witt; see Section 26.5 in Volume II.

2. Under the assumption that K contains a primitive n-th root of unity, we have given in F1 and F2 a description of the extensions E/K obtainable from K by adjoining an n-th root of an element of K; now we will investigate what happens when we adjoin not one, but a whole series of n-th roots of elements of K. But in order to do that we must go over some basic facts about finite abelian groups.

First let G be *any abelian group* (written multiplicatively). Consider the set

$$(15) \qquad\qquad G^* = \operatorname{Hom}(G, \mathbb{C}^\times)$$

of all homomorphisms $\chi : G \to \mathbb{C}^\times$ from G into the multiplicative group $\mathbb{C}^\times$ of the field of complex numbers. This has the natural group structure, given by (pointwise) multiplication of functions; it is called the *character group* or *dual* of G, and its elements are called the *characters* of G. (We have already encountered an interesting example of a character, the Legendre symbol of Chapter 11.)

We say that G is a *group of exponent m* if $\sigma^m = 1$ for every σ in G. If G is of exponent m and $\chi \in G^*$ is a character of G, the equalities $\chi(\sigma)^m = \chi(\sigma^m) = \chi(1) = 1$ show that the values of χ are all m-th roots of unity, so χ can also be viewed as a homomorphism

$$(16) \qquad\qquad \chi : G \to W_m(\mathbb{C})$$

from G into the group $W_m(\mathbb{C})$ of m-th roots of unity in $\mathbb{C}$.

If $G = G_1 \times G_2$ is a direct product, there is clearly a natural isomorphism

$$(17) \qquad\qquad (G_1 \times G_2)^* \simeq G_1^* \times G_2^*$$

given by $\chi \mapsto (\chi|_{G_1}, \chi|_{G_2})$.

F3. *For every finite abelian group G there is a (noncanonical) isomorphism*

$$(18) \qquad\qquad G \simeq G^*.$$

Proof. We use the fact (whose proof can be looked up in the next section) that G is a direct product of cyclic groups, $G = C_1 \times C_2 \times \cdots \times C_r$. In view of (17), it is enough to verify the existence of an isomorphism (18) in the case that G is cyclic.

So let $G = \langle \sigma \rangle$, with σ of order $n = |G|$. Then $G \simeq W_n(\mathbb{C})$, and we will be done if we show that the homomorphism

$$G^* \to W_n(\mathbb{C})$$
$$\chi \mapsto \chi(\sigma)$$

is an isomorphism. Injectivity is clear. Let ζ denote a generator of $W_n(\mathbb{C})$, that is, a root of unity of order n in $\mathbb{C}$. The equation $\chi(\sigma^k) = \zeta^k$ then yields a *well defined* character χ of G with $\chi(\sigma) = \zeta$; this proves surjectivity.   □

F4. *Let G, H be abelian groups and*

$$\varphi : G \times H \to \mathbb{C}^\times$$

*a bilinear (bimultiplicative) map, also known as a **pairing**. Then φ naturally gives rise to homomorphisms*

$$\varphi_1 : G \to H^* \quad and \quad \varphi_2 : H \to G^*$$
$$\sigma \mapsto \varphi(\sigma, \cdot) \qquad\qquad \tau \mapsto \varphi(\cdot, \tau)$$

If φ is nondegenerate (equivalently, if φ_1 and φ_2 are injective), G is finite if and only if H is; in this case φ_1 and φ_2 are isomorphisms, and so in particular

$$G \simeq H^* \quad and \quad H \simeq G^*.$$

Proof. Suppose φ is nondegenerate and H, say, is finite. Using F3 we successively get

$$|G| \leq |H^*| = |H| \leq |G^*| = |G|,$$

so G is also finite. The other assertions follow.   □

F5. *Let G be a finite abelian group. For every subgroup A of G the sequence*

$$(19) \qquad 1 \longrightarrow (G/A)^* \xrightarrow{\text{inf}} G^* \xrightarrow{\text{res}} A^* \longrightarrow 1$$

is exact, that is, the image of each map equals the kernel of the next.

Proof. For $\chi \in G^*$, the character res χ by definition takes the same values as χ on every $a \in A$. For $\psi \in (G/A)^*$, the character inf ψ is defined by $(\text{inf } \psi)(\sigma) = \psi(\sigma \bmod A)$ for every $\sigma \in G$. The exactness of (19) is then obvious everywhere except at A^*, where it amounts to the surjectivity of res $: G^* \to A^*$. But by the fundamental homomorphism theorem and F3 we can write

$$|\text{res } G^*| = \frac{|G^*|}{|(G/A)^*|} = \frac{|G|}{|G/A|} = |A|.$$ □

F6. *For any finite abelian group G the pairing*

$$G^* \times G \to \mathbb{C}^\times$$
$$(\chi, \sigma) \mapsto \chi(\sigma)$$

is nondegenerate, and so (by F4) it provides a natural isomorphism

(20) $$G \simeq (G^*)^* =: G^{**}.$$

Proof. If $\chi(\sigma) = 1$ for every σ we have $\chi = 1$ by definition. To show that our pairing is also nondegenerate in the second variable, let σ be a nontrivial element of G; we must check that there exists $\chi \in G^*$ such that $\chi(\sigma) \neq 1$.

Consider the subgroup $A = \langle \sigma \rangle$ of G. By F3, $A \neq 1$ implies $A^* \neq 1$. Because of the surjectivity of $\mathrm{res} : G^* \to A^*$ (see F5), there does exist some $\chi \in G^*$ that is nontrivial on A; that is, the condition $\chi(\sigma) \neq 1$ is satisfied. $\square$

We now come to the previously announced field-theoretical applications of these results. We need:

Notations and assumptions. Let K be a field, C an algebraic closure of K and n a natural number. *We assume that K contains a primitive n-th root of unity.*

Given a subset A of $K^\times$, we define the set of n-th roots of elements of A (in the chosen algebraic closure C of K) by

$$\sqrt[n]{A} = \{\alpha \in C \mid \alpha^n \in A\}.$$

Consider the extension $E = K(\sqrt[n]{A})$ in C, which is clearly *Galois*. If $A_1 = \langle A \rangle$ is the subgroup of $K^\times$ generated by A, we obviously have $K(\sqrt[n]{A}) = K(\sqrt[n]{A_1})$. Thus from now on we may assume that A is a *subgroup* of $K^\times$. Denoting by $K^{\times n}$ the subgroup of n-th powers of elements of $K^\times$, we can also assume that

$$K^{\times n} \subseteq A,$$

if necessary after replacing A by the subgroup $AK^{\times n}$ of $K^\times$.

We speak of an *abelian field extension L/K* when L/K is Galois and its Galois group is abelian.

F7. *Suppose K contains a primitive n-th root of unity and let A be a subgroup of $K^\times$ such that $K^{\times n} \subseteq A$. Then the extension $K(\sqrt[n]{A})/K$ is abelian and its Galois group G is a group of exponent n. There is a canonical pairing*

(21)
$$G \times A \to W_n(K)$$
$$(\sigma, a) \mapsto \frac{\sigma\alpha}{\alpha}, \quad \text{where } \alpha^n = a.$$

This in turn gives rise to a nondegenerate pairing

(22) $$G \times A/K^{\times n} \longrightarrow W_n(K).$$

$K(\sqrt[n]{A})/K$ *is finite if and only if the group* $A/K^{\times n}$ *is finite, and in this case there are natural isomorphisms*

$$(23) \qquad\qquad G \simeq (A/K^{\times n})^*, \quad G^* \simeq A/K^{\times n}.$$

Here, for an abelian group G of exponent n, the dual is considered to be the group $G^* = \operatorname{Hom}(G, W_n(K))$. Otherwise the isomorphisms in (23) are not canonical; indeed we have $W_n(K) \simeq W_n(\mathbb{C})$, but not, in general, canonically.

Proof of F7. We set $E := K(\sqrt[n]{A})$. For $\sigma \in G$ and $\alpha \in E^\times$ such that $\alpha^n = a \in A$ we have $(\sigma\alpha)^n = a$, so there is a unique n-th root of unity $\zeta = \zeta(\sigma, \alpha)$ such that

$$(24) \qquad\qquad \frac{\sigma\alpha}{\alpha} = \zeta(\sigma, \alpha).$$

By assumption, $\zeta(\sigma, \alpha)$ is in K. If some $\alpha' \in E$ also satisfies $\alpha'^n = \alpha^n = a$, there is $\eta \in W_n(K)$ such that $\alpha = \eta\alpha'$; there follows

$$\frac{\sigma\alpha}{\alpha} = \frac{\sigma(\eta\alpha')}{\alpha} = \frac{\eta\sigma(\alpha')}{\alpha} = \frac{\sigma(\alpha')}{\alpha'}.$$

Thus the map (21) is well defined. It is multiplicative in the second variable, since obviously

$$\frac{\sigma(\alpha\beta)}{\alpha\beta} = \frac{\sigma(\alpha)}{\alpha}\frac{\sigma(\beta)}{\beta}.$$

Since $\zeta(\sigma, \alpha) \in K$, applying some $\rho \in G$ to both sides of (24) immediately gives $\rho\sigma\alpha = \zeta(\sigma, \alpha)\rho\alpha = \zeta(\sigma, \alpha)\zeta(\rho, \alpha)\alpha$, and hence

$$\zeta(\rho\sigma, \alpha) = \zeta(\rho, \alpha)\zeta(\sigma, \alpha).$$

Thus (21) is multiplicative also in the first variable. Moreover we see that the Galois group G is abelian. Take $\sigma \in G$. If $\sigma\alpha/\alpha = 1$ for any α such that $\alpha^n \in A$, we obviously get $\sigma = 1$. Thus the pairing (21) is nondegenerate with respect to the first variable. On the other hand, let $\alpha^n = a \in A$ be given; then a necessary and sufficient condition for $\sigma\alpha/\alpha$ to equal 1 for all $\sigma \in G$ is that α belong to $K^\times$, which is to say that a lie in $K^{\times n}$. Thus one gets from (21) the *nondegenerate* pairing (22). The remaining assertions now follow immediately from F4, with the replacement of $\mathbb{C}^\times$ by $W_n(K)$. $\qquad\square$

Remark. From F7 we recover F1 by taking $A = \langle \gamma \rangle K^{\times n}$.

Definition. A Galois extension E/K is called *abelian of exponent n* if $G(E/K)$ is an abelian group of exponent n. An abelian extension E/K where K contains a primitive n-th root of unity is also called a *Kummer extension*, in honor of E. Kummer (1810–1893), who earned immortal fame for his profound work in number theory. Kummer himself actually considered only certain special extensions that arose in his number-theoretic studies; the more general, purely algebraic laws of what is now called Kummer theory came later. Their usefulness in number theory is astonishing, given how simple it is to derive them.

Theorem 4 (Kummer theory). *Under the assumption that K has a primitive n-th root of unity, and keeping the earlier notations, the map*

$$(25) \qquad A \mapsto K(\sqrt[n]{A}) =: E_A$$

is a bijection between the set of subgroups A of $K^{\times}$ such that $K^{\times n} \subseteq A$ and the set of subfields E of C for which E/K is abelian of exponent n. Moreover E_A/K is finite if and only if the group $A/K^{\times n}$ is finite, and then there are natural isomorphisms

$$(26) \qquad G^* \simeq A/K^{\times n}, \quad G \simeq (A/K^{\times n})^*,$$

where $G = G(E_A/K)$ (and the convention flagged immediately after the statement of F7 is maintained).

Proof. We must show that the map (25) is one-to-one and onto; everything else then follows from F7.

Surjectivity: Let E/K be abelian of exponent n. Setting

$$A := E^{\times n} \cap K^{\times}$$

we get $K(\sqrt[n]{A}) \subseteq E$. Suppose that E is not contained in $K(\sqrt[n]{A}) = E_A$. Then there exists a *finite* subextension F/K of E/K such that $F \not\subseteq E_A$. This extension is abelian of exponent n because E/K is. The finite abelian group $G(F/K)$ is a direct product of finite cyclic subgroups. By Galois theory, then, F/K is a composite of cyclic subextensions of F/K; therefore we could just as well have chosen F so that F/K it cyclic, and we assume we did so. By Theorem 1, we have F is generated over K by some α such that $\alpha^n \in K^{\times}$. By definition, then, α belongs to E_A. But then $F \subseteq E_A$, contradicting our assumption.

Injectivity: In view of the preceding discussion, what is left to prove is: If $E = K(\sqrt[n]{A})$ for some subgroup A of $K^{\times}$ containing $K^{\times n}$, then A must coincide with the subgroup

$$(27) \qquad A_E = E^{\times n} \cap K^{\times}.$$

The inclusion $A \subseteq A_E$ is clear. Conversely, let a be any element of A_E. Since $K(\sqrt[n]{A_E}) = E = K(\sqrt[n]{A})$, there exist elements $a_1, \dots, a_r$ of A such that $K(\sqrt[n]{a}) \subseteq K(\sqrt[n]{a_1}, \dots, \sqrt[n]{a_r})$, which means that

$$(28) \qquad K(\sqrt[n]{a}, \sqrt[n]{a_1}, \dots, \sqrt[n]{a_r}) = K(\sqrt[n]{a_1}, \dots, \sqrt[n]{a_r}).$$

It suffices to prove that a is an element of the group $A' = \langle a_1, \dots, a_r, K^{\times n} \rangle$. To do this, consider the field

$$E' = K(\sqrt[n]{a_1}, \dots, \sqrt[n]{a_r}) = K(\sqrt[n]{A'}).$$

Because of (28), $a \in A_{E'}$. We see therefore that the extension E/K may be assumed *finite* with loss of generality. But taking into account (23) in F7, we have

$$A : K^{\times n} = A_E : K^{\times n} < \infty,$$

from which we get the desired equality $A = A_E$ because $A \subseteq A_E$. $\qquad \square$

Remark. Let $E/K = K(\sqrt[n]{A})/K$ be any *Kummer extension of exponent n* as above, with Galois group G. Then, even in the case $E : K = \infty$, the nondegenerate pairing (22) gives rise to an isomorphism

$$G \simeq (A/K^{\times n})^*,$$

as can be deduced easily from the finite case, effecting the regress via projective limits. Similarly, (22) gives in the infinite case an isomorphism

$$A/K^{\times n} \simeq G^*,$$

if by G^* we understand the group of *continuous* homomorphisms from the *topological group* $G = G(E/K)$, with the Krull topology, into the discrete group $W_n(K)$.

3. The preceding field-theoretic considerations are a good motivation for spending some time at this point on the *structure of finite abelian groups*. Although this is elementary material that you may have already studied, a brief and self-contained exposition of it geared toward our needs is useful. Actually the treatment of abelian groups is not at all out of place here; N. H. Abel (1802–1829), in his investigations of algebraic equations, made patent right from the beginning the close connections between equations and commutative groups.

Definition. A group G is called a *torsion group* if for every $x \in G$ there exists $n \in \mathbb{N}$ such that $x^n = 1$.

If p is a prime number, G is a *p-group* if for every $x \in G$ there is a p-power p^k such that $x^{p^k} = 1$.

F8. *A finite group G is a p-group if and only if its order $|G|$ is a p-power.*

Proof. Let $n = |G|$. We know that $x^n = 1$ for every $x \in G$. Thus, if n is a p-power, G is a p-group. If n is not a p-power, it has distinct prime factors $q \neq p$. By *Cauchy's Theorem* (page 101) there exists $x \in G$ such that $\operatorname{ord} x = q$. Therefore G is not a p-group; if it were, the order of x would be a factor of a p-power. $\qquad\square$

Remark. In the case of an *abelian* group G we need not resort to Cauchy's Theorem: Let G be an abelian p-group of order $n > 1$. For any $x \neq 1$ in G the order of x is a p-power p^k, that is, the subgroup $H = \langle x \rangle$ has order p^k. By induction we can assume that the order of the quotient group G/H is also a p-power. This gives the result, since $|G| = |G/H| \cdot |H|$.

Convention. In the sequel the group operation for an *abelian group* A will mostly be denoted *additively*. For $x \in A$ and $n \in \mathbb{Z}$ we then write nx instead of x^n.

Remark. Every abelian group is a $\mathbb{Z}$-module (and vice versa). The point of this observation is that the main results about to be stated for abelian groups carry over with only minor terminological changes to the context of modules over an arbitrary *principal ideal domain R* instead of $\mathbb{Z}$. We will formulate them for abelian groups only, but give the proofs in such a way that they can be generalized without difficulty to modules over a PID.

Definition. Let A be any *abelian* group, and let p be a prime number. Set

$$A_p = \{x \in A \mid \text{there exists } k \in \mathbb{N} \text{ such that } p^k x = 0\}.$$

Since A is abelian, A_p is a *subgroup* of A. By definition, A_p is a p-group. We call A_p the *p-component* of A.

F9. *Every **abelian torsion group** A is the direct sum of its p-components; that is, the map*

$$\varphi : \bigoplus_p A_p \to A$$
$$(x_p)_p \mapsto \sum_p x_p$$

is an isomorphism.

Proof. (i) Clearly φ is a (well defined) homomorphism of groups.

(ii) Take $x = (x_p)_p \in \bigoplus_p A_p$, and suppose $\varphi(x) = 0$. Let $p_1, \ldots, p_r$ be (distinct) indices such that $x_p = 0$ for all $p \neq p_i$, and consider

$$(29) \qquad n = p_1^{k_1} \ldots p_r^{k_r} \quad \text{with } p_i^{k_i} = \text{ord } x_{p_i}.$$

Set $n_i = n/p_i^{k_i}$ for $1 \leq i \leq r$. Because $n_1, \ldots, n_r$ are relatively prime, there is a relation

$$(30) \qquad c_1 n_1 + c_2 n_2 + \cdots + c_r n_r = 1 \quad \text{with } c_i \in \mathbb{Z}.$$

Set $x_i = x_{p_i}$. Then $0 = n_i \varphi(x) = n_i \sum_k x_k = n_i x_i$ and hence $x_i = 1 x_i = c_i n_i x_i = 0$ for all $1 \leq i \leq r$. Therefore $x = 0$, and we have shown that φ is injective.

(iii) Take any $y \in A$. Since A is a torsion group, there exists $n \in \mathbb{N}$ such that

$$(31) \qquad\qquad ny = 0.$$

Let (29) be the prime factorization of n, define numbers n_i as in part (ii) and choose coefficients c_i such that (30) holds. Then

$$(32) \qquad y = 1y = \sum_{i=1}^{r} c_i n_i y = \sum_{i=1}^{r} x_i,$$

with $x_i = c_i n_i y$. Since $p_i^{k_i} x_i = c_i p_i^{k_i} n_i y = c_i ny = 0$, each x_i belongs to A_{p_i}. Define $x = (x_p)_p \in \bigoplus_p A_p$ by setting $x_p = x_i$ if $p = p_i$ and $x_p = 0$ otherwise; then $\varphi(x) = \sum x_i = y$, by (32). Therefore φ is surjective as well. $\qquad\square$

Definition. Let G be a group. We say that G is a *group of exponent m* if

$$(33) \qquad\qquad x^m = 1 \quad \text{for every } x \in G$$

(in the abelian case, $mx = 0$).

Remarks. (1) If G is a group of exponent m and m divides m', then G is also a group of exponent m'. So our terminology is perhaps a bit questionable (but see the next definition). It is nonetheless very practical, particularly as regards the formulation of Kummer theory results such as the ones above.

(2) If G is a finite group of order n we have $x^n = 1$ for every $x \in G$, so G is a group of exponent n.

Definition. Let G be a group, and suppose there exists $m \in \mathbb{N}$ such that (33) holds. Among all such m let e be the smallest. We call this number *the* exponent of G and denote it by

$$e(G) := e.$$

If no $m \in \mathbb{N}$ satisfies (33), we set $e(G) = 0$. Clearly,

$$G \text{ is finite} \;\Rightarrow\; e(G) \neq 0 \;\Rightarrow\; G \text{ is a torsion group}$$

(and neither implication is reversible, as the examples $W(\mathbb{C}) \simeq \mathbb{Q}/\mathbb{Z}$ and $\mathbb{F}_p^{(\mathbb{N})}$ show).

F10. *Let G be a group such that $e(G) \neq 0$. Any $m \in \mathbb{N}$ for which (33) holds must be a multiple of $e(G)$:*

$$(34) \qquad\qquad\qquad e(G) \mid m.$$

Thus $e(G)$ is the least common multiple of the orders of elements of G:

$$(35) \qquad\qquad\qquad e(G) = \text{lcm} \, \{\text{ord } x \mid x \in G\}.$$

In particular,

$$(36) \qquad\qquad\qquad \text{ord } x \mid e(G) \quad \text{for all } x \in G.$$

Proof. Set $e = e(G)$ and write $m = qe + r$ by applying division with remainder. For every x in G we have $1 = x^m = x^{eq}x^r$, so $x^r = 1$. If r were not 0 we would get the contradiction $e \leq r < e$. This proves (34). $\qquad\qquad$ □

Remarks. (i) If G is finite we have $e(G) \neq 0$, and

$$(37) \qquad\qquad\qquad e(G) \text{ divides } |G|.$$

(ii) For $G = S_3$ we have $e(G) = |G|$.

(iii) For any finite *abelian* group G we only have $e(G) = |G|$ when G is *cyclic*. This quite nontrivial fact will become apparent as a consequence of F12 below.

(iv) We show that (iii) easily implies the following theorem, first encountered in Section 9: *If G is a finite subgroup of the multiplicative group $K^\times$ of a field K, then G is cyclic.* For let n be the order of G and $e = e(G)$ its exponent. Every $x \in G$ satisfies $x^e = 1$ and is thus a root of the polynomial $X^e - 1 \in K[X]$. There can be at most e such roots in K, so $n \leq e$. There follows $e = n$, and then (iii) implies that G is cyclic.

(v) Consider the direct product $G = \prod_{i \in I} G_i$ of given groups G_i, for $i \in I$. Set $e = e(G)$ and $e_i = e(G_i)$. Then obviously

$$(38) \qquad\qquad e = \operatorname{lcm}\{e_i \mid i \in I\}.$$

(vi) Assume $e(G) \neq 0$. If G is a p-group, $e(G)$ is a p-power.

F11. *Let A be an abelian group with $e(A) \neq 0$, and let $p_1, \ldots, p_r$ be the distinct prime factors of $e := e(A)$. Setting $A_i = A_{p_i}$ and $e_i = e(A_i)$, the map*

$$(39) \qquad \begin{array}{c} A_1 \times A_2 \times \cdots \times A_r \to A \\ (x_1, x_2, \ldots, x_r) \mapsto \sum_i x_i \end{array}$$

is an isomorphism, and

$$(40) \qquad\qquad e = e_1 e_2 \ldots e_r$$

is the prime factorization of e.

Proof. A_p is zero for every prime p that does not come into the prime factorization of e. By F9, then, (39) is indeed an isomorphism. The rest follows from remarks (v) and (vi) above. $\qquad\square$

Remark. Let A be a *finite* abelian group of order n and let $n = p_1^{\nu_1} \ldots p_r^{\nu_r}$ be the prime factorization of n. Then $p_1, \ldots, p_r$ are exactly the distinct prime factors of $e = e(A)$. The orders of the groups A_i of F11 are given by

$$(41) \qquad\qquad |A_i| = p_i^{\nu_i};$$

in other words, A_i is the p_i-*Sylow group* of A, for each $1 \leq i \leq r$.

Proof. With the notation of F11, we have $A \simeq A_1 \times \cdots \times A_r$, so $|A| = \prod_{i=1}^r |A_i|$. Being a p_i-group, A_i has order a power of p_i (see F8). The result follows, because the prime factorization of n is unique. $\qquad\square$

F12. *Let A be an abelian group with $e(A) \neq 0$. There is an element a in A such that*

$$(42) \qquad\qquad \operatorname{ord} a = e(A).$$

Proof. (i) We consider first the special case where $e := e(A)$ is a p-power for some prime p. For every $x \in G$, the order of x divides e. Since e is a p-power, we have

$$\operatorname{ord} x = p^{\nu(x)},$$

with a well defined $\nu(x) \in \mathbb{N} \cup 0$. Now, by (35),

$$e = \operatorname{lcm}\{p^{\nu(x)} \mid x \in A\} = p^{\max\{\nu(x) \mid x \in A\}}.$$

Given $a \in A$ with $\nu(a) = \max\{\nu(x) \mid x \in A\}$, then, we have $e = p^{\nu(a)} = \operatorname{ord} a$.

(ii) For the general case we invoke F11. Using the same notation as in that statement, we see from part (i) that, since $e_i = e(A_i)$ is a power of p_i for each i, there exist a_i such that

(43) $\operatorname{ord} a_i = e_i$ for $1 \le i \le r$.

Consider the element $\tilde{a} = (a_1, \ldots, a_r)$ in $A_1 \times \cdots \times A_r$. Since (39) is an isomorphism, it is enough to show that $\operatorname{ord} \tilde{a} = e(A)$. By (40) and (43) we have, on the one hand, $e(A) = e_1 e_2 \ldots e_r = \operatorname{ord} a_1 \operatorname{ord} a_2 \ldots \operatorname{ord} a_r$; on the other, each element $\tilde{a} = (a_1, \ldots, a_r)$ of the direct product $A_1 \times \cdots \times A_r$ obviously satisfies $\operatorname{ord} \tilde{a} = \operatorname{lcm}(\operatorname{ord} a_1, \ldots, \operatorname{ord} a_r)$. But because all the a_i have orders relatively prime to one another, we have $\operatorname{lcm}(\operatorname{ord} a_1, \ldots, \operatorname{ord} a_r) = \operatorname{ord} a_1 \operatorname{ord} a_2 \ldots \operatorname{ord} a_r$.   □

Definition. Let G be a group and M a subset of G. Denote by

$$\langle M \rangle$$

the intersection of all subgroups of G containing M; equivalently, $\langle M \rangle$ is the least subgroup of G containing M. When $M = \{a_1, \ldots, a_r\}$ is a finite set we also write $\langle a_1, \ldots, a_r \rangle$ for $\langle M \rangle$.

When $\langle M \rangle = G$ we say that M *generates* G (or that the elements of M do so). A group G is called *finitely generated* if there are finite many elements $a_1, \ldots, a_r \in G$ such that $G = \langle a_1, \ldots, a_r \rangle$.

Let A be a *finitely generated abelian* group. By assumption, there are elements $a_1, \ldots, a_r$ in A such that

$$A = \left\{ \sum_{i=1}^{r} c_i a_i \mid c_i \in \mathbb{Z} \right\} = \mathbb{Z} a_1 + \cdots + \mathbb{Z} a_r = \langle a_1 \rangle + \cdots + \langle a_r \rangle.$$

First we will assume that A is a *torsion group*. (Then A is necessarily *finite*; this is easy to check, but we will not make use of finiteness, so that our considerations can serve also for the more general case of modules over a PID.) Taking F11 into account, we can reduce to the case of a p-group. For that case we have:

Theorem 5. *Let $A \ne 0$ be a finitely generated abelian p-group. Then A is a finite direct product of cyclic p-groups, that is,*

(44) $A \simeq \mathbb{Z}/p^{\nu_1} \times \mathbb{Z}/p^{\nu_2} \times \cdots \times \mathbb{Z}/p^{\nu_s},$

where the ν_i are natural numbers. The numbers $s, \nu_1, \nu_2, \ldots, \nu_s$ are uniquely determined if we impose the condition

(45) $\nu_1 \ge \nu_2 \ge \ldots \ge \nu_s \ge 1.$

Proof. Suppose that A is generated by $a_1, \ldots, a_r$, where $r = r(A)$ is the minimal number of generators of A (i.e., the smallest cardinality of a set that generates A). To prove the theorem's first assertion, we work by induction on $r(A)$, and we show that (44) holds with $s = r(A)$.

For $r(A) = 1$ everything is clear. Let $r(A) > 1$ and denote by $e = e(A)$ the exponent of A. Clearly $e \neq 0$. Since A is a p-group, one sees immediately that one at least of $a_1, \ldots, a_r$ must have order e; suppose without loss of generality that it is a_1:

$$\text{(46)} \qquad \text{ord } a_1 = e.$$

Now consider $\bar{A} = A/\langle a_1 \rangle$, with quotient homomorphism $x \mapsto \bar{x}$. Obviously $\bar{A} = \langle \bar{a}_2, \ldots, \bar{a}_r \rangle$, so $r(\bar{A}) < r(A)$. By the induction hypothesis this means that

$$\text{(47)} \qquad \bar{A} = \langle \bar{b}_2 \rangle \times \cdots \times \langle \bar{b}_s \rangle$$

for certain $b_i \in A$ (and $s \leq r$). Moreover each b_i is replaceable by representative of the same class modulo $\langle a_1 \rangle$. By (47), we have in any case

$$\text{(48)} \qquad A = \mathbb{Z}a_1 + \mathbb{Z}b_2 + \cdots + \mathbb{Z}b_s = \langle a_1, b_2, \ldots, b_s \rangle.$$

(This means that $r(A) \leq s$, so in fact $r = s$.) We now show that one can choose the b_i in such a way that the sum (48) is *direct*, which will obviously imply the existence part of the theorem. Let $e_i = \text{ord } \bar{b}_i$. Then we have

$$\text{(49)} \qquad e_i b_i = k_i a_1 \quad \text{with } k_i \in \mathbb{Z}.$$

Since $e_i = \text{ord } \bar{b}_i \,|\, \text{ord } b_i \,|\, e$, we have $e/e_i \in \mathbb{N}$. Multiplication of (49) by e/e_i yields

$$(e/e_i)k_i a_1 = e b_i = 0.$$

Using (46) it follows that $(e/e_i)k_i$ is divisible by e, and hence that k_i/e_i must be an integer. Now set $b_i' = b_i - (k_i/e_i)a_1$; then $\bar{b}_i' = \bar{b}_i$, and $e_i b_i' = 0$ because of (49). Therefore

$$\text{ord } b_i' \,|\, e_i = \text{ord } \bar{b}_i = \text{ord } \bar{b}_i' \,|\, \text{ord } b_i',$$

so ord $b_i' = e_i$. Thus we can assume without loss of generality that

$$\text{(50)} \qquad \text{ord } b_i = e_i = \text{ord } \bar{b}_i \quad \text{for all } 2 \leq i \leq s.$$

But in this case (48) is a *direct sum*, because an equality of the form

$$\text{(51)} \qquad \sum_{i=1}^{s} c_i b_i = 0, \quad \text{with } c_i \in \mathbb{Z} \text{ (and } b_1 := a_1)$$

implies first $\sum_{i=2}^{s} c_i \bar{b}_i = 0$ (since $\bar{b}_1 = 0$) and then, using (47),

$$\text{ord } \bar{b}_i \,|\, c_i \quad \text{for } 2 \leq i \leq s.$$

Thus we conclude from (50) that ord b_i divides c_i, and hence

$$c_i b_i = 0 \quad \text{for } 2 \leq i \leq s.$$

But then $c_1 b_1$ also vanishes, so all summands in (51) are zero.

We must still show the uniqueness of the representation in Theorem 5. Assume that, besides (44) and (45), there holds

$$A \simeq \mathbb{Z}/p^{\mu_1} \times \mathbb{Z}/p^{\mu_2} \times \cdots \times \mathbb{Z}/p^{\mu_t},$$

where $\mu_1 \geq \mu_2 \geq \ldots \geq \mu_t \geq 1$ are natural numbers. Clearly $p^{\mu_1} = e(A) = p^{\nu_1}$, so certainly $\mu_1 = \nu_1$. We now apply induction on $e(A)$. Suppose $e(A) > p$. The subgroup pA of A obviously satisfies

$$pA \simeq \mathbb{Z}/p^{\nu_1-1} \times \cdots \times \mathbb{Z}/p^{\nu_s-1} \simeq \mathbb{Z}/p^{\mu_1-1} \times \cdots \times \mathbb{Z}/p^{\mu_t-1}.$$

The induction hypothesis then implies immediately that $\nu = (\nu_1, \ldots, \nu_s)$ and $\mu = (\mu_1, \ldots, \mu_t)$ already coincide at all entries except perhaps those of value 1; in other words, there exists $d \geq 1$ such that $\nu_d > 1$ and that

$$\nu = (\nu_1, \ldots, \nu_d, \underbrace{1, \ldots, 1}_{n \text{ entries}}), \quad \mu = (\nu_1, \ldots, \nu_d, \underbrace{1, \ldots, 1}_{m \text{ entries}}).$$

To show that the number of 1's is the same for μ and ν, consider the subgroup $A(p) = \{a \in A \mid pa = 0\}$ of A. We have

$$A(p) \simeq (\mathbb{Z}/p)^d \times (\mathbb{Z}/p)^m \simeq (\mathbb{Z}/p)^d \times (\mathbb{Z}/p)^n.$$

This implies $m = n$, because we are talking of vector spaces over the field $\mathbb{Z}/p$. The same argument also serves to start the induction, in the case $e(A) = p^1$. □

(The last step of the proof can be simplified by considering the order of the group; but again we refrain from using this argument so the proof applies more generally.)

Theorem 5′. *Let $A \neq 0$ be a finitely generated abelian torsion group (that is, a finite abelian group). Then A is a finite direct product of cyclic groups. More precisely, there exist natural numbers $e_1, e_2, \ldots, e_s > 1$ such that*

$$(52) \qquad\qquad e_{j+1} \mid e_j \quad \text{for } 1 \leq j \leq s-1$$

and that the isomorphism

$$(53) \qquad\qquad A \simeq \mathbb{Z}/e_1 \times \mathbb{Z}/e_2 \times \cdots \times \mathbb{Z}/e_s$$

holds. The numbers $s, e_1, \ldots, e_s$ are uniquely determined.

Proof. If $p_1, \ldots, p_r$ are the distinct prime factors of $e = e(A) \neq 0$, we have by F11

$$(54) \qquad\qquad A = A_1 \times \cdots \times A_r \quad \text{with } A_i = A_{p_i}.$$

Now decompose the A_i according to Theorem 5:

$$(55) \qquad\qquad A_i \simeq \prod_{j=1}^{s} \mathbb{Z}/p_i^{\nu(i,j)},$$

where we pad as needed with $v(i, j) = 0$, in order to be able to have the same number s of factors for each i. Multiply together the highest prime powers, one for each i, then the second highest, and so on. This yields a sequence

$$(56) \qquad e_j = p_1^{v(1,j)} p_2^{v(2,j)} \ldots p_r^{v(r,j)}$$

of natural numbers satisfying (52) and $e_j > 1$. Now, because of (56), we have

$$\mathbb{Z}/e_j \simeq \prod_{i=1}^{r} \mathbb{Z}/p_i^{v(i,j)}.$$

Thus by combining the factors in the same pattern as the prime powers, one obtains the isomorphism (53).

The uniqueness part of Theorem 5′ is true because the $A_i = A_{p_i}$ are canonically determined and their decompositions according to Theorem 5 are also unique (for each type). □

Theorem 5′ wholly explains the structure of finitely generated abelian groups that are *torsion groups*. To deal with the general case, we first look at the diametric opposite of torsion groups:

Definition. A group G is called *torsionfree* if, apart from the identity, no element of G has finite order.

F13. *Let A be any **abelian** group. Then*

$$A_T = \{x \in A \mid \text{there exists } n \in \mathbb{N} \text{ such that } nx = 0\}$$

is a subgroup of A. By definition, A_T is a torsion group. The quotient group A/A_T is torsionfree.

The easy proof is left as an exercise.

Theorem 6. *If $A \neq 0$ is a torsionfree finitely generated abelian group, then*

$$A \simeq \mathbb{Z} \times \cdots \times \mathbb{Z} = \mathbb{Z}^r,$$

where $r := r(A)$ is the minimal number of generators of A.

Proof. We wish to show that A is *free abelian in r generators*, which by definition means that $A \simeq \mathbb{Z}^r$. We first establish that A is isomorphic to a subgroup of $\mathbb{Z}^r$, then show that any such subgroup is isomorphic to $\mathbb{Z}^d$ for some $d \leq r$. That settles the question, since $r \leq d$ as well (obviously $\mathbb{Z}^d$ has a generating set with d elements).

Suppose $A = \mathbb{Z}a_1 + \mathbb{Z}a_2 + \cdots + \mathbb{Z}a_r$. After renumbering, let $\{a_1, \ldots, a_n\}$ be a *maximal linearly independent* subset of $\{a_1, \ldots, a_r\}$. (*Linearly independent* here means that if $\sum c_i a_i = 0$ with $c_i \in \mathbb{Z}$ then $c_1 = c_2 = \cdots = c_n = 0$.) In view of the definition of $a_1, \ldots, a_n$, there is for each i with $n < i \leq r$ a relation of the form

$$m_i a_i = \sum_{j=1}^{n} c_j a_j \quad \text{for some nonzero } m_i \in \mathbb{Z}.$$

Taking the lcm we see that there exists $m \in \mathbb{Z} \smallsetminus \{0\}$ such that

$$ma_i \in \mathbb{Z}a_1 + \cdots + \mathbb{Z}a_n \quad \text{for every } i.$$

Setting $F := \mathbb{Z}a_1 + \cdots + \mathbb{Z}a_n$, we rewrite this as

$$(57) \qquad\qquad mA \subseteq F.$$

Now F, by its definition and the linear independence of $a_1, \ldots, a_n$, is free abelian in n generators: an isomorphism $F \simeq \mathbb{Z}^n$ is given by $(x_1, \ldots, x_n) \mapsto \sum x_i a_i$. At the same time A and mA are isomorphic (by the map $a \mapsto ma$, since A is torsionfree). So, as announced,

$$(58) \qquad\qquad A \simeq mA \subseteq F \simeq \mathbb{Z}^n \quad \text{with } n \leq r.$$

Our theorem will be completely proved after we show that *any subgroup of $\mathbb{Z}^n$ is isomorphic to $\mathbb{Z}^d$ for some $d \leq n$* (where we allow $d = 0$ to represent the zero group).

We do this by induction on n. The case $n = 1$ is clear: every nonzero subgroup of $\mathbb{Z}$ is of the form $m\mathbb{Z} \simeq \mathbb{Z}$. Suppose $n > 1$ and denote by $\pi : \mathbb{Z}^n \to \mathbb{Z}$ the projection onto the n-th factor: $\pi(x_1, \ldots, x_n) = x_n$. Let $B \subset \mathbb{Z}^n$ be our subgroup and take $B' = B \cap \ker \pi$; this is isomorphic to a subgroup of $\mathbb{Z}^{n-1}$, so the induction assumption implies that

$$B' \simeq \mathbb{Z}^k, \quad \text{with } k \leq n - 1.$$

If $B = B'$ we are done. If instead $B \neq B'$, it is easy to check that

$$(59) \qquad\qquad B = B' \times \mathbb{Z}a,$$

where $a \in B$ is such that πa generates the subgroup πB. Thus $B \simeq \mathbb{Z}^k \times \mathbb{Z} \simeq \mathbb{Z}^{k+1}$ with $k + 1 \leq n$, and our assertion is proved. $\qquad\qquad\square$

Remarks. (1) Should anyone need convincing that $\mathbb{Z}^m$ and $\mathbb{Z}^n$ can only be isomorphic if $m = n$, here is one reason: An isomorphism between $\mathbb{Z}^m$ and $\mathbb{Z}^n$ implies, for any prime p, an isomorphism between $p\mathbb{Z}^m$ and $p\mathbb{Z}^n$, and so also an isomorphism between $\mathbb{Z}^m/p\mathbb{Z}^m \simeq (\mathbb{Z}/p)^m$ and $\mathbb{Z}^n/p\mathbb{Z}^n \simeq (\mathbb{Z}/p)^n$. But then $m = n$ follows by comparing dimensions.

(2) The existence of the direct-sum decomposition (59) falls under the following principle: *Given an exact sequence*

$$0 \to B \xrightarrow{\;i\;} A \xrightarrow{\;\pi\;} C \to 0$$

of abelian groups, if there is a homomorphism $\sigma : C \to A$ such that $\pi \circ \sigma = \mathrm{id}_C$, then A is the direct sum of iB and σC, and hence $A \simeq B \times C$. The simple proof is left to the reader; note that the endomorphism $\varepsilon := \sigma \circ \pi$ of A satisfies $\varepsilon^2 = \varepsilon$.

Theorem 7 (Classification of finitely generated abelian groups). *Let A be a finitely generated abelian group. There exist integers $c_1, c_2, \ldots, c_s \geq 0$, all distinct and different from 1, satisfying*

$$(60) \qquad\qquad c_j \mid c_{j+1} \quad \text{for } 1 \leq j \leq s-1,$$

and such that

$$(61) \qquad\qquad A \simeq \mathbb{Z}/c_1 \times \mathbb{Z}/c_2 \times \cdots \times \mathbb{Z}/c_s.$$

The numbers $s, c_1, \ldots, c_s$ are uniquely determined; $c_1, \ldots, c_s$ are called the invariants of A.

Proof. Let A_T be the torsion component of A and set $\bar{A} = A/A_T$ (see F13). Being a homomorphic image of A, the group $\bar{A}$ is finitely generated. It is also torsionfree, so Theorem 6 shows that

$$\bar{A} \simeq \mathbb{Z}^r = \mathbb{Z}/0 \times \cdots \times \mathbb{Z}/0$$

for some $r \geq 0$ (with the obvious interpretation in the case $r = 0$). Denote by $\pi : a \to \bar{a}$ the canonical homomorphism from A to $\bar{A}$, and let $\bar{a}_1, \ldots, \bar{a}_r$ form a $\mathbb{Z}$-basis of $\bar{A}$. There exists exactly one homomorphism $\sigma : \bar{A} \to A$ such that $\sigma(\bar{a}_i) = a_i$ for all $1 \leq i \leq r$; it satisfies $\pi \circ \sigma = \mathrm{id}_{\bar{A}}$. By Remark 2 after Theorem 6, then,

$$A \simeq A_T \times \bar{A};$$

in particular A_T is also a homomorphic image of A, and so finitely generated. By Theorem 5′, then, there is a (possibly empty) family of natural numbers $c_1, c_2, \ldots, c_q > 1$ such that

$$A_T \simeq \mathbb{Z}/c_1 \times \cdots \times \mathbb{Z}/c_q \quad \text{and} \quad c_j \mid c_{j+1}.$$

Setting $c_{q+1} = \cdots = c_{q+r} = 0$, we obtain (60) and (61).

The uniqueness part, which is very important, comes out easily from the uniqueness statements in Theorem 5′ and Remark 1 to Theorem 6. $\qquad\square$

Remark. Still in the situation of Theorem 7, *the number s equals $r(A)$, the minimal number of generators of A.* For obviously $r(A) \leq s$. Suppose $s \geq 1$. If p is a prime divisor of c_1 we have $A/pA \simeq (\mathbb{Z}/p)^s$, so $s = r(A/pA) \leq r(A)$.

In the framework of the question we are pursuing, Theorem 7 represents a fully satisfying outcome (and one which, as mentioned, extends to modules over any principal ideal domain R instead of $\mathbb{Z}$). In some other situations, however, one often needs a stronger result:

Theorem 8. *Let $F \simeq R^n$ be a free module in n generators over a principal ideal domain R, and let N be a submodule of F. There exists a basis $b_1, \ldots, b_n$ of F and elements $c_1, \ldots, c_m$ (where $m \leq n$) in $R \setminus \{0\}$ such that $c_i \mid c_{i+1}$ and that $c_1 b_1, \ldots, c_m b_m$ form a basis of N. The number m and the elements $c_1, \ldots, c_m$ are uniquely determined apart from multiplication by units.*

Proof. If we take existence as proved, we get an isomorphism

$$F/N \simeq R/c_1 \times \cdots \times R/c_m \times R^{n-m};$$

then uniqueness follows immediately from Theorem 7 (but note that here the early c_i's can be units in R).

One might think that the existence part can also be plucked somehow from Theorem 7, but that impression is deceptive. We are better off resorting to the methods of *linear algebra*. Both R-modules, F and N, possess R-bases, one with n and the other with m elements, where $m \leq n$. (Regarding N see the discussion following (58) on page 162.) The assertion now follows directly from a well known result:

Invariant Factor Theorem. *Let A be an $n \times m$ matrix over a principal ideal domain R, with $m \leq n$. Then A is equivalent over R to a matrix obtained from an $m \times m$ diagonal matrix by adding $n-m$ rows of zeros and where the diagonal entries $c_1, \ldots, c_m$ satisfy $c_i \mid c_{i+1}$.*

This is proved in LA II, p. 148, in the case where R is a Euclidean ring (also with the assumption $m = n$, but that makes scant difference.) If R is a PID but not Euclidean, it's not possible to get by with only the usual *elementary operations*, and the reader might ponder what to do instead. Note that certain other manipulations of two rows or columns are also permissible, whereby the leading entries a, b are replaced by $d, 0$, with $d = \gcd(a, b)$.   □

Apart from uniqueness, Theorem 7 is an easy consequence of Theorem 8: Let M be a finite generated module over a principal ideal domain R and n the minimal number of generators of M. Then $M \simeq R^n/N$. Theorem 8 gives

$$M \simeq R/c_1 \times \cdots \times R/c_m \times R^{n-m}.$$

By the minimality of n, none of the c_i is a unit in R. Hence we get a decomposition of the desired form, setting $c_i := 0$ for $m < i \leq n$.

15

Solvability of Equations

1. Let K be a field. The roots of a quadratic polynomial

$$(1) \qquad f(X) = X^2 + pX + q$$

can be represented, in a splitting field E of f over K, as

$$(2) \qquad -\frac{p}{2} \pm \sqrt{d}, \quad \text{where } d = \left(\frac{p}{2}\right)^2 - q;$$

here $\sqrt{d}$ denotes an element of E whose square is d, and we have assumed that char $K \neq 2$.[1] Moreover,

$$(3) \qquad 4d = p^2 - 4q$$

is the *discriminant* of f in the sense of Section 8.3.

In the early sixteenth century Italian mathematicians (Scipione Ferro, Nicolo Tartaglia, Girolamo Cardano) found out that the roots of a real cubic polynomial

$$(4) \qquad f(X) = X^3 + pX + q$$

too can be expressed using appropriate radicals: namely, in the form

$$(5) \qquad \sqrt[3]{-\frac{q}{2} + \sqrt{\left(\frac{q}{2}\right)^2 + \left(\frac{p}{3}\right)^3}} + \sqrt[3]{-\frac{q}{2} - \sqrt{\left(\frac{q}{2}\right)^2 + \left(\frac{p}{3}\right)^3}},$$

where the cube roots must be suitably interpreted (see F10). The same formula finds the roots of $f \in K[X]$ for any field K, provided of course that char $K \neq 2, 3$. Under this assumption, moreover, any cubic polynomial $g(X) = X^3 + aX^2 + bX + c$ can be reduced to a polynomial of the form (4), via the transformation $f(X) := g(X - \frac{1}{3}a)$.

---

[1] If char $K = 2$, the polynomial f in (1) is inseparable if and only if $p = 0$. If this is the case, f has as its only zero the unique square root of q. If $p \neq 0$, the roots of f can be expressed in terms of a solution α of the equation $Y^2 - Y = q/p^2$: they are $p\alpha$ and $p(\alpha+1)$.

The question whether similar formulas exist for the roots of polynomials of *any degree* occupied mathematicians for a long time. It was finally answered in the negative by *Abel* in 1826, and it provided a decisive impetus for the work of *Galois* (1831).

This problem of *solvability of equations by means of radicals* is close in spirit to the problem of constructibility with ruler and compass, discussed in Chapter 1. We will proceed in a similar way as we did then, with the difference that we now have at our fingertips a well-developed conceptual apparatus.

2. First we have to make precise what is meant by solving an equation by radicals.

Definition 1.   (i) Let F/K be a field extension. We say that F *arises from K by the successive adjunction of radicals* (*of exponents $n_1, \ldots, n_r$ respectively*) if there is a finite chain

(6) $$K = K_0 \subseteq K_1 \subseteq \cdots \subseteq K_r = F$$

of intermediate fields K_i of F/K such that each K_i is obtained from K_{i-1} by the adjunction of an n_i-th root (also called a *radical of exponent n_i*). In this case we call F/K a *radical extension*.

(ii) An extension E/K is called *solvable by radicals* (*of exponents $n_1, \ldots, n_r$ successively*) if there is a field extension F/K such that $E \subseteq F$ and F arises from K by the successive adjunction of radicals (of exponents $n_1, \ldots, n_r$ respectively).

(iii) A polynomial $f(X) \in K[X]$ is called *solvable by radicals* (*of exponents $n_1, \ldots, n_r$ successively*) if there is an extension E/K that is solvable by radicals (*of exponents $n_1, \ldots, n_r$ successively*) and such that f splits into linear factors over E.

Remarks.   (a) Obviously, any radical extension can be obtained by successively adjoining radicals of *prime* exponents.

(b) Let F_1, F_2 be intermediate fields of an extension C/K and let $F_1 F_2$ be their composite in C. Clearly, if F_1/K is a radical extension, so is $F_1 F_2/F_2$; and if $F_1 F_2/F_2$ and F_2/K are radical extensions, so is $F_1 F_2/K$. Putting it together we see that if F_1/K and F_2/K are both radical extensions, so is $F_1 F_2/K$. Corresponding statements follow for extensions solvable by radicals.

(c) If F/K is a radical extension and F' is the normal closure of F/K, the extension F'/K is also radical. This follows from (b), because F' is the composite over K of the fields conjugate to F (inside some algebraic closure C of F). The corresponding statement for extensions solvable by radicals follows.

(d) From (c) we derive: If $f \in K[X]$ is *irreducible* and E/K is an extension solvable by radicals such that f has a root in E, then f is solvable by radicals.

(e) Any finite and *purely separable* extension E/K is a radical extension; see F15 in Chapter 7.

Next we consider a *radical extension* F/K, assumed to be *Galois*, and we investigate what consequences can be derived regarding the Galois group $G = G(F/K)$. Take a chain of subfields (6) with all terms different and such that the adjoined radicals all have prime exponent; that is, assume that

(7) $$K_i = K_{i-1}(\alpha_i) \quad \text{with } \alpha_i \notin K_{i-1}, \ \alpha_i^{p_i} \in K_{i-1}, \ p_i \text{ prime}$$

for all i. Then $p_i \neq \operatorname{char} K$ for all i; otherwise K_i/K_{i-1} would be purely inseparable of degree p_i, contradicting the separability of F/K (see F3 in Section 7). If we set

$$n = p_1 p_2 \ldots p_r,$$

there is a primitive n-th root of unity ζ in the algebraic closure C of F. We now make the additional assumption that

$$\zeta \in K.$$

Then, by F1 in Chapter 14, each extension

(8) $$K_i/K_{i-1} \text{ is } cyclic \text{ of degree } p_i.$$

Now, by Galois theory, the chain (6) of intermediate fields of the extension F/K corresponds to a chain

$$G = H_0 \supseteq H_1 \supseteq \cdots \supseteq H_r = 1$$

of subgroups $H_i = G(F/K_i)$ of $G = G(F/K)$, and because of (8), each H_i is a normal subgroup of H_{i-1}, with quotient group H_{i-1}/H_i cyclic of order p_i. Thus we are led to:

Definition 2. A group H is called *solvable* (or *metacyclic*) if there exists a chain

(9) $$G = H_0 \supseteq H_1 \supseteq \cdots \supseteq H_r = 1$$

of subgroups H_i of G such that

(10) $$H_i \trianglelefteq H_{i-1} \text{ and } H_{i-1}/H_i \text{ is cyclic of prime order.}$$

(The H_i need not be normal subgroups of G.)

We next assemble a list of simple facts about solvable groups:

F1. *Let G be a finite group.*

  (a) *If G is solvable, so is any subgroup of G.*

  (b) *If G is solvable, so is any quotient group G/N of G.*

  (c) *Suppose $N \trianglelefteq G$. If N and G/N are solvable, so is G.*

  (d) *If G is abelian, G is solvable.*

  (e) *If G is a p-group, G is solvable.*

We postpone the proof for a bit, so we can get to something meatier right away:

Theorem 1. *If E/K is an extension solvable by radicals and E' is the normal closure of E/K, the automorphism group $G(E'/K)$ is solvable.*

Proof. By Remark (c) following Definition 1, E'/K is solvable by radicals if E/K is. Thus we may as well assume that E/K is *normal: $E = E'$.* By assumption there is a radical extension F/K such that E is a subfield of F. Again by Remark (c) we can assume that F/K is normal. Now, the natural map $G(F/K) \to G(E/K)$ is surjective; thus by part (b) of F1, it suffices to show that $G(F/K)$ is solvable. Altogether this shows that we can assume that E/K is both a *radical* and a *normal* extension; we further assume that E arises from K by the successive adjunction of radicals of *prime* exponents $p_1, p_2, \ldots, p_r$.

Now let n be the product of all the p_i such that $p_i \neq \operatorname{char} K$, and fix a primitive n-th root of unity ζ (in the algebraic closure of E). Consider the diagram of fields

(11)

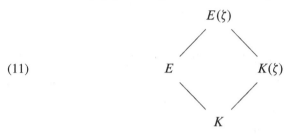

Clearly, $E(\zeta)/K$ is normal. Moreover $E(\zeta)/K(\zeta)$ is a radical extension with exponents $p_1, \ldots, p_r$, since E/K is one. By F9 in Chapter 9, the group $G(K(\zeta)/K)$ is abelian, and thus solvable, by part (d) of F1. Thus, if we take the solvability of $G(E(\zeta)/K(\zeta))$ as granted, part (c) of F1 shows that $G(E(\zeta)/K)$ is solvable and thus also $G(E/K)$, by part (b) of F1. What is left to prove, then, is that $G(E(\zeta)/K(\zeta))$ is solvable, or, otherwise put, that $G(E/K)$ is solvable under the additional assumption that $\zeta \in K$.

If E/K is Galois, there is nothing left to show, in view of what we said before Definition 2. The general case is completed by induction on r. For $r = 0$ there is nothing to show. Thus take $r > 0$ and assume without loss of generality that $K_1 \neq K$. If $p_1 = \operatorname{char} K$, the extension K_1/K is purely inseparable, so $G(K_1/K) = 1$; if $p_1 \neq \operatorname{char} K$, on the other hand, $G(K_1/K)$ is solvable. The induction hypothesis applied to the normal radical extension E/K_1 says that $G(E/K_1)$ is solvable. Because $G(K_1/K)$ and $G(E/K_1)$ are solvable, so is $G(E/K)$, by part (c) of F1.   □

The reader is encouraged to work out the proof again under the assumption that $\operatorname{char} K = 0$, which makes it less fussy.

Proof of F1. (b) Let $\pi : G \to \bar{G}$ be a surjective homomorphism. Given a chain (9) of subgroups of G, where each H_i is a normal subgroup of H_{i-1}, the application of π yields a chain

$$\bar{G} = \pi G = \pi H_0 \supseteq \pi H_1 \supseteq \cdots \supseteq \pi H_r = 1,$$

where again each πH_i is a normal subgroup of πH_{i-1}. Moreover π gives rise to surjective homomorphisms

$$\pi_i : H_{i-1}/H_i \to \pi H_{i-1}/\pi H_i.$$

If H_{i-1}/H_i has prime order, the Fundamental Homomorphism Theorem for groups and the Euler–Lagrange Theorem imply that either $\pi H_{i-1} = \pi H_i$ or $\pi H_{i-1}/\pi H_i \simeq H_{i-1}/H_i$. This proves that $\overline{G}$ is solvable.

(c) We can start with a chain

$$G/N = U_0 \supseteq U_1 \supseteq \cdots \supseteq U_m = 1$$

of subgroups G/N, where each U_i is a normal subgroup of U_{i-1} of prime index, and a similar chain

$$N = H_m \supseteq H_{m+1} \supseteq \ldots \supseteq H_n = 1$$

for N. Let $\pi : G \to G/N$ be the quotient homomorphism, and consider the subgroups $\pi^{-1}(U_i)$ of G. We have $\pi^{-1}(U_m) = \pi^{-1}(1) = N = H_m$, so there is no ambiguity in notation if we set

$$H_i := \pi^{-1}(U_i) \text{ for } i = 0, 1, \ldots, m.$$

Of course H_i is a normal subgroup of H_{i-1}; moreover π yields an isomorphism

$$H_{i-1}/H_i \simeq U_{i-1}/U_i.$$

Thus we obtain a chain

$$G = H_0 \supseteq H_1 \supseteq \cdots \supseteq H_m \supseteq H_{m+1} \supseteq \cdots \supseteq H_n = 1,$$

where each H_i is a normal subgroup of H_{i-1} with quotient group H_{i-1}/H_i cyclic of prime order. Therefore G is solvable.

(d) Let G be a finite abelian group. We prove that G is solvable by induction on the order of G. Leaving aside the trivial case $G \neq 1$, take an element $\sigma \in G$ whose order is a prime p. Let N be the subgroup generated by σ. Since G is abelian, N is trivially normal in G. By the induction assumption, G/N is solvable. Since N too is solvable, being a group of prime order, part (c) above shows that G is solvable.

Moreover using (d) and (c) we see by induction that, if a finite group G admits a chain of subgroups

$$G = H_0 \trianglerighteq H_1 \trianglerighteq \cdots \trianglerighteq H_r = 1$$

for which all the quotient groups H_{i-1}/H_i are *abelian*, G is solvable. Thus we can forgo the stipulation that the H_{i-1}/H_i in (10) have prime order. On the other hand, we remark that any solvable group G admits a chain $G = N_0 \supseteq N_1 \supseteq \cdots \supseteq N_r = 1$ where each N_i is a *normal subgroup of G*, so that the quotient groups N_{i-1}/N_i are all abelian (see §15.2 in the Appendix).

(e) The solvability of p-groups is guaranteed by the (stronger) statement of F8 in Chapter 10, again taking into account that a group of prime order is automatically cyclic.

(a) Let H be a subgroup of G. Starting from a chain (9) of subgroups of G having the properties listed in Definition 2, we obtain by intersection with H a chain

$$H = H_0 \cap H \supseteq H_1 \cap H \supseteq \cdots \supseteq H_r \cap H = 1.$$

For each $i = 1, 2, \ldots, r$, consider the restriction π_i of the quotient homomorphism $H_{i-1} \to H_{i-1}/H_i$ to the subgroup $H_{i-1} \cap H$ of H_{i-1}:

$$\pi_i : H_{i-1} \cap H \to H_{i-1}/H_i.$$

Its kernel is $(H_{i-1} \cap H) \cap H_i = H_i \cap H$. Thus $H_i \cap H$ is a normal subgroup of $H_{i-1} \cap H$, and π_i yields an injective homomorphism

$$H_{i-1} \cap H / H_i \cap H \to H_{i-1}/H_i.$$

Thus, either $H_{i-1} \cap H$ coincides with $H_i \cap H$ or $H_{i-1} \cap H / H_i \cap H \simeq H_{i-1}/H_i$ is cyclic of prime order. The solvability H follows. $\square$

We take the opportunity afforded by the preceding considerations to formulate the relevant isomorphism theorem:

F2 (Noether's isomorphism theorem for groups). *Let N be a normal subgroup of a group G, with quotient map $\pi : G \to G/N$. For every subgroup H of G, the restriction of π to H gives rise to an isomorphism*

$$H/H \cap N \simeq \pi(H);$$

moreover $\pi^{-1}(\pi(H)) = HN$, *so* π *also gives rise to an isomorphism* $HN/N \simeq \pi(H)$. *Altogether one gets a natural isomorphism*

(12) $$HN/N \simeq H/H \cap N.$$

Getting back to the problem of departure: we now show that, with the right stipulation regarding the characteristic, Theorem 1 admits a converse. (The reader might consider first the case char $K = 0$, where certain technical complications do not arise.)

Theorem 2. *Let E/K be a finite field extension and let E' be the normal closure of E/K. If the group $G(E'/K)$ is solvable and its order is not divisible by the characteristic of K, the extension E/K is solvable by radicals.*

Proof. Clearly we may as well assume that E/K is *normal* ($E' = E$). Let E_s be the separable closure of K in E. The extension E/E_s is radical (see Remark (e) after Definition 1), so it suffices to show that E_s/K is solvable by radicals. However,

E_s/K is Galois (because the normality of E/K obviously implies that of E_s/K); since $G(E_s/K) \simeq G(E/K)$ we can therefore assume that E/K is *Galois*.

Let G be the Galois group of E/K, of order $n = E : K$. Since char K does not divide n, the algebraic closure C of E contains a primitive n-th root of unity ζ. Now consider again the diagram (11). Trivially, $K(\zeta)/K$ is a radical extension, since ζ is an n-th root of 1. Therefore it is enough to show that $E(\zeta)/K(\zeta)$ is solvable by radicals (because then so is $E(\zeta)/K$, and *a fortiori* E/K). By the *Translation Theorem* of Galois theory (Chapter 12, Theorem 1), the group $G(E(\zeta)/K(\zeta))$ is isomorphic to a subgroup of $G = G(E/K)$; and since G is solvable, so is $G(E(\zeta)/K(\zeta))$, by F1. Thus we see that we can assume without loss of generality that the ground field K contains a primitive n-th root of unity.

That $G = G(E/K)$ is solvable means there is a chain

$$G = H_0 \supseteq H_1 \supseteq \cdots \supseteq H_r = 1$$

of subgroups H_i of G such that each H_i (where $1 \le i \le r$) is a normal subgroup of H_{i-1}, with cyclic quotient H_{i-1}/H_i of prime order p_i. By Galois theory we have an associated chain

$$K = K_0 \subseteq K_1 \subseteq \cdots \subseteq K_r = E$$

of intermediate fields K_i of E/K, where each extension K_i/K_{i-1} (for $1 \le i \le r$) is Galois, with Galois group isomorphic to H_{i-1}/H_i. Thus

(13) $$G(K_i/K_{i-1})$$

is *cyclic* of prime order p_i. Since K_{i-1} contains a primitive p_i-th root of unity (since p_i divides n), Theorem 1 of Chapter 14 says that K_i arises from K_{i-1} by adjunction of a p_i-th root. This completes the proof of Theorem 2. $\qquad \square$

Remark 1. We can extend the notion of a radical extension given in Definition 1 (and likewise that of an extension solvable by radicals) by allowing the possibility that, if char $K = p > 0$, some elements K_i in the chain (6) be obtained from K_{i-1} by the adjunction of a root of a polynomial of the form $X^p - X - \gamma \in K_{i-1}[X]$ (compare Theorem 3 in Chapter 14). Then the statement of Theorem 1 remains valid, and Theorem 2 is valid *without any condition on the characteristic*. The proofs are analogous to the ones given earlier, with recourse to Theorem 3 of Chapter 14 when required.

Remark 2. Let F/K be a field extension. We will say that F *arises from K by the successive adjunction of **irreducible** radicals of prime exponents* $p_1, \ldots, p_r$ if there is a finite chain

$$K = K_0 \subseteq K_1 \subseteq \cdots \subseteq K_r = F$$

of intermediate fields K_i of F/K such that each K_i arises from K_{i-1} by the adjunction of a root of an *irreducible* polynomial of the form

$$X^{p_i} - \gamma_i \in K_{i-1}[X], \quad \text{with } p_i \text{ prime.}$$

An extension E/K is called *solvable by irreducible radicals* if E is a subfield of a field F that can be obtained from K by the successive adjunction of *irreducible* radicals.

If E/K is such an extension and E' is the normal closure of E/K, the extension $G(E'/K)$ is solvable, by Theorem 1. For the converse we have to strengthen a bit our assumption on the ground field K:

Theorem 2'. *Let E/K be a finite extension, with normal closure E'. If the group $G(E'/K)$ is solvable and the characteristic of K is either zero or greater than all primes dividing the order of $G(E'/K)$, the extension E/K is solvable by **irreducible** radicals.*

The proof follows the same pattern as that of Theorem 2: First we can assume that E/K is *normal*. Since any purely inseparable extension is always solvable by *irreducible* radicals, we can also assume that E/K is *Galois*. Now everything goes through as in the proof of Theorem 2, assuming the following result to have been proved:

Lemma 1. *Let K be a field and n a natural number. Let K' be the field of n-th roots of unity over K. If the characteristic of K is either 0 or greater than all prime factors of n, the extension K'/K is solvable by **irreducible** radicals.*

Proof. We work by induction on n. The cases $n = 1, 2$ being trivial, assume $n > 2$. There is a primitive $\varphi(n)$-th root of unity η in the algebraic closure C of K', because no prime factor of $\varphi(n)$ is greater than all prime factors of n (formula (29) in Chapter 9). Now consider the diagram

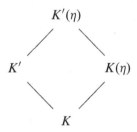

Since $\varphi(n) < n$, the extension $K(\eta)/K$ is solvable by irreducible radicals, by the induction hypothesis. We thus see that we can henceforth assume that the ground field K contains a primitive $\varphi(n)$-th root of unity. Now, $G(K'/K)$ is isomorphic to a subgroup of the group of prime residue classes modulo n (Chapter 9, F9). Thus $G = G(K'/K)$ is certainly abelian, and its order divides $\varphi(n)$. Every prime factor q of $|G|$ is therefore a factor of $\varphi(n)$. Since abelian groups are solvable, G has a chain of subgroups

$$G = H_0 \supseteq H_1 \supseteq \cdots \supseteq H_r = 1$$

where each H_i (for $1 \leq i \leq r$) is a normal subgroup of H_{i-1} or prime index q_i. By the Galois correspondence we get a corresponding chain

$$K = K_0 \subseteq K_1 \subseteq \cdots \subseteq K_r = K'$$

of intermediate fields K_i of K'/K, where each extension K_i/K_{i-1} is Galois with Galois group of prime order q_i. Each q_i divides $\varphi(n)$; thus K and therefore also K_{i-1} contain a primitive q_i-th root of unity. By Theorem 1 in Chapter 14, the field K_i arises from K_{i-1} by the adjunction of an irreducible radical. Therefore K' arises from K by the adjunction of irreducible radicals. □

Let $f \in K[X]$ be a polynomial over a field K, and let E be a splitting field of f over K. The automorphism group $G = G(E/K)$ is also called the *group of f* (or *of the equation $f = 0$) over K*. If all prime factors of f are separable in $K[X]$, the extension E/K is Galois; in this case G is called the *Galois group of f over K* (compare Chapter 8, Definition 3). An arbitrary polynomial $f \in K[X]$ is solvable by radicals if and only if E/K is solvable by radicals; see Definition 1(iii). With this, Theorems 1 and 2 imply:

Theorem 3. *Let f be a polynomial over a field K. If f is solvable by radicals, the group of f over K is solvable. Conversely, if the group of f over K is solvable and its order is not a multiple of the characteristic of K, the polynomial f is solvable by radicals.*

Remarks. (1) In characteristic 0, Theorem 3 becomes: *A polynomial $f \in K[X]$ is solvable by radicals if and only if its group over K is solvable.*

(2) If the characteristic of K is 0 or greater than all prime factors of $|G|$, Theorems 3 and 2' together imply: *A polynomial $f \in K[X]$ is solvable by radicals if and only if it is solvable by irreducible radicals.*

(3) If we use the extended notion of radicals introduced in Remark 1 to Theorem 2 (namely allowing roots of polynomials $X^p - X - \gamma$ where $p = \operatorname{char} K > 0$), the italicized statements in parts (1) and (2) hold regardless of the characteristic.

3. The foregoing considerations raise two obvious questions:

A. Are there *any* finite groups that are not solvable?

B. What finite groups occur as Galois groups of Galois extensions?

We first take up question B and show that, for any n, the *symmetric group S_n* is the Galois group of an appropriate Galois extension. To this effect we consider, over some arbitrary ground field k, the *field $k(X_1, X_2, \ldots, X_n)$ of rational functions in the n variables $X_1, \ldots, X_n$* (that is, the fraction field of the polynomial ring $k[X_1, \ldots, X_n]$ in the n variables $X_1, \ldots, X_n$ over k). For every $r \in k(X_1, \ldots, X_n)$ there exist polynomials $g_1, g_2 \in k[X_1, \ldots, X_n]$ such that

$$(14) \qquad r = \frac{g_1}{g_2}.$$

Of course this representation is not unique, but we have

$$(15) \qquad r = \frac{g_1}{g_2} = \frac{h_1}{h_2} \iff g_1 h_2 = h_1 g_2.$$

Take $\alpha_1, \ldots, \alpha_n$ in some extension k' of k. The expression

(16)
$$r(\alpha_1, \ldots, \alpha_n) := \frac{g_1(\alpha_1, \ldots, \alpha_n)}{g_2(\alpha_1, \ldots, \alpha_n)}.$$

is well-defined when it is defined at all, that is, when in the representation (14) we can choose g_2 so that $g_2(\alpha_1, \ldots, \alpha_n) \neq 0$. This can be easily seen from (15).

We now study the polynomial

(17) $$f(X) = \prod_{i=1}^{n} (X - X_i) = X^n - s_1 X^{n-1} + s_2 X^{n-2} + \cdots + (-1)^n s_n$$

over $k[X_1, \ldots, X_n]$ and over $k(X_1, \ldots, X_n)$. Note that $s_i \in k[X_1, \ldots, X_n]$, so we write $s_i = s_i(X_1, \ldots, X_n)$. We call s_i the i-th *elementary symmetric function*. We have

$$s_1 = X_1 + X_2 + \cdots + X_n,$$
$$s_2 = X_1 X_2 + \cdots + X_1 X_n + \cdots + X_{n-1} X_n,$$
$$\vdots$$
$$s_n = X_1 X_2 \ldots X_n.$$

The symmetric group S_n acts in a natural way on $k[X_1, \ldots, X_n]$ and $k(X_1, \ldots, X_n)$: Given $\sigma \in S_n$, we set

$$\sigma h(X_1, \ldots, X_n) := h(X_{\sigma(1)}, X_{\sigma(2)}, \ldots, X_{\sigma(n)});$$

that is, the map $h \mapsto \sigma h$ of $k[X_1, \ldots, X_n]$ into itself is the unique homomorphism of k-algebras

(18) $$k[X_1, \ldots, X_n] \to k[X_1, \ldots, X_n] \quad \text{such that } X_i \mapsto X_{\sigma(i)}.$$

It is clearly an isomorphism. Next, the map $h \mapsto \sigma h$ on $k(X_1, \ldots, X_n)$ is the unique extension of (18) to the fraction field. In this way each $\sigma \in S_n$ defines a k-automorphism of $k(X_1, \ldots, X_n)$. In the sequel we will regard S_n as a subgroup of the automorphism group of the extension $k(X_1, \ldots, X_n)/k$:

$$S_n \subseteq \text{Aut}(k(X_1, \ldots, X_n)/k) \subseteq G(k(X_1, \ldots, X_n)/k)$$

(see page 66 for notation). For conciseness we set $F = k(X_1, \ldots, X_n)$. Then the elements $\sigma \in S_n$ act (coefficientwise) on the ring $F[X]$ of polynomials in one variable X over F. The polynomial f in (17) satisfies

$$\sigma f = \prod_{i=1}^{n} (X - \sigma(X_i)) = \prod_{i=1}^{n} (X - X_{\sigma(i)}) = \prod_{i=1}^{n} (X - X_i) = f.$$

Thus the coefficients s_i of f lie in the fixed field of S_n in $k(X_1, \ldots, X_n)$. But clearly $k(X_1, \ldots, X_n)$ is a splitting field of the *separable* polynomial $f(X) = \prod_i (X - X_i)$ over the field

$$k(s_1, s_2, \ldots, s_n).$$

Therefore $k(X_1, \ldots, X_n)/k(s_1, \ldots, s_n)$ is a *Galois* extension. Every automorphism of this extension permutes the roots $X_1, \ldots, X_n$ of f, and as just seen every permutation of $X_1, \ldots, X_n$ is obtained in this way. Thus

$$(19) \qquad G\big(k(X_1, \ldots, X_n)/k(s_1, \ldots, s_n)\big) = S_n.$$

With this we have exhibited S_n as the Galois group of a field extension.

Definition 3. An element $r \in k(X_1, \ldots, X_n)$ is called *symmetric* if

$$\sigma r = r \quad \text{for all } \sigma \in S_n,$$

that is, $r = r(X_1, \ldots, X_n)$ remains unchanged by any permutation of the variables.

Thus the *symmetric functions* of $k(X_1, \ldots, X_n)$ are precisely the elements of the fixed field of S_n in $k(X_1, \ldots, X_n)$. Taking (19) into account we see by Galois theory (Chapter 8) that the field of symmetric functions of $k(X_1, \ldots, X_n)$ coincides with the field $k(s_1, \ldots, s_n)$. To summarize:

F3. *Consider over the field of rational functions $k(X_1, \ldots, X_n)$ the polynomial*

$$f(X) = \prod_{i=1}^{n} (X - X_i) = X^n - s_1 X^{n-1} + s_2 X^{n-2} + \cdots + (-1)^n s_n$$

whose coefficients $s_1, \ldots, s_n$ are the elementary symmetric functions in $X_1, \ldots, X_n$. The extension $k(X_1, \ldots, X_n)/k(s_1, \ldots, s_n)$ is Galois, and its Galois group is naturally identified with the symmetric group S_n. A rational function $r \in k(X_1, \ldots, X_n)$ is symmetric if and only if it lies in $k(s_1, \ldots, s_n)$; that is, if and only if it can be expressed as a rational function in the elementary symmetric functions $s_1, \ldots, s_n$. The polynomial $f(X)$ is irreducible over $k(s_1, \ldots, s_n)$.

Proof. Only the last statement has not yet been proved. S_n is the Galois group of f over $k(s_1, \ldots, s_n)$. Since S_n acts transitively on the roots $X_1, \ldots, X_n$ of f, the irreducibility of f is guaranteed by F7 in Chapter 8.   □

We will now look at things from a different angle. As before, let k be an arbitrary ground field. Let $K = k(u_1, \ldots, u_n)$ be the field of rational functions in the n variables $u_1, u_2, \ldots, u_n$ over k. In the polynomial ring $K[X]$ over K, consider the polynomial

$$(20) \qquad g(X) = X^n - u_1 X^{n-1} + u_2 X^{n-2} + \cdots + (-1)^n u_n.$$

This is called the *general polynomial of degree n over k*.

Theorem 4. *The general polynomial of degree n over k is separable, and it is irreducible over the field $K = k(u_1, \ldots, u_n)$ of its coefficients. Its Galois group over K (also known as the "Galois group of the general equation of degree n over k") is isomorphic to the symmetric group S_n.*

Proof. Let E be the splitting field of g over $K = k(u_1, \ldots, u_n)$. Over E we have

(21) $$g(X) = \prod_{i=1}^{n} (X - x_i), \quad \text{with } x_i \in E.$$

Then $E = K(x_1, \ldots, x_n) = k(x_1, \ldots, x_n)$. Is E a field of rational functions in n variables over k? To answer the question, consider the polynomial ring $k[X_1, \ldots, X_n]$ in the n variables $X_1, \ldots, X_n$ over k, and as before take the polynomial

(22) $$f(X) := \prod_{i=1}^{n} (X - X_i).$$

Then let

(23) $$k[X_1, \ldots, X_n] \to k[x_1, \ldots, x_n]$$

be the unique homomorphism of k-algebras taking each X_i to x_i. This map obviously satisfies

$$s_i = s_i(X_1, \ldots, X_n) \mapsto s_i(x_1, \ldots, x_n) = u_i.$$

Therefore (23) gives rise to a homomorphism of k-algebras

(24) $$k[s_1, \ldots, s_n] \to k[u_1, \ldots, u_n] \quad \text{such that } s_i \mapsto u_i.$$

This is an isomorphism: indeed, since $k[u_1, \ldots, u_n]$ is the polynomial ring in the variables $u_1, \ldots, u_n$, there is a homomorphism of k-algebras

(25) $$k[u_1, \ldots, u_n] \to k[s_1, \ldots, s_n] \quad \text{such that } u_i \mapsto s_i,$$

and the maps (25) and (24) are clearly inverse to each other. The isomorphism (24) has a unique extension to the fraction field:

(26) $$K' = k(s_1, \ldots, s_n) \to K = k(u_1, \ldots, u_n).$$

We now claim that (23) too is an isomorphism. In particular, $x_1, \ldots, x_n$ are pairwise distinct and hence g is separable. The surjectivity of (23) is clear; we must prove its injectivity. Let $h(X_1, \ldots, X_n) \in k[X_1, \ldots, X_n]$ satisfy $h(x_1, \ldots, x_n) = 0$, and consider

$$N(h) = \prod_{\sigma \in S_n} \sigma h = h \cdot \prod_{\sigma \neq 1} \sigma h \in k[X_1, \ldots, X_n].$$

Clearly $N(h)$ lies in the fixed field of S_n, so $N(h) \in k(s_1, \ldots, s_n)$. Also $N(h)$ lies in the kernel of (23) because h does. But the map (23) coincides with the isomorphism (26) on $k(s_1, \ldots, s_n) \cap k[X_1, \ldots, X_n]$. Thus $N(h) = 0$ and hence $h = 0$.

We now can extend (23) in unique fashion to an isomorphism $k(X_1, \ldots, X_n) \to k(x_1, \ldots, x_n)$ of fraction fields, and taking this together with (26) we get the commutative diagram

(27)

$$
\begin{array}{ccc}
k(X_1, \ldots, X_n) & \xrightarrow{\simeq} & k(x_1, \ldots, x_n) \\
\uparrow & & \uparrow \\
K' & \xrightarrow{\simeq} & K
\end{array}
$$

The isomorphism (26) gives rise to an isomorphism

$$K'[X] \to K[X]$$

of polynomial rings; this isomorphism maps f to g. Because f is irreducible in $K'[X]$ we then deduce that g is irreducible in $K[X]$. Finally, (27) yields

$$G\big(k(x_1,\ldots,x_n)/K\big) \simeq G\big(k(X_1,\ldots,X_n)/K'\big) \simeq S_n$$

(see F3). This completes the proof of Theorem 4.   □

The proof also showed that (24) is an isomorphism, and hence that the field $k(s_1,\ldots,s_n)$ of symmetric functions can be seen as the field of rational functions in the n variables $s_1,\ldots,s_n$. Therefore:

F4. *Let* $r \in k(X_1,\ldots,X_n)$ *be symmetric. There exists a **unique***

$$g \in k(X_1,\ldots,X_n) \quad \text{such that } r = g(s_1,\ldots,s_n).$$

In fact we have more:

Fundamental Theorem on Symmetric Functions. *Every symmetric **polynomial** h in $k[X_1,\ldots,X_n]$ has one and only expression in the form*

$$h = g(s_1,\ldots,s_n)$$

*for g a **polynomial** in $k[X_1,\ldots,X_n]$.*

This theorem can be proved directly with some effort; see, for example, van der Waerden, *Algebra I*. But it can also be shown to follow from F4 if one has the right conceptual tools; we do this in Chapter 16.[2]

F5. *Given any finite group G, there exists Galois extensions E/F such that $G(E/F) \simeq G$.*

Proof. Set $n = |G|$. Then G is isomorphic to a subgroup U of S_n (Chapter 10, F2). By F3 (or Theorem 4) there exists a Galois extension E/K such that $G(E/K) = S_n$. Let F be the fixed field of U in E. By Galois theory, $G(E/F) = U \simeq G$.

$$
\left(
\begin{array}{c}
E \\
\mid \\
F
\end{array}
\;U \simeq G
\right)
$$

$$
\begin{array}{c}
S_n \\
\\
\mid \\
K
\end{array}
$$

     □

---

[2] For a delightful little exercise, show that, conversely, F4 follows from the theorem just stated; this is simple once you see how. On the other hand, F4 implies: *If r is symmetric and $r = g/h$ for $g, h \in k[X_1,\ldots,X_n]$ relatively prime, then g and h are symmetric.* To see this, show first that if two elements of $k[s_1,\ldots,s_n]$ are relatively prime in $k[s_1,\ldots,s_n]$, they are also relatively prime in $k[X_1,\ldots,X_n]$.

Remarks. With F5 we have been able to give a surprisingly simple answer to our Question B (page 173). Another, much harder question is this: *Once a ground field K is fixed*, for what finite groups G is there an extension E such that E/K is Galois with $G(E/K) \simeq G$? In particular, taking the case $K = \mathbb{Q}$, one would like to know whether *every* finite group G occurs as the Galois group of a Galois extension $E/\mathbb{Q}$. This is a central problem of *inverse Galois theory*. It can be shown using more or less elementary methods that for every natural number n there exists $f \in \mathbb{Q}[X]$ such that the Galois group of f over $\mathbb{Q}$ is isomorphic to S_n (see van der Waerden, *Algebra I*). A much deeper result of Scholz and Shafarevitch says that every *solvable* group G occurs as the Galois group of a Galois extension $E/\mathbb{Q}$.

As to our Question A on whether there exist nonsolvable finite groups, the following statement will be proved in the next section:

F6. *The symmetric group S_n is not solvable for $n \geq 5$, and it is solvable for $n \leq 4$.*

As a corollary we obtain:

Theorem 5 (Abel, Ruffini). *The general polynomial of degree n over k is not solvable by radicals if $n \geq 5$.*

Proof. Let g be as in (20) the general polynomial of degree n over k and let $K = k(u_1, \ldots, u_n)$ be its coefficient field. By Theorem 4, the Galois group of g over K is isomorphic to the symmetric group S_n. If the polynomial $g \in K[X]$ were solvable by radicals, therefore, S_n would have to be a solvable group, by Theorem 3; and F6 says this is not the case for $n \geq 5$.   □

F7. *Let K be a field with char $K \neq 2, 3$. Every $f \in K[X]$ of degree at most 4 is solvable by irreducible radicals.*

Proof. Let G be the group of f over K. Then G is isomorphic to a subgroup of S_n, where n is the degree of f. In view of the second sentence of F6, together with part (a) of F1, G is solvable. The assertion then follows from Theorem 2′.   □

4. We now must prove F6; and we take the opportunity to talk a bit about permutations. In this section M will denote a set with n elements, say $M = \{1, 2, \ldots, n\}$, and $S = S(M) \simeq S_n$ will denote the group of all permutations of M. The group S acts on M via

$$(\sigma, a) \mapsto \sigma a = \sigma(a).$$

For a given $\sigma \in S$ we consider in particular the action of the *cyclic group* $H = \langle \sigma \rangle$ on M. The orbit of $a \in M$ under H is also called the *orbit of a under σ*. Let $H_a = \{\rho \in H \mid \rho a = a\}$ be the stabilizer of a under H, and $d = H : H_a$ its index in H. Then $\sigma^i a = \sigma^j a \iff \sigma^{i-j} a = a \iff \sigma^{i-j} \in H_a \iff d \mid i - j \iff i \equiv j \bmod d$, so the orbit

$$Ha = \{a, \sigma a, \sigma^2 a, \ldots, \sigma^{d-1} a\} \quad \text{has } d \text{ elements.}$$

Now set $a_i := \sigma^{i-1} a$, so that

$$\sigma a_1 = a_2, \qquad \sigma a_2 = a_3, \qquad \ldots, \qquad \sigma a_d = a_1.$$

Definition 4. An element $\rho \in S(M)$ is called a *cycle of length d* if there exist d distinct elements $a_1, a_2, \ldots, a_d$ in M such that

$$\rho a_i = a_{i+1} \text{ for } i < d; \quad \rho a_d = a_1; \quad \rho a = a \text{ for all } a \in M \smallsetminus \{a_1, \ldots, a_d\}.$$

Remarks. (a) A cycle of length 1 is the identity: $\rho = 1$.

(b) Given d distinct elements $a_1, a_2, \ldots, a_d \in M$ there is obviously a unique $\rho \in S(M)$ satisfying the conditions in Definition 4. This cycle ρ of length d is denoted by

$$\rho = (a_1 a_2 \ldots a_d).$$

Note that $(a_1 a_2 \ldots a_d) = (a_2 a_3 \ldots a_d a_1) = \cdots = (a_d a_1 \ldots a_{d-1})$. Moreover, $(a) = 1$ for any $a \in M$.

(c) A cycle of length d obviously has order d.

(d) Cycles transform elegantly under *inner automorphisms*: for any $\tau \in S(M)$,

$$(28) \qquad \tau(a_1 a_2 \ldots a_d)\tau^{-1} = (\tau a_1 \, \tau a_2 \ldots \tau a_d).$$

Given $\sigma \in S(M)$, define the set $W(\sigma) = \{a \in M \mid \sigma a \neq a\}$; this is the union of orbits of σ (i.e., orbits under $\langle \sigma \rangle$) that have length greater than 1. Two permutations $\sigma, \tau \in S(M)$ are called *disjoint* if $W(\sigma) \cap W(\tau) = \emptyset$. Obviously, in this case the two permutations commute: $\sigma \tau = \tau \sigma$.

Again let $\sigma \in S(M)$ be given and let $C_1, \ldots, C_r$ be the distinct orbits of σ. As we saw above, there exists for each C_j a cycle ρ_j such that $\rho_j a = \sigma a$ for all $a \in C_j$ and $W(\rho_j) \subseteq C_j$. By definition, $\rho_1, \rho_2, \ldots, \rho_r$ are pairwise disjoint. We claim that

$$\sigma = \rho_1 \rho_2 \ldots \rho_r.$$

Indeed, any given $a \in M$ lies in precisely one C_j, and then

$$\rho_1 \rho_2 \ldots \rho_r a = \rho_j a = \sigma a.$$

We thus get the first statement in the following result:

F8. *Every $\sigma \in S(M)$ can be represented as a product*

$$(29) \qquad \sigma = \rho_1 \rho_2 \ldots \rho_r$$

of pairwise disjoint cycles with

$$(30) \qquad \sum_i \text{length } \rho_i = n.$$

This representation is unique apart from the order of the factors. In (29) one can of course omit cycles of length 1, but then (30) is no longer satisfied.

Proof. What remains to be proved is the uniqueness. Take $\sigma \neq 1$ and let $\sigma = \gamma_1 \gamma_2 \ldots \gamma_s$ be another decomposition into pairwise disjoint cycles $\gamma_1, \ldots, \gamma_s$ of length greater than 1. Take $a \in W(\gamma_1)$. There is precisely one j such that $a \in W(\rho_j)$; we may as well assume it is $j = 1$. Then

$$\gamma_1 a = \sigma a = \rho_1 a, \quad \gamma_1 \sigma = \sigma \gamma_1, \quad \rho_1 \sigma = \sigma \rho_1.$$

It follows that $\gamma_1(\gamma_1^{k-1} a) = \gamma_1^k a = \sigma^k a = \rho_1^k a = \rho_1(\rho_1^{k-1} a)$. Thus the orbits $W(\gamma_1)$ and $W(\rho_1)$ of a under γ_1 and ρ_1 coincide, and because of this we get $\gamma_1 x = \sigma x = \rho_1 x$ for all x in $W(\gamma_1) = W(\rho_1)$. Outside of this set both permutations act trivially. Altogether we obtain $\gamma_1 = \rho_1$, and by cancellation $\rho_2 \ldots \rho_r = \gamma_2 \ldots \gamma_s$. By induction we are done. □

F9. (i) *Let $\sigma \in S(M)$ have the decomposition $\sigma = \rho_1 \rho_2 \ldots \rho_r$ into pairwise disjoint cycles. The order of σ is the least common multiple of the lengths of the ρ_i.*

(ii) *Two elements σ, σ' of $S(M)$ are conjugate in $S(M)$ if and only if their decompositions into cycles have the same type* (the *type* is defined in the proof).

Proof. (i) Since all the ρ_i commute, we have

$$(31) \qquad\qquad \sigma^m = \rho_1^m \rho_2^m \ldots \rho_r^m \quad \text{for each } m.$$

In particular, if $m = v$ is the lcm of the lengths (which is to say, the orders) of the cycles ρ_i, equation (31) implies that $\operatorname{ord}(\sigma) \,|\, v$. From (31) we get the cycle decomposition of σ^m. Thus from $\sigma^m = 1$ we get $\rho_i^m = 1$ for all i. Therefore $\operatorname{ord}(\rho_i)$ divides $\operatorname{ord}(\sigma)$ for all i, and hence so does v.

(ii) A decomposition

$$\sigma = \rho_1 \rho_2 \ldots \rho_r$$

into pairwise disjoint cycles is called *normalized* if it satisfies condition (30) in F8. We say that $\sigma \in S_n$ has *type* $c_1, c_2, \ldots, c_n$ if the normalized decomposition of σ contains precisely c_j cycles of length j. We have $n = \sum_j j c_j$. For an arbitrary $\tau \in S_n$, equality (28) says that

$$\tau \sigma \tau^{-1} = (\tau \rho_1 \tau^{-1})(\tau \rho_2 \tau^{-1}) \ldots (\tau \rho_r \tau^{-1})$$

is the normalized cycle decomposition of $\tau \sigma \tau^{-1}$; this of course has the same type as the normalized decomposition of σ.

Conversely, suppose $\sigma, \sigma' \in S(M)$ have the same type and let

$$\sigma = \rho_1 \ldots \rho_r, \quad \sigma' = \rho_1' \ldots \rho_{r'}'$$

be their normalized decompositions. Then $r = r'$, and after renumbering we can assume that

$$\text{length } \rho_i = \text{length } \rho_i' \quad \text{for } 1 \le i \le r.$$

For fixed i set

$$\rho_i = (a_1 \ldots a_d), \quad \rho_i' = (a_1' \ldots a_d'),$$

and let φ_i be the bijection from $M_i = \{a_1, \ldots, a_d\}$ onto $M_i' = \{a_1', \ldots, a_d'\}$ defined by $\varphi_i(a_j) = a_j'$. Because the $\rho_1, \ldots, \rho_r$ are disjoint we can form a map $\tau : M \to M$ coinciding with φ_i on each M_i, and this is a bijection because the $\rho_1', \ldots, \rho_r'$ are also disjoint. Again from (28) we get

$$\tau \sigma \tau^{-1} = (\tau \rho_1 \tau^{-1}) \ldots (\tau \rho_r \tau^{-1}) = \rho_1' \ldots \rho_r' = \sigma',$$

so σ and σ' are conjugate in $S(M)$. $\qquad\square$

We denote by

$$(32) \qquad \qquad \mathrm{sgn} : S_n \to \{+1, -1\}$$

the well known *signature map* or *parity map*, which assigns to each permutation $\sigma \in S_n$ its sign or parity $\mathrm{sgn}(\sigma)$ (see for instance LA I, p. 160). A permutation σ is called *even* if $\mathrm{sgn}(\sigma) = 1$; otherwise it is called *odd*. If

$$(33) \qquad \qquad \sigma = \tau_1 \tau_2 \ldots \tau_s$$

is a representation of σ as a product of *transpositions* (cycles of length two), we have

$$\mathrm{sgn}(\sigma) = (-1)^s.$$

Note that the $\tau_1, \ldots, \tau_s$ in (33) are generally not pairwise disjoint, nor is a representation in the form (33) unique; only the parity of s is determined by σ.

For *cycles* σ of length d, one clearly has

$$\sigma = (a_1 a_2 \ldots a_d) = (a_1 a_2)(a_2 a_3) \ldots (a_{d-1} a_d),$$

so

$$(34) \qquad \qquad \mathrm{sgn}(\sigma) = (-1)^{d-1}.$$

Since the map (32) is a homomorphism, the set

$$A_n := \{\sigma \in S_n \mid \mathrm{sgn}(\sigma) = 1\} = \ker \mathrm{sgn}$$

is a normal subgroup of S_n, of index 2 if $n \geq 2$. It is called the *alternating group* of degree n, and it consists of all *even* permutations in S_n. By (34), cycles of odd length are even and cycles of even length are odd.

We now can easily show: S_n *is solvable for* $n \leq 4$. For $S_1 = 1$ and $S_2 \simeq \mathbb{Z}/2\mathbb{Z}$ this is clear; for S_3 we have the chain

$$(35) \qquad \quad S_3 \trianglerighteq A_3 \trianglerighteq 1 \quad \text{with } S_3/A_3 \simeq \mathbb{Z}/2\mathbb{Z}, \ A_3 \simeq \mathbb{Z}/3\mathbb{Z}.$$

In S_4, consider the set

$$(36) \qquad \qquad V_4 := \{1, (12)(34), (13)(24), (14)(23)\},$$

consisting of the identity and of all the *double transpositions* in S_4. Since

$$(12)(34) \cdot (13)(24) = (14)(23) = (13)(24) \cdot (12)(34),$$

this is a subgroup of S_4, isomorphic to $\mathbb{Z}/2\mathbb{Z} \times \mathbb{Z}/2\mathbb{Z}$. It is called the *Klein four-group*. In view of (28), it is a *normal subgroup of* S_4. Clearly V_4 is contained in A_4, and by considering the orders we see that $A_4/V_4 \simeq \mathbb{Z}/3\mathbb{Z}$. Now the chain

$$(37) \qquad\qquad\qquad S_4 \trianglerighteq A_4 \trianglerighteq V_4 \trianglerighteq 1$$

shows that S_4 is solvable. Moreover one easily sees that V_4 is the only nontrivial normal subgroup of A_4, and that $S_4/V_4 \simeq S_3$.

Lemma. *Let $n \geq 5$ and let G be a subgroup of S_n containing all three-cycles (that is, cycles of length three). If N is a normal subgroup of G with abelian quotient G/N, then N contains all three-cycles.*

Proof. Let (abc) be a three-cycle. Since $n \geq 5$, there exist d, e such that $d, e \in M \smallsetminus \{a, b, c\}$ and $e \neq d$. Set $\sigma = (ace)$ and $\rho = (abd)$. By (28) we have $\rho\sigma\rho^{-1} = (\rho a\ \rho c\ \rho e) = (bce)$, so

$$\rho\sigma\rho^{-1}\sigma^{-1} = (bce)(eca) = (abc).$$

But N contains all commutators of elements of G since G/N was assumed abelian; in particular N contains (abc). $\qquad\qquad\qquad\qquad\qquad\qquad\qquad\qquad\qquad\quad\square$

With this lemma we can now easily show that S_n *is not solvable for* $n \geq 5$. For otherwise there would be a chain $S_n = G_0 \trianglerighteq G_1 \trianglerighteq G_2 \trianglerighteq \cdots \trianglerighteq G_m = 1$ with all factors G_{i-1}/G_i abelian. By induction, it would follow from the lemma that every G_i contains all three-cycles. But this is impossible since $G_m = 1$.

The unsolvability of S_n for $n \geq 5$ has had interesting consequences for us from the field-theoretic point of view; but in fact a much more encompassing result is true:

Theorem 6. *The alternating group A_n is simple for $n \geq 5$.*

Recall that a group $G \neq 1$ is called *simple* if it has no normal subgroup apart from itself and 1. A proof of Theorem 6 is outlined in §15.13 and §15.16 in the Appendix.

5. Returning to our earlier line of investigation, we now wish to find an explicit solution by radicals for *cubic polynomials* f over a field K of characteristic distinct from 2 and 3. As remarked on page 165, we may as well assume that f has the form

$$(38) \qquad\qquad\qquad f(X) = X^3 + pX + q;$$

we also assume that f is irreducible in $K[X]$. Let E be a splitting field of f over K and let $\alpha_1, \alpha_2, \alpha_3$ be the roots of f in E. We now proceed according to the theory. First we adjoin a primitive third root of unity ζ, and so form the diagram

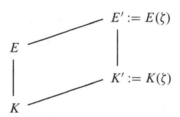

Now let $F' := K'(\sqrt{D})$, where D denotes the *discriminant* of f (Chapter 8, Definition 4). Since $\sqrt{D}$ is not preserved by any transposition of roots, we have $E' : K'(\sqrt{D}) \leq 3$. At the same time, since f is irreducible, $E : K$ is divisible by 3, and hence so is $E' : K'$, since $K' : K \leq 2$. Putting it together we get

$$E' : F' = 3.$$

Thus the extension E'/F' is *cyclic of degree* 3. By the material in Chapter 14 — see in particular equation (6) there — we have $E' = F'(v)$, where v, a *Lagrange resolvent*, is defined by

$$v = \beta + \zeta^2 \sigma(\beta) + \zeta \sigma^2(\beta),$$

where $\beta \in E'$ is to be chosen so that $v \neq 0$ but is otherwise arbitrary. We first try $\beta = \alpha_1$, so that

(39) $v = \alpha_1 + \zeta^2 \alpha_2 + \zeta \alpha_3,$

and we consider at the same time the resolvent

(40) $u = \alpha_1 + \zeta \alpha_2 + \zeta^2 \alpha_3.$

We have $u + v = 2\alpha_1 + (\zeta + \zeta^2)\alpha_2 + (\zeta + \zeta^2)\alpha_3 = 3\alpha_1$ since $1 + \zeta + \zeta^2 = 0$ and $\alpha_1 + \alpha_2 + \alpha_3 = 0$; and observing that multiplication by ζ or ζ^2 in (39) and (40) induces a cyclic permutation on the indices, we likewise get the last two equations in the following trio:

(41) $u + v = 3\alpha_1, \quad \zeta^2 u + \zeta v = 3\alpha_2, \quad \zeta u + \zeta^2 v = 3\alpha_3.$

By construction, u^3 and v^3 lie in F'; one sees on conjugacy grounds that these are the roots of a quadratic equation over K', and we can actually write down this equation explicitly by computing the sum and product of u^3 and v^3. First we have

$$u^3 + v^3 = (u + v)(u + \zeta v)(u + \zeta^2 v) = 3\alpha_1 \cdot 3\alpha_3 \cdot 3\alpha_2,$$

where the second equality comes from (41). Therefore

(42) $u^3 + v^3 = -27q.$

Next, from (39) and (40) we get

$$(43) \quad uv = \alpha_1^2 + \alpha_2^2 + \alpha_3^2 + \zeta^2(\alpha_1\alpha_2 + \alpha_2\alpha_3 + \alpha_1\alpha_3) + \zeta(\alpha_1\alpha_2 + \alpha_2\alpha_3 + \alpha_1\alpha_3)$$
$$= \alpha_1^2 + \alpha_2^2 + \alpha_3^2 + (\zeta^2 + \zeta)p = \alpha_1^2 + \alpha_2^2 + \alpha_3^2 - p$$
$$= (\alpha_1 + \alpha_2 + \alpha_3)^2 - 3p = -3p,$$

so

$$(44) \qquad\qquad u^3 v^3 = -27p^3.$$

Thus u^3 and v^3 are the roots of the quadratic polynomial

$$(45) \qquad\qquad X^2 + 27qX - 27p^3,$$

whose coefficients actually lie in K. There follows

$$\left(\frac{u}{3}\right)^3 = -\frac{q}{2} + \sqrt{\left(\frac{q}{2}\right)^2 + \left(\frac{p}{3}\right)^3}, \qquad \left(\frac{v}{3}\right)^3 = -\frac{q}{2} - \sqrt{\left(\frac{q}{2}\right)^2 + \left(\frac{p}{3}\right)^3}.$$

Together with (41) we then get:

F10. *Let K have characteristic distinct from 2 and 3. The roots of the polynomial $f(X) = X^3 + pX + q \in K[X]$ are given by*

$$(46) \qquad \sqrt[3]{-\frac{q}{2} + \sqrt{\left(\frac{q}{2}\right)^2 + \left(\frac{p}{3}\right)^3}} + \sqrt[3]{-\frac{q}{2} - \sqrt{\left(\frac{q}{2}\right)^2 + \left(\frac{p}{3}\right)^3}},$$

where one cube root γ can be chosen at will, but the other cube root γ' must be chosen so that $3\gamma\gamma' = -p$. (The choice of a square root is arbitrary but must be the same in both terms.) The discriminant D of $f(X)$ is

$$(47) \qquad\qquad D = -4p^3 - 27q^2.$$

Proof. Regardless of whether f is irreducible, we have seen above that, if u and v are defined by (39) and (40), they satisfy the relations (41), and u^3, v^3 are the roots of the quadratic equation (45). It follows that every root of f must have the form (46). To be safe we must show that for any f (even if inseparable), the numbers given by (46) are indeed zeroes of f. The cube roots $\gamma = u/3$ and $\gamma' = v/3$ that occur in (46) satisfy $\gamma^3 + \gamma'^3 = -q$ and $27\gamma^3\gamma'^3 = -p^3$ (since u^3, v^3 are solutions of the quadratic equation $X^2 + 27qX - 27p^3 = 0$), and by assumption they are normalized so that $3\gamma\gamma' = -p$. It follows that $(\gamma+\gamma')^3 + p(\gamma+\gamma') + q = \gamma^3 + \gamma'^3 + 3\gamma^2\gamma' + 3\gamma\gamma'^2 + p(\gamma+\gamma') + q = -q + 3\gamma\gamma'(\gamma+\gamma') + p(\gamma+\gamma') + q = 0$.

Finally we have to verify the formula (47) for the discriminant. Setting $\Delta = (\alpha_1 - \alpha_2)(\alpha_1 - \alpha_3)(\alpha_2 - \alpha_3)$, we have

$$
\begin{aligned}
27\Delta &= (3\alpha_1 - 3\alpha_2)(3\alpha_1 - 3\alpha_3)(3\alpha_2 - 3\alpha_3) \\
&= (u + v - \zeta^2 u - \zeta v)(u + v - \zeta u - \zeta^2 v)(\zeta^2 u + \zeta v - \zeta u - \zeta^2 v) \\
&= \big((1-\zeta)^2 u + (1-\zeta)v\big)\big((1-\zeta)u + (1-\zeta^2)v\big)\big((\zeta^2-\zeta)u + (\zeta-\zeta^2)v\big) \\
&= (1-\zeta)^2\big((1+\zeta)u + v\big)\big(u + (1+\zeta)v\big)\big((u-v)(\zeta^2-\zeta)\big) \\
&= (\zeta-1)^3 \zeta(-\zeta^2 u + v)(u - \zeta^2 v)(u - v) \\
&= (1-\zeta)^3 (u - \zeta v)(u - \zeta^2 v)(u - v) = (1-\zeta)^3 (u^3 - v^3);
\end{aligned}
$$

setting $D = \Delta^2$ and using the fact that $(\zeta - 1)^2 = \zeta^2 - 2\zeta + 1 = -3\zeta$ we get

$$
27^2 D = (\zeta - 1)^6 (u^3 - v^3)^2 = -27(u^3 - v^3)^2,
$$

so $-27D$ coincides with the discriminant of the quadratic polynomial (45):

$$
-27D = 27^2 q^2 + 4 \cdot 27 p^3.
$$

This implies (47).   □

Note that (47) can be rewritten as

$$
(48) \qquad D = -4 \cdot 27 \left(\left(\frac{p}{3} \right)^3 + \left(\frac{q}{2} \right)^2 \right),
$$

so that the *square root* appearing in the Cardano formula (46) is related to the discriminant by

$$
(49) \qquad \sqrt{-3D} = 2 \cdot 9 \sqrt{ \left(\frac{q}{2} \right)^2 + \left(\frac{p}{3} \right)^3 }.
$$

Remarks. (1) From the introductory remarks of this section we see that, *if the cubic polynomial f is irreducible, the extension $E/K(\sqrt{D})$ is cyclic of degree* 3. Moreover, since f has degree 3, the irreducibility of f in $K[X]$ is equivalent to there being no roots of f in K. In the sequel we will identify the Galois group G of f over K with a subgroup of S_3, by fixing the numbering of the roots $\alpha_1, \alpha_2, \alpha_3$ of f.

If f is irreducible we therefore have:

$$
(50) \qquad G = \begin{cases} A_3 & \text{if } D \text{ is a square in } G, \\ S_3 & \text{if } D \text{ is not a square in } G. \end{cases}
$$

An example of the first alternative with $K = \mathbb{Q}$ — that is, a cubic polynomial over $\mathbb{Q}$ whose Galois group G over $\mathbb{Q}$ is cyclic of order 3 — is given by

$$
(51) \qquad X^3 - 3X + 1.
$$

For this is clearly irreducible (Chapter 5, F8), and it has discriminant $D = 81$ by (47). Examples of the second alternative, illustrating the (generic) case $G = S_3$, are given by $X^3 - X + 1$, $X^3 - 2$, and $X^3 - 4X + 1$, with discriminants -23, -108, 229.

(2) Assume that K is a subfield of $\mathbb{R}$. Then, for any cubic polynomial f over K,

(52) $D \geq 0 \iff$ all the roots of f lie in $\mathbb{R}$.

To see this, note first that f, being a polynomial of odd degree, has at least one real root. Either all roots $\alpha_1, \alpha_2, \alpha_3$ of f are real or, say, α_1 is real and $\alpha_3 = \bar{\alpha}_2 \notin \mathbb{R}$. In the former case $\Delta = (\alpha_1 - \alpha_2)(\alpha_1 - \alpha_3)(\alpha_2 - \alpha_3)$ is real, so $D = \Delta^2 \geq 0$; in the latter case, $\Delta = (\alpha_1 - \alpha_2)(\alpha_1 - \bar{\alpha}_2)(\alpha_2 - \bar{\alpha}_2)$ is purely imaginary, so $D = \Delta^2 < 0$.

Suppose in addition that $D \neq 0$, so $\alpha_1, \alpha_2, \alpha_3$ are all distinct. By (48) we have

$$D > 0 \iff \sqrt{\left(\frac{q}{2}\right)^2 + \left(\frac{p}{3}\right)^3} \notin \mathbb{R}.$$

Thus, for f separable, equation (52) says that the Cardano formula (46) yields three real roots if and only if the *square root* appearing in the formula is nonreal ("casus irreducibilis").

6. In this section we will deal with *solvable equations of prime degree*. We will need to have at hand the following group-theoretic result.

Lemma. *Let p be a prime number and let G be a subgroup of the full permutation group of a set M with p elements. Assume that G acts **transitively** on M.*

(a) *Any normal subgroup $N \neq 1$ of G also acts transitively on M.*

(b) *If G is solvable, it contains a unique subgroup H of order p (which is then necessarily normal in G).*

Proof. (a) Take $a \in M$. For any σ in G we have

$$\sigma N a = N \sigma a.$$

Since G acts transitively on M, all orbits of N have the same length, say m. Thus m divides $p = |M|$ (see F1' in Chapter 10). Since $N \neq 1$ we have $m > 1$. It follows that $m = p$, that is, N acts transitively on M.

(b) Since G acts transitively on the p-element set M, the order of G is divisible by p (Chapter 10, F4). Assume $|G| > p$ (otherwise there is nothing to prove). Because G is solvable, it contains a normal subgroup N distinct from G and from 1. By part (a), N acts transitively on M. By induction on the order of the group, we can therefore assume that N has *exactly one* subgroup H of order p. Then H is a *characteristic subgroup* of the group N; that is, for each automorphism α of N we have $\sigma H = H$. In particular, H is a normal subgroup *of G* (because every inner automorphism of G gives rise to an automorphism of N, since N is normal in G). So G contains a *normal* subgroup H of order p. Let H' be any subgroup of order p and assume that H' is distinct from H. Then $H \cap H' = 1$, and the quotient $H'H/H$ satisfies

$$H'H/H \simeq H'/H' \cap H \simeq H',$$

so the subgroup $H'H$ of G has order p^2 (see F2). It follows that p^2 divides $|G|$, and hence $p!$, contradicting the primality of p. $\square$

Now let $f \in K[X]$ be an *irreducible* polynomial of *prime degree p*. We assume that f is *separable* (that is, if $p = $ char K, the polynomial does not have the form $c(X^p - a)$; see Chapter 7, F12). The Galois group G of f over K acts *transitively* on the p-element set M of roots of f in a splitting field E of f over K (Chapter 8, F7).

Assume that the Galois group G of f over K is *solvable*. By part (b) of the preceding lemma it follows, first, that G contains an element ρ of order p. Regarded as a permutation of M, the element ρ is then necessarily a *cycle of length p*, that is, for some numbering $\alpha_1, \alpha_2, \ldots, \alpha_p$ of the roots of f in E, we have

$$(53) \qquad \rho\alpha_i = \alpha_{i+1} \quad \text{for } i < p, \qquad \rho\alpha_p = \alpha_1.$$

We now identify $M = \{\alpha_1, \alpha_2, \ldots, \alpha_p\}$ with the p-element field $\mathbb{F}_p$ via the map $\alpha_i \mapsto i$, thus also identifying G with a subgroup of $S(\mathbb{F}_p)$. Then (53) becomes

$$(54) \qquad \rho x = x + 1 \quad \text{for } all \ x \in \mathbb{F}_p.$$

In other words, the action of ρ on elements of $\mathbb{F}_p$ is simply *translation* by 1. But more is true: Let σ be any element of G. By the lemma, the subgroup generated by ρ is a *normal subgroup* of G. Therefore

$$\sigma\rho\sigma^{-1} = \rho^a,$$

for some natural number $a < p$. Thus $\sigma\rho x = \rho^a \sigma x$ for all $x \in \mathbb{F}_p$; hence, by (54),

$$\sigma(x + 1) = \sigma x + a.$$

Setting $b := \sigma(0)$, we get $\sigma(1) = b+a$, $\sigma(2) = b+a+a$, ..., and in general

$$\sigma x = ax + b \quad \text{for all } x \in \mathbb{F}_p.$$

Definition 5. A permutation $\sigma \in S(\mathbb{F}_p)$ is called *affine* if there exist elements $a \in \mathbb{F}_p^\times$ and $b \in \mathbb{F}_p$ such that

$$(55) \qquad \sigma x = ax + b \text{ for all } x \in \mathbb{F}_p.$$

A subgroup G of $S(\mathbb{F}_p)$ is called an *affine subgroup of $S(\mathbb{F}_p)$* if every element of G is affine.

Theorem 7 (Galois).   (I) *Let $f \in K[X]$ be irreducible of prime degree p; also assume that f is separable (that is, not of the form $c(X^p - a)$ if $p = $ char K). As before we regard the Galois group G of f over K as a subgroup of $S(\mathbb{F}_p)$. Then, if G is solvable, it is an affine subgroup of $S(\mathbb{F}_p)$.*

(II) *Every affine subgroup of $S(\mathbb{F}_p)$ is solvable.*

Proof. Part (I) has been proved above. For part (II), let's denote an element $\sigma \in S(\mathbb{F}_p)$ of the form (55) by $\sigma_{a,b}$; note that $b \in \mathbb{F}_p$ and $a \in \mathbb{F}_p^\times$ are uniquely determined by σ. A simple calculation shows that

$$\sigma_{a,b} \circ \sigma_{a',b'} = \sigma_{aa',b+ab'}.$$

188 15 Solvability of Equations

Let G denote the set of all affine elements of $S(\mathbb{F}_p)$. Then G is a subgroup of $S(\mathbb{F}_p)$, and the map $\sigma = \sigma_{a,b} \mapsto a$ is a homomorphism form G onto the multiplicative group $\mathbb{F}_p^{\times}$ of $\mathbb{F}_p$. Let N be its kernel. Obviously, N consists of all elements of the form $\sigma_{1,b}$, that is, of all *translations*. The map $\sigma_{1,b} \mapsto b$ is then an isomorphism of N onto the additive group $\mathbb{Z}/p\mathbb{Z}$ of $\mathbb{F}_p$. Thus N is cyclic (of order p). The quotient G/N is isomorphic to the multiplicative group $\mathbb{F}_p^{\times}$, by the foregoing. Therefore G/N is also cyclic, by Chapter 9, Theorem 2. Putting it all together we conclude that G is solvable (see F1). Since a subgroup of a solvable group is solvable, we are done. $\qquad\square$

The lovely result just proved is Proposition VII of Galois's "Mémoire sur les conditions de résolubilité des équations par radicaux", and can be found in *Écrits et mémoires mathématiques d'Evariste Galois*, Gauthier-Villars, Paris, 1962.

Remark. The group $S(\mathbb{F}_p)$ itself is not affine for $p \geq 5$, since it has precisely $p(p-1)$ elements of the form (55). By Theorem 7, therefore, S_p is not solvable (since up to isomorphism, S_p occurs as the Galois group of the general equation of degree p: see Theorem 4). In particular, S_5 is not solvable. And since for $n \geq 5$ the symmetric group S_n clearly has a subgroup isomorphic to S_5, we have proved again that S_n cannot be solvable for $n \geq 5$.

As an application of Theorem 7 we obtain another result of Galois:

F11. *Let $f \in K[X]$ be irreducible of prime degree p, and assume f is separable. Let E be a splitting field of f over K and G the Galois group of E/K.*

If G is solvable, E arises by adjunction of any two distinct roots of f.

Conversely, if there exist two roots α, β of f such that $E = K(\alpha, \beta)$, the group G is solvable.

Proof. Let G be solvable and let α and $\beta \neq \alpha$ be roots of f in E. We must show that $E = K(\alpha, \beta)$, or equivalently, by Galois theory, that $G(E/K(\alpha, \beta)) = 1$. Assume for a contradiction that $G(E/K(\alpha, \beta))$ contains an element $\sigma \neq 1$. Then α and β are both left fixed by σ. But a map of the form (55) distinct from the identity obviously has either exactly one fixed point or none.

For the converse, suppose that $E = K(\alpha, \beta)$, where α, β are roots of f. Hence

$$(56) \qquad\qquad |G| = E : K \leq p(p-1),$$

because $K(\alpha) : K = p$ and $K(\alpha, \beta) : K(\alpha) \leq p-1$ no matter what. Also since $K(\alpha) : K = p$, there is an element ρ of order p in G. If $H = \langle \rho \rangle$ is normal in G, it follows, as we saw above, that G is isomorphic to an affine subgroup of $S(\mathbb{F}_p)$. Thus G is solvable, by Theorem 7. On the other hand, if H is not normal in G, we can find $\sigma \in G$ such that $H' := \sigma^{-1} H \sigma$ is distinct from H. Because p is prime, $H \cap H' = 1$. Therefore HH' (though it need not be a subgroup of G) contains p^2 elements, in contradiction with (56). $\qquad\square$

The following consequence of F11 was first stated by Kronecker:

F12. *Let K be a subfield of the field $\mathbb{R}$ of real numbers, and let $f \in K[X]$ be an irreducible polynomial of prime degree $p > 2$. If the Galois group of f over K is solvable, f has either exactly one root in $\mathbb{R}$ or all its roots in $\mathbb{R}$.*

Proof. Being a polynomial of odd degree, f certainly has at least one root α in $\mathbb{R}$. Assume there is another one, $\beta \in \mathbb{R}$, with $\beta \neq \alpha$. Using F11 we conclude that the subfield $K(\alpha, \beta)$ of $\mathbb{R}$ must be a splitting field of f over K. But then all the roots of f lie in $\mathbb{R}$. □

Remarks. Using F12 it is not hard to find polynomials of prime degree with rational coefficients that are not solvable by radicals over $\mathbb{R}$. For example:

(i) *For every prime $p \geq 5$, the polynomial*

$$(57) \qquad\qquad f(X) = X^p - 4X + 2$$

*is **not** solvable by radicals over $\mathbb{Q}$.*

Proof. First, by Eisenstein's criterion, f is *irreducible* over $\mathbb{Q}$. If f were solvable by radicals over $\mathbb{Q}$, it would have, by F12, either exactly one real root or exactly p real roots. But a simple analytic argument shows that in fact f has exactly three real roots: If k is the number of real roots of f, the derivative $f'(X) = pX^{p-1} - 4$ has at least $k - 1$ roots in $\mathbb{R}$, by Rolle's Theorem. But clearly $pX^{p-1} - 4$ has exactly two real roots, so $k \leq 3$. On the other hand, the Intermediate Value Theorem gives $k \geq 3$, since $f(-2) < 0$, $f(0) > 0$, $f(1) < 0$, $f(2) > 0$. □

For $p = 5$ more can be proved:

(ii) *The Galois group of $X^5 - 4X + 2$ over $\mathbb{Q}$ is isomorphic to the full symmetric group S_5.*

Proof. As already seen, the polynomial (57) in the case $p = 5$ has precisely two nonreal roots, which we can view as elements of $\mathbb{C}$, by the Fundamental Theorem of Algebra (see Remark to Definition 2 in Chapter 6). The permutation of the roots of f determined by complex conjugation, $z \mapsto \bar{z}$, is therefore a *transposition*. Hence, if we regard the Galois group G of f over $\mathbb{Q}$ as a subgroup of S_5, this group contains a transposition. Since G also contains a cycle of length 5 (any element of order 5), it must coincide with S_5, as follows form the following general fact, whose proof is left as an exercise. □

(iii) *If a subgroup G of S_n contains both a transposition and a cycle of length n, and n is a prime number, then $G = S_n$.* (The primality of n is an unavoidable assumption; for instance, in S_4 the elements (1234) and (24) generate a subgroup of order 8.)

The converse of the statement of F12 is not true; just consider the polynomial

$$(58) \qquad\qquad f(X) = X^5 - X - 1,$$

which has a single real root (again by calculus), but whose Galois group is isomorphic to S_5. Justifying this last assertion is not so simple, but it's easy to see at least that f is *irreducible* over $\mathbb{Q}$: it suffices to show that f is irreducible as a polynomial over some field $\mathbb{F}_p$ (see F9 and F6 in Chapter 5), which is the case for instance when $p = 5$ (Theorem 3 in Chapter 14). The irreducibility of f implies, if nothing else, that the Galois group G has an element of order 5, that is, a cycle of length 5. To prove the equality $G = S_5$, one can for example — in view of (iii) above — check that G contains a *transposition*. One way to do this is to look at f over $\mathbb{F}_2$; here we have the *prime factorization*

$$X^5 - X - 1 = (X^2 + X + 1)(X^3 + X^2 + 1) \text{ in } \mathbb{F}_2[X].$$

Hence the Galois group $\overline{G}$ of $X^5 - X - 1$ over $\mathbb{F}_2$ certainly contains a transposition. Now by a general principle due to *Dedekind*, this allows one to conclude that G, too, contains a transposition; see F13 in Chapter 16. (Another way of proving that $G = S_5$ is outlined in §15.25 in the Appendix.)

7. To conclude this chapter we will cite a beautiful — and fairly deep — theorem of *David Hilbert* (1862–1943), without undertaking to prove it here.

Hilbert Irreducibility Theorem. *Let k stand for the field $\mathbb{Q}$ or for any finitely generated extension of $\mathbb{Q}$. Then k has the following property:*

*If $f = X^m + a_{m-1}(t_1, \ldots, t_n)X^{m-1} + \cdots + a_0(t_1, \ldots, t_n)$, with $m, n \geq 1$, is an **irreducible** polynomial in the ring $k[t_1, \ldots, t_n, X]$ of polynomials in the $n+1$ variables $t_1, \ldots, t_n, X$ over k, there exist infinitely many $(c_1, \ldots, c_n) \in k^n$ such that the polynomial $f(c_1, \ldots, c_n, X) \in k[X]$ is also irreducible.*

Surely this theorem already speaks for itself; but its relevance to Galois theory will be put in sharp focus by a later result (Chapter 16, F14).

Remarks.   (a) A field having the property stated for k in the conclusion of the theorem is called *Hilbertian*. It turns out that a field is already Hilbertian if the property holds with $n = 1$.

(b) Let k be Hilbertian. For $n > 1$ a stronger version of the property holds, namely: If in addition to the irreducible polynomial $f \in k[t_1, \ldots, t_n, X]$ we are given a nonzero polynomial g in $k[t_1, \ldots, t_n]$, there exist infinitely many $c \in k^n$ such that $f(c, X)$ is irreducible in $k[X]$ *and such that $g(c) \neq 0$*. See Lang, *Diophantine geometry*, Chapter 8. This further implies the following fact, by F12 in Chapter 7: *If an irreducible polynomial $f \in k[t_1, \ldots, t_n, X]$ is separable as a polynomial over $k(t_1, \ldots, t_n)$, there exist infinitely many $c \in k^n$ for which $f(c, X)$ is irreducible and separable over k.*

16

Integral Ring Extensions
with Applications to Galois Theory

1. We now explain how the notion of an *algebraic field extension* can be generalized to *rings* in the appropriate way. Let A be a ring with unity and R a subring of A containing the unity of A. Suppose also that R is *central*, that is to say, each of its elements commutes with all elements of A. (In particular, R is commutative.) In this situation we say that

$$A/R \text{ is a } \textit{ring extension}.$$

If A/R is a ring extension, we can regard A in a natural way as an R-*algebra*. Conversely, if A is an algebra (with unity element 1) over a commutative ring R with unity, there is a natural ring homomorphism

$$R \to A$$
$$a \mapsto a1$$

whose image R' is a subring of the center of A and contains the unity of A. Then A/R' is a ring extension in the sense just defined.

Definition 1. Let A be an algebra over a commutative ring R with unity. An element α of A is called *integral over* R if there exists a *normalized* polynomial

(1) $$f(X) = X^n + a_{n-1}X^{n-1} + \cdots + a_0 \in R[X]$$

of degree $n \geq 1$ over R such that

(2) $$f(\alpha) = \alpha^n + a_{n-1}\alpha^{n-1} + \cdots + a_0\alpha^0 = 0.$$

An equation of the form (2) is called an *integrality equation* for α over R.

Examples. (1) For a *field extension* E/K, an element of E is integral over K if and only if it is algebraic over K.

(2) Let k be a field and $K = k(X_1, \ldots, X_n)$ the field of rational functions in n variables $X_1, \ldots, X_n$ over k. Let $s_1, \ldots, s_n$ be the elementary symmetric functions in $X_1, \ldots, X_n$. Then each X_i is integral over $k[s_1, \ldots, s_n]$; see (17) in Chapter 15.

(3) The elements of $\mathbb{C}$ that are integral over $\mathbb{Z}$ are called *algebraic integers*. The following complex numbers, for example, are algebraic integers: 5, $\sqrt{2}$, $1+i$, $e^{2\pi i/n}$ for $n \in \mathbb{N}$, and $\frac{1}{2}(-1+\sqrt{5})$.

F1. *Let A be an algebra over a commutative ring R with unity. If A is finitely generated as an R-module, every element of A is integral over R.*

Proof. This is a direct generalization of the fact that every finite field extension is algebraic (F4 in Chapter 2). Admittedly, the proof is harder. Since we still lack certain conceptual tools, we fall back on the following "classical" argument:

By assumption, A possesses a finite set of generators $\beta_1, \ldots, \beta_n$ over R. We can assume that $\beta_1 = 1$ (by adding this extra generator if needed). We can also assume that R is a subring of A. Let $\alpha \in A$ be given. For each $1 \le j \le n$ we have

$$\alpha\beta_j = \sum_{k=1}^{n} a_{kj}\beta_k \quad \text{for some } a_{kj} \in R.$$

Otherwise stated, there are relations

$$(3) \qquad \sum_{k=1}^{n} (a_{kj} - \delta_{jk}\alpha)\beta_k = 0 \quad \text{for } 1 \le j \le n,$$

where δ_{jk} is the Kronecker delta. Make the abbreviation

$$c_{jk} := a_{kj} - \delta_{jk}\alpha$$

and denote by $C = (c_{jk})_{j,k}$ the corresponding $n \times n$ matrix over the commutative ring $R[\alpha]$. The *adjoint matrix* $\tilde{C} = (\tilde{c}_{jk})_{j,k}$ of C satisfies

$$\tilde{C}C = \det(C)\,E_n,$$

where E_n is the $n \times n$ identity matrix (see for example LA I, p. 148). From (3) there follows, for all $1 \le i \le n$,

$$0 = \sum_{j=1}^{n}\tilde{c}_{ij}\left(\sum_{k=1}^{n} c_{jk}\beta_k\right) = \sum_{k=1}^{n}\sum_{j=1}^{n}\tilde{c}_{ij}c_{jk}\beta_k = \sum_{k=1}^{n}\det(C)\,\delta_{ik}\beta_k = \det(C)\beta_i.$$

Since $\beta_1 = 1$ this implies

$$\det(C) = \det\left((a_{jk} - \delta_{jk}\alpha)_{j,k}\right) = 0.$$

Thus α is a root of the polynomial

$$f(X) = \det\left((\delta_{jk}X - a_{jk})\right) \in R[X].$$

But $f(X)$ is normalized of degree $n \ge 1$, so α is indeed integral over R. □

Definition 2. A ring extension A/R is called *integral* if every element of A is integral over R.

A ring extension A/R is called *finite* if A is finitely generated as an R-module.

Given this definition, F1 can be rephrased very simply:

F1′. *Every finite ring extension A/R is integral.*

The following integrality criterion for elements can then be stated:

F2. *Let A be an algebra over a commutative ring R with unity. Given an element α in A, there is equivalence between:*

  (i) *α is integral over R.*

  (ii) *The subalgebra $R[\alpha]$ of A is finitely generated as an R-module.*

  (iii) *There exists a subalgebra A' of A that contains α and is finitely generated as an R-module.*

Proof. (i) $\Rightarrow$ (ii): By definition, $R[\alpha] = \{g(\alpha) \mid g \in R[X]\}$. Let $f(\alpha) = 0$ be an integrality equation for α over R. Since f is normalized, division with rest yields for any $g \in R[X]$ a representation

$$g(X) = h(X)f(X) + r(X),$$

with $h(X), r(X) \in R[X]$ and $\deg r < \deg f =: n$. Since $g(\alpha) = r(\alpha)$ it follows that $1, \alpha, \dots, \alpha^{n-1}$ generate the R-module $R[\alpha]$.

The implication (ii) $\Rightarrow$ (iii) is trivial, and (iii) $\Rightarrow$ (i) follows from F1.   □

*In the sequel we will assume all rings to be **commutative with unity**.*

Lemma 1. *If A/R and B/A are finite ring extensions, B/R is also finite.*

Proof. The relations $A = Re_1 + \cdots + Re_m$ and $B = Af_1 + \cdots + Af_n$ obviously imply $B = Re_1 f_1 + \cdots + Re_m f_n$.   □

F3. *For a ring extension A/R, there is equivalence between:*

  (i) *There exist finitely many elements $\alpha_1, \dots, \alpha_m$ of A integral over R and such that $A = R[\alpha_1, \dots, \alpha_m]$.*

  (ii) *A/R is finite.*

Proof. (ii) $\Rightarrow$ (i) is clear: We can choose for $\alpha_1, \dots, \alpha_m$ a set of elements that generate A as an R-module. Then we actually have $A = R\alpha_1 + \cdots + R\alpha_m$, and F1′ says that the α_i are integral over R.

(i) $\Rightarrow$ (ii) is proved by induction on m. For $m = 0$ there is nothing to prove. Suppose (i) holds with some $m \geq 1$ and set $A' = R[\alpha_1, \dots, \alpha_{m-1}]$. Then $A = A'[\alpha_m]$, and α_m is integral over A'. Therefore, by F2, A/A' is finite. A'/R is also finite, by the induction hypothesis. The finiteness of A/R follows from Lemma 1.   □

F4 and Definition 3. *Let A/R be a ring extension. The subset*

$$C = \{\alpha \in A \mid \alpha \text{ is integral over } R\}$$

*is a **subring** of A containing R. We call C the **integral closure** of R in A.*

Proof. Clearly $R \subseteq C$, since every $a \in R$ is a root of the normalized polynomial $X - a$ over F. Now let α, β be elements of C, and take the subalgebra $R[\alpha, \beta]$ of the R-algebra A. By F3, the extension $R[\alpha, \beta]/R$ is finite. Thus, by F1′, all elements of $R[\alpha, \beta]$ are integral over R, so

$$R[\alpha, \beta] \subseteq C.$$

In particular, $\alpha + \beta$, $\alpha - \beta$ and $\alpha\beta$ all belong to C. This completes the proof. □

Definition 4. Let A/R be a ring extension. We say that R is *integrally closed* in A if R coincides with its integral closure in A.

F5. *For every ring extension A/R, the integral closure C of R in A is integrally closed in A.*

Proof. Let $\alpha \in A$ be integral over C. Then

$$\alpha^n + \alpha_{n-1}\alpha^{n-1} + \cdots + \alpha_0 = 0, \quad \text{with } \alpha_i \in C, \ n \geq 1.$$

Clearly α is then integral over the R-subalgebra $A' = R[\alpha_0, \alpha_1, \ldots, \alpha_{n-1}]$ of C as well. By F3 the extension A'/R is finite; $A'[\alpha]/A'$ is also finite, by F2. Therefore $A'[\alpha]/R$ is finite. But then, by F1′, the element α must be integral over R (since it belongs to $A'[\alpha]$). This shows that $\alpha \in C$. □

F6. *Let A/R and B/A be ring extensions. If A/R and B/A are integral, so is B/R (and conversely).*

Proof. Let C be the integral closure of R in B. Since A/R was assumed integral, we have $A \subseteq C$. Since B/A is also integral by assumption, B/C is integral. But now F5 says that C is integrally closed in B, so $B = C$. □

2. We now turn our attention to subrings of fields.

F7. *Let E/R be a ring extension, and assume that E is a field. If $\alpha \in E$ is algebraic (equivalently, integral) over the field K of fractions of R in E, there exists a nonzero $c \in R$ such that $c\alpha$ is integral over R.*

Proof. Let

(4) $$f(\alpha) = \alpha^n + a_{n-1}\alpha^{n-1} + \cdots + a_0 = 0$$

be an algebraic equation for α over K. There certainly exists a nonzero $c \in R$ such that

$$ca_i \in R \quad \text{for all } 0 \leq i \leq n - 1.$$

Multiplying (4) by c^n we get

$$(c\alpha)^n + ca_{n-1}(c\alpha)^{n-1} + \cdots + a_0 c^n = 0,$$

which is an integrality equation for $c\alpha$ over R. □

Definition 5. An integral domain R is called *integrally closed*, or *normal*, if R is integrally closed in its fraction field.

F8. *Every unique factorization domain R is integrally closed.*

Proof. This has already been stated in different words in Chapter 5, F8 — and proved using Gauss's Lemma! One can also justify the statement as follows: Any element α of $K = \operatorname{Frac} R$ has the form $\alpha = a/b$, with $a, b \in R$; since R is a UFD, one can also assume that a, b are relatively prime. Now let α be integral over R, satisfying, say, $\alpha^n + a_{n-1}\alpha^{n-1} + \cdots + a_0 = 0$ with each a_i in R. Multiplication by b^n yields

$$a^n + ba_{n-1}a^{n-1} + \cdots + a_0 b^n = 0,$$

so b divides a^n. But because a, b are relatively prime and R is a UFD, this can only happen if b is a unit of R. Then $\alpha = a/b$ lies in R. □

Remark. We can now show how the Fundamental Theorem on Symmetric Functions (page 177) can be derived from F4 in Chapter 15: Let $h(X_1, \ldots, X_n)$ be a symmetric polynomial in $k[X_1, \ldots, X_n]$. We already know that h lies in the subfield $k(s_1, \ldots, s_n)$ of $k(X_1, \ldots, X_n)$ generated by the elementary symmetric functions $s_1, \ldots, s_n$. This subfield is the fraction field of the ring $R := k[s_1, \ldots, s_n]$. We also know that each X_i is *integral* over $k[s_1, \ldots, s_n]$; thus, by F4, so is each element of $k[X_1, \ldots, X_n]$. In particular, h is integral over R. But $R = k[s_1, \ldots, s_n]$ is a polynomial ring in n variables over a field k; therefore R is a UFD, by Gauss's Theorem (page 46), and hence integrally closed, by F8. Thus h does lie in $R = k[s_1, \ldots, s_n]$ as desired.

Lemma 2. *Let A/R be a ring extension and $\sigma : A \to B$ a ring homomorphism. If $\alpha \in A$ is integral over R, then $\sigma(\alpha)$ is integral over $\sigma(R)$.*

Proof. This is clear. □

F9. *Let E/K be a finite field extension and assume K is the fraction field of an integral domain R. If R is integrally closed, the minimal polynomial of an element α of E integral over R has all its coefficients in R. In particular, $S_{E/K}(\alpha)$ and $N_{E/K}(\alpha)$ lie in R.*

Proof. Over an algebraic closure C of E, let $g := \operatorname{MiPo}_K(\alpha)$ have the factorization

$$g(X) = \prod_{i=1}^{n}(X - \alpha_i), \quad \text{with } \alpha_1 = \alpha.$$

There exist K-homomorphisms $\sigma_i : K(\alpha) \to C$ such that $\sigma_i(\alpha) = \alpha_i$. By Lemma 2, all the α_i are integral over R. The coefficients of g are polynomial expressions $s_j(\alpha_1, \ldots, \alpha_n)$ in the α_i, and thus, by F4, also integral over R. But R was assumed to be integrally closed, and g lies in $K[X]$, so we obtain $g \in R[X]$ as required. □

Another justification for F9 is provided by the next result, which in view of F8 represents a generalization of Gauss's Lemma (F7 in Chapter 5):

F10. *Let R be an integrally closed integral domain, with fraction field K. Let $f, g, h \in K[X]$ be **normalized** polynomials over K, with*

$$(5) \qquad\qquad f = gh.$$

If all the coefficients of f lie in R, so do the coefficients of g and h.

Proof. Let E be a splitting field of f over K. Over E we have

$$f(X) = \prod_{k=1}^{n} (X - \alpha_k).$$

Since f is normalized, all the α_k are integral over R. In view of (5), there exist $I, J \subseteq \{1, 2, \ldots, n\}$ such that

$$g(X) = \prod_{i \in I} (X - \alpha_i), \quad h(X) = \prod_{j \in J} (X - \alpha_j).$$

Being polynomial expressions in the integral elements α_k, the coefficients of g and h are also integral over R. Moreover they lie in K, and R is integrally closed; thus indeed $g, h \in R[X]$. $\square$

F11. *Let E/R be an **integral** ring extension. If E is a field, so is R.*

Proof. Take $\alpha \in R \smallsetminus \{0\}$. The element $1/\alpha$ of E satisfies by assumption an equation of the form
$$(1/\alpha)^n + a_{n-1}(1/\alpha)^{n-1} + \cdots + a_0 = 0$$
over R. Multiply out by α^{n-1} to get

$$1/\alpha = -a_{n-1} - a_{n-2}\alpha - \cdots - a_0\alpha^{n-1} \in R.$$

Thus R is a field. $\square$

The next result is useful in various contexts:

F12. *Let A be a subring of $\mathbb{C}$ obtained from $\mathbb{Z}$ by adjoining algebraic integers. For a given prime p the natural homomorphism $\mathbb{Z} \to \mathbb{F}_p$ can be extended to a ring homomorphism from A into an algebraic closure of $\mathbb{F}_p$.*

Proof. First note that the principal ideal pA of A is distinct from A. Otherwise there would be a relation $1 = p\alpha$ with $\alpha \in A$, and then we would have $1/p \in A \cap \mathbb{Q} = \mathbb{Z}$.

Next, since $pA \neq A$, there is a *maximal ideal* $\mathfrak{P}$ of A such that $pA \subseteq \mathfrak{P}$ (see Chapter 6, F12). The inclusion $\mathbb{Z} \subseteq A$ then yields a natural homomorphism

$$(6) \qquad\qquad \mathbb{Z}/p\mathbb{Z} \to A/\mathfrak{P}.$$

This map is injective, since $\mathbb{Z}/p\mathbb{Z}$ is a field. Because $\mathfrak{P}$ is maximal, $A/\mathfrak{P}$ is also a field. We can then view $\bar{A} := A/\mathfrak{P}$ as an extension of $\mathbb{F}_p = \mathbb{Z}/p\mathbb{Z}$, via the map (6). Since A is integral over $\mathbb{Z}$, the field $\bar{A}$ is algebraic over $\mathbb{F}_p$ (see Lemma 2), and so is contained in an algebraic closure of $\mathbb{F}_p$. Thus the natural map $A \to A/\mathfrak{P} = \bar{A}$ extends the map $\mathbb{Z} \to \mathbb{Z}/p\mathbb{Z} = \mathbb{F}_p$ and yields a homomorphism of A into an algebraic closure of $\mathbb{F}_p$. $\qquad\square$

Remark. In addition one sees easily that if $A/\mathbb{Z}$ is finite, say with $A = \mathbb{Z}[\alpha_1, \ldots, \alpha_n]$, there can only be finitely many extensions with the properties stated in the conclusion of F12.

We would not want to pass up the chance to point out that F12 is a special case of a result that is quite general:

F12*. *Let A/R be an **integral** ring extension (of arbitrary commutative rings). Every homomorphism from R into an **algebraically closed** field F can be extended to a homomorphism from A into F. In other words*: *For each prime ideal $\mathfrak{p}$ of R there is at least one prime ideal $\mathfrak{P}$ of A such that*

$$(7) \qquad\qquad \mathfrak{p} = \mathfrak{P} \cap R.$$

We will not prove F12* here; see §16.12 in the Appendix.

3. Now we would like to show how the basic results about ring extensions presented in the last two sections are useful, for instance, when one is investigating the Galois group of a given equation. Naturally enough, we keep in mind first the case where the ground field is $K = \mathbb{Q}$.

Suppose then that we are given a *normalized* polynomial $f \in \mathbb{Q}[X]$ of degree $n \geq 1$. First we *get rid of multiple roots*, by taking the gcd of f and f' and dividing f by it. We can also arrange for *all the coefficients of f to be in* $\mathbb{Z}$, by making a substitution $X \mapsto X/c$ for a judiciously chosen integer c (much as in the proof of F7) and dividing by the leading coefficient to keep the polynomial normalized. These changes do not affect the Galois group G of f over $\mathbb{Q}$.

It is now natural to look at the reduction modulo some appropriate prime p. We cannot hope that this will still leave the Galois group unaltered — consider that only *cyclic* Galois groups occur over a finite field — but we can expect to obtain some partial information about G. We denote by $\bar{f} = \bar{f}(X)$ the canonical image of f in $\mathbb{F}_p[X]$ and we assume moreover that $\bar{f}$ has no multiple roots. Over a splitting field E of f over $\mathbb{Q}$ we have

$$(8) \qquad\qquad f(X) = (X - \alpha_1)(X - \alpha_2)\ldots(X - \alpha_n).$$

If A denotes the integral closure of $\mathbb{Z}$ in E, one can extend the natural map $\mathbb{Z} \to \mathbb{F}_p$ into a homomorphism φ from A into an algebraic closure of $\mathbb{F}_p$ (see F12); over this algebraic closure we have

$$(9) \qquad\qquad \bar{f}(X) = (X - \bar{\alpha}_1)(X - \bar{\alpha}_2)\ldots(X - \bar{\alpha}_n),$$

with $\bar{\alpha}_i = \varphi(\alpha_i)$. We can then state the following law:

Theorem 1. *We maintain the preceding notation and the assumption that $\bar{f}$ has no multiple roots. As a group of permutations of the roots $\bar{\alpha}_1, \bar{\alpha}_2, \ldots, \bar{\alpha}_n$ of $\bar{f}$, the Galois group $G(\bar{f})$ of $\bar{f}$ over $\mathbb{F}_p$ is isomorphic to a subgroup of the Galois group $G(f)$ of f over $\mathbb{Q}$, likewise regarded as a group of permutations of the roots $\alpha_1, \alpha_2, \ldots, \alpha_n$ of f.*

We postpone the proof of this theorem a bit in order to illustrate its application to the case of the polynomial

$$(10) \qquad f(X) = X^5 - X - 1.$$

For $p = 2$, the prime factorization of $\bar{f}$ in $\mathbb{F}_p[X]$ is

$$(11) \qquad \bar{f}(X) = (X^2 + X + 1)(X^3 + X^2 + 1).$$

From this we see immediately that $G(\bar{f})$ contains a transposition (as well as a three-cycle). Because of Theorem 1, the Galois group $G(f)$ over $\mathbb{Q}$ must also contain a transposition (as well as a three-cycle).

If we now examine f modulo the prime $p = 5$ as well, we can in fact conclude, as outlined on page 190, that $G(f)$ is isomorphic to the full symmetric group S_5.

An important ingredient in the application of Theorem 1, of course, is that $G(\bar{f})$ is isomorphic to a subgroup of $G(f)$ not just as an abstract group, but rather *with preservation of the permutation structure*; otherwise in the $p = 2$ example one would not be able to conclude that $G(f)$ contains a *transposition*, only some element of order 2. □

The proof of Theorem 1 will be given in a more general framework. Instead of $\mathbb{Z}$ we will consider an arbitrary *integrally closed* ring R with fraction field K. We will start from a normalized polynomial $f(X) \in R[X]$ of degree $n \geq 1$. Let E be a splitting field of f over K and let A be the *integral closure* of R in E. Also suppose given a *maximal ideal* $\mathfrak{p}$ of R with quotient field $\bar{R} = R/\mathfrak{p}$. We assume that the natural map $R \to \bar{R}$ can be extended to a homomorphism

$$\varphi : A \to F$$

from A into an algebraic closure F of $\bar{R}$. (This does not represent a restriction; for $R = \mathbb{Z}$ we have seen why in F12, and for the general case one would resort to F12*.) In general there are many ways to extend $R \to \bar{R}$ to a homomorphism from A into a given algebraic closure F of $\bar{R}$, but we imagine having chosen such an extension once and for all, and denote it by $\alpha \mapsto \bar{\alpha}$; let its image be $\bar{A}$ and its kernel $\mathfrak{P}$. An obvious idea is to form the set

$$(12) \qquad G_{\mathfrak{P}} = \{\sigma \in G \mid \sigma\mathfrak{P} = \mathfrak{P}\}$$

of all those elements σ of the Galois group $G = G(f) = G(E/K)$ of f over K that map the kernel $\mathfrak{P}$ of φ into itself. Clearly $G_{\mathfrak{P}}$ is a subgroup of G, and by the definition of $G_\mathfrak{P}$ each element $\sigma \in G_\mathfrak{P}$ gives rise to an automorphism of the algebraic field extension $\bar{A}/\bar{R}$ which is *well defined* by the condition

$$(13) \qquad\qquad \bar{\sigma}(\bar{\alpha}) = \overline{\sigma(\alpha)}.$$

Note that the extension $\bar{A}/\bar{R}$ is *normal*, which can be seen as follows: Let $\bar{\beta} \in \bar{A}$ be arbitrary and take $g = \mathrm{MiPo}_K(\beta)$. Since the normalized polynomial $g \in R[X]$ splits into linear factors over E, one sees by applying φ that $\bar{g}$ also splits into linear factors over $\bar{A}$. Since $\bar{g}(\bar{\beta}) = 0$ the assertion follows; moreover one gets $\bar{R}(\bar{\beta}) : \bar{R} \le E : K$, on account of which the largest separable subextension of $\bar{A}/\bar{R}$ must be *finite*. As we have already seen, the map

$$(14) \qquad\qquad \begin{array}{c} G_\mathfrak{P} \to G(\bar{A}/\bar{R}) \\ \sigma \mapsto \bar{\sigma} \end{array}$$

affords a natural homomorphism between the subgroup $G_\mathfrak{P}$ of G and the group $G(\bar{A}/\bar{R})$ of the normal field extension $\bar{A}/\bar{R}$. As before, let $\bar{f}$ be the canonical image of the given polynomial f in $\bar{R}[X]$. If (8) is the factorization of f over E, (9) is the factorization of $\bar{f}$ over $\bar{A}$. Since $\bar{R}[\bar{\alpha}_1, \ldots, \bar{\alpha}_n]$ is contained in $\bar{A}$, the map (14) yields a homomorphism

$$(15) \qquad\qquad G_\mathfrak{P} \to G(\bar{f})$$

from $G_\mathfrak{P}$ into the group of $\bar{f}$ over $\bar{R}$.

From now on we assume that $\bar{f}$ has no multiple roots, so the $\bar{\alpha}_1, \bar{\alpha}_2, \ldots, \bar{\alpha}_n$ in (9) are all distinct. Then the map (15) is obviously *injective*: For if $\sigma \alpha_i = \alpha_j$ with $\sigma \in G_\mathfrak{P}$ and $j \ne i$, we get $\bar{\sigma} \bar{\alpha}_i = \overline{\sigma \alpha_i} = \bar{\alpha}_j$, that is, $\bar{\sigma} \ne 1$. And what's more, we have the following theorem (which encompasses the statement of Theorem 1):

Theorem 2. *In the situation above, assuming that $\bar{f}$ has no multiple roots, the maps* (14) *and* (15) *are isomorphisms.*

Proof. Let's first assume that (14) is already known to be surjective. Because (15) is injective, as we have just seen, it follows that $G(\bar{f}) = G(\bar{A}/\bar{R})$, and we are done.

The proof that (14) is surjective is carried out in two steps:

(1) Let Z be the fixed field of $G_\mathfrak{P}$ in E and let $S = A \cap Z$ be the integral closure of R in Z. We wish to show that $\bar{S} = \bar{R}$. For this we take elements $1, \sigma_2, \ldots, \sigma_r$ of G that give rise to the distinct K-homomorphisms from Z into E; here $r = Z : K$. By definition, then, $\mathfrak{P} \ne \sigma_i \mathfrak{P}$ and $\sigma_i^{-1} \mathfrak{P} \ne \mathfrak{P}$. We claim that

$$(16) \qquad\qquad \sigma_i^{-1} \mathfrak{P} \cap S \ne \mathfrak{P} \cap S$$

as well. For, given $x \in \sigma_i^{-1} \mathfrak{P}$ not lying in $\mathfrak{P}$, the element

$$N_{E/Z}(x) = \prod_{\sigma \in G_\mathfrak{P}} \sigma x$$

lies in $\sigma_i^{-1}\mathfrak{P} \cap S$, but not in $\mathfrak{P}$. Since the natural map $S/\mathfrak{P} \cap S \to A/\mathfrak{P}$ is injective and $\mathfrak{P}$ is a maximal ideal of A, the intersection $\mathfrak{P} \cap S$ is a maximal ideal of S (see F11). Now let s be any element of S. An application of the *Chinese Remainder Theorem* (see Section 4.5, Lemma and F16) now yields, in view of (16), an element $z \in S$ such that

$$z \equiv s \bmod \mathfrak{P}, \qquad z \equiv 1 \bmod \sigma_i^{-1}\mathfrak{P} \quad \text{for } i = 2, \dots, r.$$

Then, setting

$$a := N_{Z/K}(z) = z \cdot \prod_{i=2}^{r} \sigma_i z,$$

we have obtained an element of R such that $a \equiv s \bmod \mathfrak{P}$. This shows that indeed $\bar{S} = \bar{R}$.

(2) To prove the surjectivity of (14) we can assume from now on that $G_{\mathfrak{P}} = G$. Let $\rho \in G(\bar{A}/\bar{R})$ be given. We will use the fact that ρ is determined by its action on a *primitive element* of the largest separable subextension of $\bar{A}/\bar{R}$. Take $\bar{\beta}$, with $\beta \in K$, to be such a primitive element. Let g be the minimal polynomial of β over K and let $g(X) = (X - \beta_1)(X - \beta_2) \dots (X - \beta_m)$ be its factorization over E. Since ρ can only take $\bar{\beta}$ to a root of $\bar{g}$, we have $\rho\bar{\beta} = \bar{\beta}_i$ for some i. By the irreducibility of g, however, there is some $\sigma \in G$ such that $\sigma\beta = \beta_i$. There follows $\bar{\sigma} = \rho$, which proves the claim. $\qquad \square$

As a consequence of Theorem 1 we now mention a fact first stated by Dedekind:

F13. *Let f be a normalized polynomial with coefficients in $\mathbb{Z}$ and let p be a prime number for which the polynomial $\bar{f}$ of $\mathbb{F}_p[X]$ determined by f has no multiple roots. Let the prime factorization of $\bar{f}$ in $\mathbb{F}_p[X]$ be*

$$(17) \qquad\qquad \bar{f} = \bar{f}_1 \bar{f}_2 \dots \bar{f}_r,$$

each $\bar{f}_i$ having degree n_i. Regarded as a group of permutations of the roots of f, the Galois group $G(f)$ of f over $\mathbb{Q}$ contains an element σ whose decomposition into cycles has the form

$$(18) \qquad\qquad \sigma = \sigma_1 \sigma_2 \dots \sigma_r, \quad \text{with length } \sigma_i = n_i.$$

(A marvelous converse was proved by Frobenius: see §11.11 in the Appendix.)

Proof. We start from the Galois group $G(\bar{f})$ of $\bar{f}$ over $\mathbb{F}_p$. The *orbits* of the action of $G(\bar{f})$ on the set of roots of $\bar{f}$ are precisely the sets of roots of the distinct irreducible factors $\bar{f}_1, \bar{f}_2, \dots, \bar{f}_r$ of $\bar{f}$ in (17). But the Galois group $G(\bar{f})$ is cyclic, and so generated by a single element $\bar{\sigma}$. Directly from the definition (Section 15.4), we conclude that the cycle decomposition of $\bar{\sigma}$ has the form $\bar{\sigma} = \bar{\sigma}_1 \bar{\sigma}_2 \dots \bar{\sigma}_r$ with cycles $\bar{\sigma}_i$ of length $n_i = \deg \bar{f}_i$. By Theorem 1, then, each $G(f)$ contains a permutation σ of the same type. $\qquad \square$

Remark. Under the assumptions underlying Theorem 2, we have: *If $\mathfrak{P}'$ is another prime ideal of A such that $\mathfrak{P}' \cap R = \mathfrak{p}$, there exists an element σ in the Galois group G of E/K such that $\sigma\mathfrak{P} = \mathfrak{P}'$.*

Proof. Assume for a contradiction that $\sigma\mathfrak{P} \neq \mathfrak{P}'$ for all $\sigma \in G$. Applying the Chinese Remainder Theorem to $\mathfrak{P}$ and the ideals $\sigma^{-1}\mathfrak{P}'$ for $\sigma \in G$, we get an $\alpha \in A$ such that

$$\alpha \equiv 0 \bmod \mathfrak{P}, \qquad \alpha \equiv 1 \bmod \sigma^{-1}\mathfrak{P}' \quad \text{for all } \sigma \in G.$$

Now the norm $N\alpha = \prod_\sigma \sigma\alpha \in R$ satisfies on the one hand $N\alpha \equiv 0 \bmod \mathfrak{P}$ and on the other $N\alpha \equiv 1 \bmod \mathfrak{P}'$. Since $\mathfrak{P} \cap R = \mathfrak{p} = \mathfrak{P}' \cap R$, this is impossible.   □

To conclude, we mention an interesting consequence of Theorem 2 to Galois Theory:

F14. *Let k be a **hilbertian** field (see Remark (a) at the end of Chapter 15). Let a subgroup G of S_n be given. G can be regarded in a natural way as a group of automorphisms of the field of rational functions $E = k(X_1, \ldots, X_n)$ in n variables over k; let $K = E^G$ denote the fixed field of G in E. Under the assumption that K is also a rational function field*

$$K = k(t_1, \ldots, t_n)$$

in n variables $t_1, \ldots, t_n$ over k, the group G can be realized as a Galois group over the field k.

Proof. By the Primitive Element Theorem there is some $\alpha \in E$ such that $E = K(\alpha)$, and we can assume that α is *integral* over $k[t_1, \ldots, t_n]$. So let $f = f(t_1, \ldots, t_n, X)$ be the minimal polynomial of α over K. Since k is hilbertian, there exist $c_1, \ldots, c_n \in k$ such that $\bar{f} = f(c_1, \ldots, c_n, X)$ is *irreducible* and *separable* in $k[X]$. It follows directly from Theorem 2, together with F12*, that the Galois group $G(\bar{f})$ of $\bar{f}$ over k is isomorphic to a subgroup of $G = G(E/K)$. Because $\bar{f}$ is irreducible we then have $|G(\bar{f})| \geq \deg \bar{f} = \deg f = |G|$, so G is isomorphic to $G(\bar{f})$, proving the assertion.   □

The relevance of F14 to *inverse Galois theory* over $\mathbb{Q}$ (page 178) stands out in view of the *Hilbert Irreducibility Theorem* (Section 15.7): A given finite group G of order n can be regarded naturally as a subgroup of S_n, and thus also as a group of automorphisms of the field of rational functions $\mathbb{Q}(X_1, \ldots, X_n)$ in n variables over $\mathbb{Q}$. If the fixed field K of G is likewise a *field of rational functions* in n variables over $\mathbb{Q}$, the given group G is isomorphic to the Galois group of a Galois extension $L/\mathbb{Q}$ with ground field $\mathbb{Q}$.

The conjecture that the field K so obtained always satisfies the condition just stated is generally attributed to *Emmy Noether*, although her 1917 work has no hint of it; nor does *Hilbert*'s foundational work of 1892 contain any intimation in this direction. In 1969 a counterexample to the conjecture was exhibited by *Swan*, for G a cyclic group of order 47 (*Invent. Math.* **7**, 148–158). Thus the central problem of inverse Galois theory is not to be put to rest so easily; and yet the methodical study of function fields remains by all means a fruitful approach. (One may also remark that the counterexamples found by Swan involve only certain cyclic groups, that is, groups whose realizability as Galois groups over $\mathbb{Q}$ is known on other grounds anyway: see §14.10 in the Appendix.)

The reader who wishes to get a glimpse of current work on inverse Galois theory is referred to *Inverse Galois theory* by G. Malle and B. H. Matzat's (Springer, 1999) and to *Generic polynomials: constructive aspects of the inverse Galois problem* by C. Jensen, A. Ledet and N. Yui (Cambridge, 2002).

The Transcendence of π

1. To prove the famous result, already stated in Chapter 2, that π is transcendent, we will frame it as a special case of a more general theorem that will be of use in other situations. Not that we shall be able to go deep into the fascinating territory of transcendental number theory, but this approach hopefully has the advantage of transparency, and in any case no shorter path to the transcendence of π is known to this author. The guiding ideas are taken from Drinfeld's booklet,[1] with some necessary minor corrections to the exposition.

Suppose that π is algebraic. Let $\beta_1 = i\pi, \beta_2, \ldots, \beta_m$ be the conjugates of $i\pi$. Since $e^{i\pi} = -1$, we have

$$(1 + e^{\beta_1})(1 + e^{\beta_2}) \ldots (1 + e^{\beta_m}) = 0.$$

Multiplying out we get

$$1 + \sum_j e^{\beta_j} + \sum_{j<k} e^{\beta_j + \beta_k} + \cdots + e^{\beta_1 + \cdots + \beta_m} = 0.$$

Denote by $\alpha_1, \alpha_2, \ldots, \alpha_n$ those exponents β_j, $\beta_j + \beta_k$, $\ldots$, $\beta_1 + \cdots + \beta_m$ that are *nonzero*, and rewrite the preceding relation as

$$(1) \qquad\qquad N + e^{\alpha_1} + e^{\alpha_2} + \cdots + e^{\alpha_n} = 0,$$

where $N \in \mathbb{N}$ and all the α_i are nonzero by assumption.

Now, a conjugacy map simply permutes the numbers $\alpha_1, \alpha_2, \ldots, \alpha_n$. But then the existence of a relation (1) is precluded by Theorem 1 on the next page.

2. As a stepping stone to Theorem 1 we state and prove an elementary approximation property of the exponential function in connection with an arbitrary polynomial

$$(2) \qquad\qquad f(X) = c_0 + c_1 X + \cdots + c_m X^m.$$

---

[1] G. I. Drinfeld, Квадратура круга и трансцендентность числа π, Vishcha shkola, Kiev, 1976; German translation: *Quadratur des Kreises und Transzendenz von π* (Mathematische Schülerbücherei, 101), VEB Deutscher Verlag der Wissenschaften, Berlin, 1980.

First we claim that *for every nonzero $x \in \mathbb{C}$ and every $j = 0, 1, 2, \ldots,$*

$$(3) \qquad j!e^x = j! + j!x + \frac{j!}{2!}x^2 + \cdots + x^j + x^{j+1}q_j(x)e^{|x|},$$

where $|q_j(x)| < 1$.

The questionable remainder of the series $j!e^x$ is $x^{j+1}\delta_j(x)$, where

$$\delta_j(x) = \frac{1}{j+1} + \frac{x}{(j+1)(j+2)} + \frac{x^2}{(j+1)(j+2)(j+3)} + \cdots.$$

Thus

$$|\delta_j(x)| \le 1 + \frac{|x|}{1 \cdot 2} + \frac{|x|^2}{1 \cdot 2 \cdot 3} + \cdots < e^{|x|}$$

for $x \ne 0$. It follows that $q_j(x) := \delta_j(x)e^{-|x|}$ does indeed satisfy $|q_j(x)| < 1$.

Now multiply (3) by c_j, for $j = 0, 1, \ldots, m$, and add together the resulting equalities. After a simple calculation, this leads to:

Lemma. *For $f(X)$ as in (2), set*

$$(4) \qquad F(X) = f(X) + f^{(1)}(X) + \cdots + f^{(m)}(X).$$

Then for every nonzero $x \in \mathbb{C}$ we have

$$(5) \qquad F(0)e^x = F(x) + e^{|x|}Q(x),$$

where

$$(6) \qquad F(0) = \sum_{j=0}^{m} c_j j!$$

and

$$(7) \qquad Q(x) = \sum_{j=0}^{m} c_j q_j(x) x^{j+1}, \quad \text{with } |q_j(x)| < 1.$$

3. We denote by $\mathbb{Q}^c$ the field of all algebraic numbers, that is, the algebraic closure of $\mathbb{Q}$ in $\mathbb{C}$.

Theorem 1. *Let $\alpha_1, \ldots, \alpha_n \in \mathbb{Q}^c \setminus \{0\}$ and $a_1, \ldots, a_n \in \mathbb{Z}$ be given, satisfying the following condition: For every automorphism σ of $\mathbb{Q}^c/\mathbb{Q}$ there is a permutation $s \in S_n$ such that $\sigma\alpha_i = \alpha_{s(i)}$ and $a_{s(i)} = a_i$ for all i. Then there exists no nonzero integer a satisfying*

$$(8) \qquad a_1 e^{\alpha_1} + a_2 e^{\alpha_2} + \cdots + a_n e^{\alpha_n} = a.$$

Proof. The α_i are roots of a polynomial $g(X) = \sum b_i X^i \in \mathbb{Z}[X]$ of degree n. For a given prime p (which will be chosen later) we consider the polynomial $f(X)$ defined by

$$(9) \qquad (p-1)! f(X) = X^{p-1} g(X)^p =: \sum_{p-1}^{m} c_j X^j \in \mathbb{Z}[X].$$

where $m = np + p - 1$. Now we form the polynomial F associated with f as in (4), and obtain, with the notations of the lemma, the equations

$$F(0)e^{\alpha_i} = F(\alpha_i) + e^{|\alpha_i|}Q(\alpha_i).$$

Multiplying by a_i and adding together, we obtain, after using (8),

(10) $aF(0) - a_1F(\alpha_1) - \cdots - a_nF(\alpha_n) = a_1e^{|\alpha_1|}Q(\alpha_1) + \cdots + a_ne^{|\alpha_n|}Q(\alpha_n).$

Now, it follows from the definitions that, for each i,

(11) $F(\alpha_i) = p \cdot h(\alpha_i)$ for some $h \in \mathbb{Z}[X]$ of degree $< np$.

(Note that α_i is a root of f of multiplicity at least p, so all the derivatives of f of order up to $p - 1$ vanish at α_i.)

Because the $\alpha_1, \ldots, \alpha_n$ are algebraic numbers, we can find $b \in \mathbb{N}$ such that all numbers $b\alpha_i^k$ with $k \leq n$ are algebraic integers. Then $b^p h(\alpha_i)$ is also an algebraic integer for every i. Multiplying by b^p we get from the left-hand side of (10) an algebraic integer; since this number must also, by assumption, be invariant under all automorphisms, we have

(12) $ab^pF(0) - a_1b^pF(\alpha_1) - \cdots - a_nb^pF(\alpha_n) \in \mathbb{Z}.$

Choose p large enough that

(13) $p \nmid ab_0b.$

The number $F(0)$, being equal to $\sum_{j=p-1}^{m} c_j j!/(p-1)!$, must be an integer, and modulo p we have

$$F(0) \equiv c_{p-1} = b_0^p \equiv b_0 \bmod p.$$

For p as in (13), therefore, the first summand in (12) is not divisible by p. In view of (11), this shows that the whole sum (12) is not divisible by p, and *a fortiori it is nonzero*. We will thus have a contradiction with (10) if we show that for large enough p,

(14) $\left| a_1b^pQ(\alpha_1)e^{|\alpha_1|} + \cdots + a_nb^pQ(\alpha_n)e^{|\alpha_n|} \right| < 1.$

By the definition of $Q(x)$ in (7), we have

$$(p-1)!|Q(x)| \leq \sum_{j=0}^{m} |c_j|\,|x|^{j+1} = |x| \sum_{j=0}^{m} |c_j|\,|x|^j$$

$$\leq |x|\,|x|^{p-1}\left(\sum_i |b_i|\,|x|^i\right)^p = |x|^p\left(\sum_i |b_i|\,|x|^i\right)^p.$$

(For the last inequality, note that the c_j arise from the b_i by a polynomial law.) From this we see that, for $i = 1, 2, \ldots, n$,

(15) $\left| a_ib^pQ(\alpha_i) \right| \leq \dfrac{M^p}{(p-1)!}$

with a constant $M > 0$ that depends only on the initial data, not on p. For large enough p, then, the right-hand side of (15) becomes as small as we please; in particular one can arrange for (14) to hold. □

Now, starting from Theorem 1, we would like to derive some more general transcendence statements. Following *Weierstrass* (see his *Werke*, vol. II), we first formulate the following result:

Lemma. *Suppose given algebraic numbers $x_1, \ldots, x_m$, all distinct. If $(a_j^{(k)})_j \in \mathbb{C}^m \setminus \{0\}$ for $k = 1, 2, \ldots, r$, form the product*

$$P = \prod_i \left(\sum_j a_j^{(i)} e^{x_j} \right) = \sum_{j_1, \ldots, j_r} a_{j_1}^{(1)} a_{j_2}^{(2)} \ldots a_{j_r}^{(r)} e^{x_{j_1} + \cdots + x_{j_r}}.$$

By collecting together terms having the same sums in the exponent, we obtain a representation

$$(16) \qquad P = \sum_{j \geq 0} c_j e^{z_j},$$

with $z_0, z_1, z_2, \ldots$ all distinct (where each c_j is a sum of products $a_{j_1}^{(1)} \ldots a_{j_r}^{(r)}$ with $x_{j_1} + \cdots + x_{j_r} = z_j$). Then at least one of the c_j in (16) is nonzero.

Proof. Since the order of the x_i is not involved, we can assume that they are ordered lexicographically as points in $\mathbb{C} = \mathbb{R}^2$: $x_1 > x_2 > \cdots > x_m$. If we define i_k, for each k, as the lowest index such that $a_{i_k}^{(k)} \neq 0$, and then set $z_i = x_{i_1} + \cdots + x_{i_r}$, the corresponding c_i is nonzero; for if

$$a_{j_1}^{(1)} \ldots a_{j_r}^{(r)} e^{x_{j_1} + \cdots + x_{j_r}}$$

is another nonzero summand, then $j_k \geq i_k$ for each k and hence $x_{j_1} + \cdots + x_{j_r} \leq x_{i_1} + \cdots + x_{i_r}$, and equality can only hold if $j_1 = i_1, \ldots, j_r = i_r$. Thus we see that $c_i = a_{i_1}^{(1)} \ldots a_{i_r}^{(r)} \neq 0$. □

Theorem 2. *If $x_1, x_2, \ldots, x_m$ are distinct algebraic numbers, then $e^{x_1}, e^{x_2}, \ldots, e^{x_m}$ are linearly independent over $\mathbb{Q}$.*

Proof. We pick a finite Galois extension $K/\mathbb{Q}$ containing all the x_i. By supplementing the $x_1, x_2, \ldots, x_m$ with conjugates as needed, we can assume that every $\sigma \in G = G(K/\mathbb{Q})$ effects a permutation of the x_i; thus there is a well defined element of S_m, still denoted by σ, such that $\sigma x_i = x_{\sigma(i)}$. Now suppose there is a nontrivial relation

$$(17) \qquad \sum_{j=1}^{m} a_j e^{x_j} = 0, \quad \text{with } a_j \in \mathbb{Q}.$$

We form the product

$$(18) \qquad \prod_\sigma \left(\sum_j a_{\sigma(j)} e^{x_j} \right) = \sum_{i=0}^{s} c_i e^{z_i},$$

where according to the lemma some coefficient on the right-hand side, say c_0, is nonzero. Clearly each σ effects a permutation of the z_i, and it is easy to see that $\sigma z_i = z_j$ always implies $c_j = c_i$. By collecting conjugates together in (18) and renumbering appropriately, we obtain

$$(19) \qquad \sum_{k=0}^{t} b_k \left(\sum_{\sigma} e^{\sigma z_k} \right) = 0, \quad \text{with } b_k \in \mathbb{Q}$$

and $b_0 \neq 0$. We next multiply (19) by $\sum_{\tau} e^{-\tau z_0}$, obtaining

$$(20) \qquad \sum_{k} b_k \left(\sum_{\sigma,\tau} e^{\sigma z_k - \tau z_0} \right) = \sum_{k} b_k \left(\sum_{\sigma,\rho} e^{\sigma(z_k - \rho z_0)} \right) = 0.$$

Let H be the subgroup of all $\rho \in G$ such that $\rho z_0 = z_0$. Then the expression $z_k - \rho z_0$ vanishes if and only if $k = 0$ and $\rho \in H$. Consider the *nonvanishing* members of the family $(z_k - \rho z_0)_{k,r}$ for all k and $\rho \in H$, and call them $y_1, y_2, \ldots$; then (20) yields

$$(21) \qquad r_0 + \sum_{i \geq 1} r_i \left(\sum_{\sigma} e^{\sigma y_i} \right) = 0,$$

with $r_i \in \mathbb{Q}$, where $r_0 = |G| \cdot |H| \cdot b_0 \neq 0$. Denoting by $\alpha_1, \alpha_2, \ldots, \alpha_n$ the members of the familiy $(sy_i)_{s,i}$, for all i and $\sigma \in G$, we finally get from (21) a relation

$$\sum_{i=1}^{n} a_i e^{\alpha_i} = a$$

of precisely the form precluded by Theorem 1. $\square$

As a consequence of Theorem 2, we obtain:

Theorem 3 (Hermite–Lindemann). *If $\alpha \neq 0$ is an algebraic number, e^α is transcendental* (*and also π, since $e^{i\pi} = -1$*).

Proof. If e^α were algebraic, there would be a nontrivial relation

$$a_0 e^0 + a_1 e^\alpha + a_2 e^{2\alpha} + \cdots + a_n e^{n\alpha} = 0 \quad \text{with } a_i \in \mathbb{Q},$$

contradicting Theorem 2. $\square$

Theorem 3 was first proved by *Lindemann*, using methods developed by *Hermite*, who had already been able to use them to show the transcendence of e.

A more general fact than Theorem 2 was stated by Lindemann and proved by Weierstrass:

Theorem 4 (Lindemann–Weierstrass). *If $x_1, \ldots, x_m$ are distinct algebraic numbers, $e^{x_1}, e^{x_2}, \ldots, e^{x_m}$ are linearly independent over $\mathbb{Q}^c$.*

Proof. Suppose

$$a_1 e^{x_1} + \cdots + a_m e^{x_m} = 0,$$

where $a_1, \ldots, a_m$ are nonvanishing algebraic numbers. Choose a finite Galois extension $K/\mathbb{Q}$ containing all the a_i. Setting $G = G(K/\mathbb{Q})$ we form the product

$$\prod_{\sigma \in G} \left(\sum_j a_j^\sigma e^{x_j} \right) = \sum_i c_i e^{z_i},$$

as in the lemma. One easily sees that each c_i is invariant under all the $\sigma \in G$; thus $c_i \in \mathbb{Q}$. In view of the lemma, this leads to a contradiction with Theorem 2.   □

As an exercise, derive from Theorem 4 the following fact: If a chord of the unit circle has as its length a (nonzero) algebraic number, the length of the corresponding arc cannot be constructed with ruler and compass, and neither can the area of the corresponding sector.

Fundamentals of Transcendental Field Extensions

1. Let E/K be a fixed field extension. Let M be a subset of E. By the *algebraic closure of M* in E we understand the algebraic closure of $K(M)$ in E (see F8 in Section 2.5). We denote this field by

$$H(M) = H_{E/K}(M).$$

We say that an element α in E is *algebraically dependent of M* (*over K*) if α lies in $H(M)$, that is, if α algebraic is algebraic over $K(M)$. Clearly

 (i) $M \subseteq H(M)$;

 (ii) $M \subseteq M' \Rightarrow H(M) \subseteq H(M')$;

 (iii) $H(H(M)) = H(M)$.

Definition 1. We say that M is *algebraically independent* (*over K*) if

$$\alpha \notin H(M \smallsetminus \{\alpha\}) \quad \text{for all } \alpha \in M,$$

that is, if every α in M is *transcendental* over $K(M \smallsetminus \{\alpha\})$. Otherwise we say that M is *algebraically dependent* (*over K*).

The formal analogy between these notions and those of *linear (in)dependence*, familiar from linear algebra, is self-evident. The following statements are also clear:

 (iv) M *algebraically dependent* if and only if there exists some $\alpha \in M$ that is algebraic over $K(M \smallsetminus \{\alpha\})$.

 (v) An element $\alpha \in M$ lies in $H(M \smallsetminus \{\alpha\})$ if and only if $H(M) = H(M \smallsetminus \{\alpha\})$.

 (vi) If α is algebraic over $K(M)$ and does not lie in M, then $M \cup \{\alpha\}$ is *algebraically dependent*.

 (vii) M is *algebraically dependent* if and only if M has a *finite subset* that is still algebraically dependent.

 (viii) M is *algebraically independent* if and only if *every finite subset* of M is algebraically independent.

F1. *M is algebraically independent if and only if, for any distinct elements* $\alpha_1, \ldots, \alpha_n$ *of M (where n is any positive integer), the canonical homomorphism of K-algebras from the polynomial ring* $K[X_1, \ldots, X_n]$ *into E defined by*

$$X_i \mapsto \alpha_i \quad \text{for } 1 \le i \le n$$

*is **injective**; in other words, if and only if there is no nontrivial algebraic relation linking the* α_i *(the meaning of this expression being that* $f(\alpha_1, \ldots, \alpha_n) = 0$ *implies* $f = 0$ *for* $f \in K[X_1, \ldots, X_n]$).

Proof. Without loss of generality, we can assume that $M = \{\alpha_1, \ldots, \alpha_n\}$ has n elements.

(1) Assume that M is algebraically dependent. Then there exists i such that α_i is algebraically dependent of $M \setminus \{\alpha_i\}$; we may as well suppose it's $i = n$. Therefore α_n is algebraic over $K(\alpha_1, \ldots, \alpha_{n-1}) = \operatorname{Frac} K[\alpha_1, \ldots, \alpha_{n-1}]$. Thus

$$\sum_{i=0}^{m} g_i(\alpha_1, \ldots, \alpha_{n-1})\alpha_n^i = 0,$$

for certain polynomials $g_i = g_i(X_1, \ldots, X_{n-1})$ in $K[X_1, \ldots, X_{n-1}]$, the last of which satisfies $g_m(\alpha_1, \ldots, \alpha_{n-1}) \ne 0$. If we set

$$f(X_1, \ldots, X_n) := \sum_{i=0}^{m} g_i(X_1, \ldots, X_{n-1})X_n^i,$$

the polynomial $f \in K[X_1, \ldots, X_n]$ satisfies $f(\alpha_1, \ldots, \alpha_n) = 0$ but $f(X_1, \ldots, X_n) \ne 0$.

(2) Conversely, assume instead that $M = \{\alpha_1, \ldots, \alpha_n\}$ is algebraically independent, and suppose that $f(\alpha_1, \ldots, \alpha_n) = 0$ for some $f \in K[X_1, \ldots, X_n]$. We must show that $f = 0$. Now, f has a representation of the form

$$(1) \qquad f(X_1, \ldots, X_n) = \sum_{i=0}^{m} g_i(X_1, \ldots, X_{n-1})X_n^i,$$

with uniquely determined polynomials $g_i(X_1, \ldots, X_{n-1}) \in K[X_1, \ldots, X_{n-1}]$. Since $f(\alpha_1, \ldots, \alpha_n) = 0$, we have

$$(2) \qquad \sum_{i=0}^{m} g_i(\alpha_1, \ldots, \alpha_{n-1})\alpha_n^i = 0.$$

This is an algebraic equation for α_n over $K(\alpha_1, \ldots, \alpha_{n-1})$. By assumption it must be trivial, that is, $g_i(\alpha_1, \ldots, \alpha_{n-1}) = 0$ for all i. By induction this implies that $g_i(X_1, \ldots, X_{n-1}) = 0$ for all i, and hence $f = 0$. $\qquad \square$

F2. *Let M be a subset of E and α any element of E.*

(a) *If M is algebraically independent but $M \cup \{\alpha\}$ is algebraically dependent, we have $\alpha \in H(M)$, that is, α depends algebraically on M.*

(b) *If B is a maximal algebraically independent subset of M, then M is contained in $H(B)$, that is, every element of M is algebraic over $K(B)$.*

Proof. (a) Under the assumption, there exist distinct elements $\alpha_1, \ldots, \alpha_{n-1}$ and $\alpha_n = \alpha$ in $M \cup \{\alpha\}$ satisfying a nontrivial algebraic relation

(3) $$f(\alpha_1, \ldots, \alpha_n) = 0 \quad \text{with } f \in K[X_1, \ldots, X_n] \text{ nonzero.}$$

Write f in the form (1); not all the $g_i(X_1, \ldots, X_{n-1})$ vanish. Because of (3), the relation (2) is satisfied. Here not all the $g_i(\alpha_1, \ldots, \alpha_{n-1})$ can vanish; for otherwise, because of the algebraic independence of $\alpha_1, \ldots, \alpha_{n-1}$ over K, all the g_i would vanish, by F1. Therefore (2) represents a nontrivial algebraic equation for α_n over $K(\alpha_1, \ldots, \alpha_{n-1})$, meaning that α_n is in fact algebraic over $K(\alpha_1, \ldots, \alpha_{n-1})$ and hence also over $K(M)$.

Part (b) is an immediate consequence of (a).   □

Definition 2. A *transcendence basis* of a field extension E/K is a set $B \subseteq E$ such that

(i) $E = H(B)$ (so the extension $E/K(B)$ is algebraic), and

(ii) B is algebraically independent (over K).

F3. *If B is a subset of E, the following conditions are equivalent:*

(i) *B is a transcendence basis of E/K.*

(ii) *If B is contained in a subset M of E such that $H(M) = E$, then B is a maximal algebraically independent subset of M.*

(iii) *There exists a subset M of E such that $H(M) = E$ and that B is a maximal algebraically independent subset of M.*

Proof. (i) $\Rightarrow$ (ii): Take $\alpha \in M \smallsetminus B$. We must show that $B \cup \{\alpha\}$ is algebraically dependent. But this is clear because $\alpha \in H(B) = E$, according to statement (vi) after Definition 1 (with B playing the role of M).

(ii) $\Rightarrow$ (iii): Take $M = E$.

(iii) $\Rightarrow$ (i): All we have to show is that $H(B) = E$. By F2(b), M is contained in $H(B)$. Therefore $E = H(M) \subseteq H(H(B)) = H(B)$.   □

Theorem 1. *Every field extension E/K has a transcendence basis. More precisely: Given a subset M of E such that $E/K(M)$ is algebraic, and given a subset C of M that is algebraically independent over K, there exists a transcendence basis B of E/K such that $C \subseteq B \subseteq M$.*

Proof. We must enlarge C to make it a maximal algebraically independent subset B of M; by F3, such a set is a transcendence basis of E/K. If M is finite, the existence of B is clear. If M is infinite, one resorts to *Zorn's Lemma*, the argument being wholly similar to the one used for the proof of Chapter 6, F11.   □

F4. *Let E/K be a field extension and M a subset of E such that $E/K(M)$ is algebraic. If C is any subset of E algebraically independent over K, there exists a subset M' of M disjoint from C and such that $C \cup M'$ is a transcendence basis of E/K.*

Proof. By Theorem 1, there exists a transcendence basis B of E/K such that $C \subseteq B \subseteq M \cup C$. Now set $M' := B \smallsetminus C$. Then M' and C are disjoint, and their union B is a transcendence basis of E/K.   □

Theorem 2. *Any two transcendence bases of a field extension E/K have the same cardinality.*

Proof. Let B and B' be transcendence bases of E/K. If B and B' are both *infinite* sets, the desired assertion follows easily on set-theoretical grounds (see §18.2 in the Appendix).

We now prove the assertion for the more interesting case, where E/K has a *finite* transcendence basis. Let B be an n-element transcendence basis of E/K and $C = \{\alpha_1, \ldots, \alpha_m\}$ an m-element algebraically independent subset of E. It suffices to show that in these circumstances m does not exceed n.

Assume for a contradiction that $m > n$. We will show by induction that for every integer k such that $0 \le k \le n$ there exist subsets

(4) $$B_0 \supseteq B_1 \supseteq \cdots \supseteq B_k$$

of B such that, for each k,

(5) $$\{\alpha_1, \ldots, \alpha_k\} \cup B_k \text{ is a transcendence basis of } E/K$$

and

(6) $$\{\alpha_1, \ldots, \alpha_k\} \cap B_k = \varnothing.$$

For $k = 0$ we take $B_0 := B$. Assume the assertion is true for $0 \le k < n$. By F4 there is a subset B_{k+1} of the set $\{\alpha_1, \ldots, \alpha_k\} \cup B_k$ that satisfies the conditions

(7) $$\{\alpha_1, \ldots, \alpha_{k+1}\} \cup B_{k+1} \text{ is a transcendence basis of } E/K$$

(8) $$\{\alpha_1, \ldots, \alpha_{k+1}\} \cap B_{k+1} = \varnothing.$$

Then B_{k+1} is necessarily contained B_k. Now, B_{k+1} and B_k cannot be equal, since otherwise $B_k \cup \{\alpha_1, \ldots, \alpha_k\} \cup \{\alpha_{k+1}\}$ would be algebraically independent according to (7), yet algebraically dependent according to (5). By virtue of (4), B_k has at most $n-k$ elements. Therefore B_n is empty. Thus $\{\alpha_1, \ldots, \alpha_n\}$ is a transcendence basis of E/K, by (5). Because $C = \{\alpha_1, \ldots, \alpha_m\}$ is algebraically independent, it cannot happen that $m > n$.   □

Definition 3. The *transcendence degree* $\mathrm{TrDeg}(E/K)$ of a field extension E/K is the cardinality of any transcendence basis of E/K.

Definition 4. A field extension E/K is called *purely transcendental* if E/K has a transcendence basis B for which $E = K(B)$.

Remarks. (1) If E/K is purely transcendental with transcendence basis B, then E is K-isomorphic to the fraction field of the polynomial ring $K[B]$ in the variables $X \in B$ over K.

(2) By Theorem 1, any field extension E/K has an intermediate field F for which F/K is purely transcendental and E/F is algebraic:

(9)

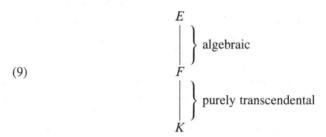

Of course, F is not unique.

Theorem 3. *Let F be an intermediate field of E/K, and let B and B' be transcendence bases for F/K and E/F, respectively. Then $B \cap B' = \varnothing$ and $B \cup B'$ is a transcendence basis for E/K. In particular,*

(10)
$$\mathrm{TrDeg}(E/K) = \mathrm{TrDeg}(E/F) + \mathrm{TrDeg}(F/K).$$

Proof. Suppose α lies in both B and B'; then α, being an element of F, is algebraic over $F(B' \smallsetminus \{\alpha\})$. But because $\alpha \in B'$, this contradicts the algebraic independence of B' over F. Thus $B \cap B' = \varnothing$.

Next we show that $E/K(B \cup B')$ is algebraic. But assumption, $F/K(B)$ is algebraic; then so is the extension

$$FK(B') \,/\, K(B)K(B') \;=\; F(B') \,/\, K(B \cup B').$$

Also by assumption, $E/F(B')$ is algebraic. Since algebraicness is transitive, the conclusion follows.

There remains to show that $B \cup B'$ is algebraically independent over K. By F4, we know there is a subset B'' of $B \cup B'$ such that $B \cap B'' = \varnothing$ and that $B \cup B''$ is a transcendence basis for E/K. Clearly B'' is contained in B', so if we prove that $B' \subseteq B''$ we are done. Suppose there exists $\alpha \in B' \smallsetminus B''$. Being an element of E, this α is algebraic over $K(B \cup B'') = K(B)(B'')$, and so also over $F(B'')$. Since $B'' \subseteq B'$ and because of our assumption, α is then algebraic over $F(B' \smallsetminus \{\alpha\})$, contradicting the algebraically independence of B' over F. $\qquad \square$

2. A fact that comes in handy on many occasions is that, for finitely generated extensions, Theorem 1 (and Remark 2 to Definition 4) can be sharpened:

Theorem 4 (Noether's Normalization Theorem). *Let A be a commutative algebra over a field K, and suppose that A is generated (as an algebra) by finitely many elements $x_1, \ldots, x_n$, meaning that*

$$(11) \qquad\qquad A = K[x_1, \ldots, x_n].$$

Then there exists some $m \leq n$ (possibly zero) and elements $u_1, \ldots, u_m$ of A with the following properties:

    (a) *The subalgebra $K[u_1, \ldots, u_m]$ of A is a polynomial algebra over K with $u_1, \ldots, u_m$ as indeterminates (if $m = 0$, by convention, $K[u_1, \ldots, u_m] = K$).*

    (b) *The ring extension $A/K[u_1, \ldots, u_m]$ is integral, and therefore, by (11), it is in fact finite.*

Proof. We may as well assume that $x_1, \ldots, x_n$ are all distinct. We use induction on n. The case $n = 1$ is clear. Suppose that $n > 1$ and that the assertion holds for $n-1$. If the $x_1, \ldots, x_n$ satisfy no nontrivial algebraic relation over K, there is nothing to show; therefore we assume instead the existence of a nonzero polynomial in $f \in K[X_1, \ldots, X_n]$ such that

$$(12) \qquad\qquad f(x_1, \ldots, x_n) = 0.$$

Let the explicit expression for f be

$$(13) \qquad\qquad f = \sum_{\nu = (\nu_1, \ldots, \nu_n)} c_\nu X_1^{\nu_1} \ldots X_n^{\nu_n}.$$

Further, let $\mu_2, \mu_3, \ldots, \mu_n$ be natural numbers (about which we will have more to say later). Setting

$$(14) \qquad\qquad y_i = x_i - x_1^{\mu_i} \quad \text{for } 2 \leq i \leq n,$$

equation (12) becomes

$$(15) \qquad\qquad f(x_1, \, y_2 + x_1^{\mu_2}, \, \ldots, \, y_n + x_1^{\mu_n}) = 0.$$

For notational simplicity we define $R = K[X_1, \ldots, X_n]$. Consider, in the polynomial ring $R[Y_2, \ldots, Y_n]$ in $n-1$ variables over R, the polynomial

$$f(X_1, Y_2 + X_1^{\mu_2}, \ldots, Y_n + X_1^{\mu_n}).$$

As a polynomial in X_1 over $K[Y_2, \ldots, Y_n]$, this has the form

$$\sum_\nu c_\nu X_1^{\nu_1 + \mu_2 \nu_2 + \cdots + \mu_n \nu_n} + g(X_1, Y_2, \ldots, Y_n),$$

where $g(X_1, Y_2, \ldots, Y_n)$ is a polynomial whose degree in X_1 is less than the degree of the polynomial on the left — it being assumed that we arrange for the summands in the sum not to cancel one another. Now we take care of the $\mu_2, \ldots, \mu_n$. First define $\mu = (1, \mu_2, \ldots, \mu_n)$ and denote by $\mu v = 1 \, v_1 + \mu_2 v_2 + \cdots + \mu_n v_n$ the usual inner product of μ with $v = (v_1, \ldots, v_n)$. Let p be a natural number such that

$$p > \deg f = \max\{v_1 + \cdots + v_n \mid c_v \neq 0\}.$$

We now choose

$$\mu = (1, p, p^2, \ldots, p^{n-1}).$$

For distinct n-tuples $v = (v_1, \ldots, v_n)$ and $v' = (v'_1, \ldots, v'_n)$ such that $c_v \neq 0$ and $c_{v'} \neq 0$ we have $\mu v \neq \mu v'$, because $v_i, v'_i < p$ for all i. (The expansion of a natural number in powers of p is unique.) Therefore

$$f(X_1, Y_2 + X_1^{\mu_2}, \ldots, Y_n + X_1^{\mu_n}) = c X_1^N + h(X_1, Y_2, \ldots, Y_n)$$

with $c \neq 0$ in K and some polynomial h of degree less than N in X_1.

Dividing this equality by c and substituting $x_1, y_2, \ldots, y_n$ for $X_1, Y_2, \ldots, Y_n$, we get using (15) an integrality equation for x_1 over $K[y_2, \ldots, y_n]$. In view of (14) and (11), then, we conclude that the ring extension

(16) $A/K[y_2, \ldots, y_n]$ is integral.

Now, by the induction hypothesis, there exist elements $u_1, \ldots, u_m$ of $K[y_2, \ldots, y_n]$ with the following properties:

(17) $K[u_1, \ldots, u_m]$ is a polynomial ring in $u_1, \ldots, u_m$ over K;

(18) $K[y_2, \ldots, y_n]/K[u_1, \ldots, u_m]$ is integral.

This proves the desired assertion, because according to (16) and (18) the extension $A/K[u_1, \ldots, u_m]$ is also integral (see F6 in Chapter 16).   □

Here is a remarkable consequence Noether's Normalization Theorem:

Theorem 5. *Let E/K be a field extension, and suppose E is **finitely generated** as a K-algebra. Then E/K is **algebraic**.*

Proof. By Theorem 4, E has as a subring some *polynomial ring* over K in finitely many indeterminates, say $F = K[u_1, \ldots, u_m]$, with the further property that E/F is integral. Since E is a field, F11 in Chapter 16 says that F must also be a field. But this is only possible for $m = 0$, because in a polynomial ring over a field K there are no invertible elements outside $K^\times$. Now, $m = 0$ implies that E/K is integral, which is to say algebraic.   □

Remark. Theorem 5 represents an important generalization of the fundamental fact, learned long ago, that an element α of a field extension over K is algebraic if and only if $K(\alpha) = K[\alpha]$ (Chapter 3, F1). At the same time, a more direct proof of Theorem 5 can be given by induction on the number of generators, as we now show.

Suppose $E = K[x_1, \ldots, x_n]$, with $n > 0$ (the case $n = 0$ being trivial). Then $E = K(x_1)[x_2, \ldots, x_n]$, and the induction hypothesis implies that each x_i is algebraic over $K(x_1)$. We set $t = x_1$. If t is algebraic over K, we are done. So suppose t is transcendental over K. Then, by F7 in Chapter 16, there is a nonzero polynomial $h = h(t) \in K[t]$ such that $h x_i$ is *integral* over $K[t]$ for all i. For any f in $E = K[x_1, x_2, \ldots, x_n]$, therefore, $h^e f$ is integral over $K[t]$ for some appropriate power h^e of h. In particular this holds for all f in $K(t)$. But $K[t]$ is integrally closed in $K(t)$, so every rational function f can be represented as a quotient g/h^e of polynomials in t, where the denominator is a power of a fixed polynomial h, independent of f. This is impossible.

Hilbert's Nullstellensatz

1. Transcendental field extensions come up naturally in the algebraic treatment of geometric problems. This chapter will serve as a appetizer for the feast that awaits the reader who wishes to delve deeper into the field of *algebraic geometry* (see for example Hartshorne's textbook of the same name).

Definition 1. In this chapter we will be working with a fixed field extension C/K and with the polynomial ring $K[X_1, \ldots, X_n]$ in n variables over K. If M is a subset of $K[X_1, \ldots, X_n]$, we set

$$\mathcal{N}(M) = \big\{ (x_1, \ldots, x_n) \in C^n \mid f(x_1, \ldots, x_n) = 0 \text{ for all } f \in M \big\}.$$

A point $x = (x_1, \ldots, x_n)$ in $\mathcal{N}(M)$ is called a *zero* of M in C^n. The set $\mathcal{N}(M)$ itself we call the *the zero set of M in C^n*. Instead of $\mathcal{N}(M)$ we sometimes use the more precise notation $\mathcal{N}_C(M)$, to exhibit the dependence on the chosen extension C of K. The subsets V of C^n of the form $V = \mathcal{N}(M)$ for some $M \subseteq K[X_1, \ldots, X_n]$ are called *affine algebraic sets of C^n defined over K*, or *algebraic K-sets of C^n* for short.
  Conversely, if N is a subset of C^n, we define

$$\mathcal{I}(N) = \big\{ f \in K[X_1, \ldots, X_n] \mid f(x_1, \ldots, x_n) = 0 \text{ for all } (x_1, \ldots, x_n) \in N \big\}.$$

If f lies in $\mathcal{I}(N)$ we say that f *vanishes on N*. Sometimes we write $\mathcal{I}_K(N)$ instead of $\mathcal{I}(N)$ to make K explicit.

F1. *The following formal properties hold*:
  (i) $M \subseteq M' \Rightarrow \mathcal{N}(M') \subseteq \mathcal{N}(M)$.
  (ii) *If* $\mathfrak{a} = (M)$ *is the ideal of* $K[X_1, \ldots, X_n]$ *generated by M, we have* $\mathcal{N}(M) = \mathcal{N}(\mathfrak{a})$.
  (iii) *For every $N \in C^n$, the set $\mathcal{I}(N)$ is an ideal of $K[X_1, \ldots, X_n]$, called the **ideal of N** in $K[X_1, \ldots, X_n]$.*
  (iv) $N \subseteq N' \Rightarrow \mathcal{I}(N') \subseteq \mathcal{I}(N)$.
  (v) $M \subseteq \mathcal{I}\mathcal{N}(M)$ *for every* $M \subseteq K[X_1, \ldots, X_n]$.

(vi) $N \subseteq \mathcal{N}\mathcal{I}(N)$ *for every* $N \subseteq C^n$.

(vii) *For any family* $(\mathfrak{a}_i)_{i \in I}$ *of ideals in* $K[X_1, \ldots, X_n]$ *we have* $\bigcap_{i \in I} \mathcal{N}(\mathfrak{a}_i) = \mathcal{N}(\sum_{i \in I} \mathfrak{a}_i)$, *where the sum* $\sum_{i \in I} \mathfrak{a}_i$ *of the given ideals* $\mathfrak{a}_i$ *is defined as the ideal generated by the union* $\bigcup_{i \in I} \mathfrak{a}_i$.

(viii) $\mathcal{N}(\mathfrak{a}) \cup \mathcal{N}(\mathfrak{b}) = \mathcal{N}(\mathfrak{a}\mathfrak{b}) = \mathcal{N}(\mathfrak{a} \cap \mathfrak{b})$ *for any ideals* $\mathfrak{a}, \mathfrak{b}$ *of* $K[X_1, \ldots, X_n]$, *where the product* $\mathfrak{a}\mathfrak{b}$ *of* $\mathfrak{a}$ *and* $\mathfrak{b}$ *is defined as the ideal of* $K[X_1, \ldots, X_n]$ *generated by all products* fg *with* $f \in \mathfrak{a}$ *and* $g \in \mathfrak{b}$.

(ix) $V = \mathcal{N}\mathcal{I}(V)$ *for any algebraic* K-set V *of* C^n.

Proof. We prove (viii) and (ix), leaving the others to the reader.

(viii) Clearly $\mathfrak{a}\mathfrak{b} \subseteq \mathfrak{a} \cap \mathfrak{b} \subseteq \mathfrak{a}, \mathfrak{b}$, so (i) yields

$$\mathcal{N}(\mathfrak{a}) \cup \mathcal{N}(\mathfrak{b}) \subseteq \mathcal{N}(\mathfrak{a} \cap \mathfrak{b}) \subseteq \mathcal{N}(\mathfrak{a}\mathfrak{b}).$$

Thus what is left to show is that $\mathcal{N}(\mathfrak{a}\mathfrak{b}) \subseteq \mathcal{N}(\mathfrak{a}) \cup \mathcal{N}(\mathfrak{b})$. Let $x \in C^n$ be a zero of $\mathfrak{a}\mathfrak{b}$ but not of $\mathfrak{a}$. Then there exists $f \in \mathfrak{a}$ such that $f(x) \neq 0$, and for every $g \in \mathfrak{b}$ we must have $g(x) = 0$, since $f(x)g(x) = fg(x) = 0$. Consequently x belongs to $\mathcal{N}(\mathfrak{b})$, as desired.

(ix) First let M be any subset of $K[X_1, \ldots, X_n]$. From (v) we get $M \subseteq \mathcal{I}\mathcal{N}(M)$, which yields the inclusion $\mathcal{N}\mathcal{I}\mathcal{N}(M) \subseteq \mathcal{N}(M)$ because of (i); on the other hand, (vi) shows that $\mathcal{N}(M) \subseteq \mathcal{N}\mathcal{I}(\mathcal{N}(M))$. Thus, for every $M \subseteq K[X_1, \ldots, X_n]$, we have

(x) $$\mathcal{N}(M) = \mathcal{N}\mathcal{I}\mathcal{N}(M).$$

But the algebraic K-sets of C^n are precisely those subsets of the form $V = \mathcal{N}(M)$, so (x) is tantamount to (ix). □

For an arbitrary ideal $\mathfrak{a}$ of $K[X_1, \ldots, X_n]$, it is generally *not* the case that $\mathfrak{a} = \mathcal{I}\mathcal{N}(\mathfrak{a})$; that is, $\mathfrak{a}$ does not necessarily coincide with the ideal of its zero set in C^n. This is just because not every ideal occurs as the ideal of some subset of C^n: if $\mathfrak{a} = \mathcal{I}(N)$ we obviously have, for every natural number m,

(1) $$f^m \in \mathfrak{a} \implies f \in \mathfrak{a}.$$

Thus, for instance, the ideal $\mathfrak{a} = (X_1^2, \ldots, X_n^2)$ of $K[X_1, \ldots, X_n]$ is not of the form $\mathfrak{a} = \mathcal{I}(N)$.

Definition 2. Let $\mathfrak{a}$ be an ideal of a *commutative* ring R. The set

$$\sqrt{\mathfrak{a}} := \{f \in R \mid \exists\, m : f^m \in \mathfrak{a}\}$$

is an ideal of R, called the *radical* of $\mathfrak{a}$. An ideal $\mathfrak{a}$ of R is called *reduced* if $\mathfrak{a} = \sqrt{\mathfrak{a}}$.

Remark. If R is a commutative ring, one can in particular consider the radical $\sqrt{0}$ of the zero ideal $0 := (0)$. By definition, this is the set of all *nilpotent* elements of R, and for this reason it is called the *nilradical* of R. If $\mathfrak{a}$ is an ideal of R, its radical $\sqrt{\mathfrak{a}}$ is the inverse image of the nilradical of the quotient ring $R/\mathfrak{a}$ under the quotient map.

It follows from these definitions that ideals of subsets of C^n must be *reduced*. But for arbitrary extensions C of K, condition (1) is generally not sufficient to ensure that an ideal $\mathfrak{a}$ of $K[X_1, \ldots, X_n]$ is an ideal of a subset. For example, take $K = C = \mathbb{R}$ and $n \geq 2$, and consider the principal ideal

$$\mathfrak{a} = (X_1^2 + X_2^2 + \cdots + X_n^2)$$

of $\mathbb{R}[X_1, \ldots, X_n]$. It is easy to see that $X_2^2 + \cdots + X_n^2$ is irreducible in $\mathbb{R}[X_1, \ldots, X_n]$. Hence $\mathfrak{a}$ is a *prime ideal* and as such it is reduced; and yet $\mathcal{N}(\mathfrak{a}) = \{0 := (0, \ldots, 0)\}$, so $\mathcal{I}(\mathcal{N}(\mathfrak{a})) = \mathcal{I}(\{0\}) = (X_1, \ldots, X_n)$. Therefore $\mathfrak{a}$ is not of the form $\mathcal{I}(N)$, since for any $N \subseteq C^n$ there is an equality dual to (x) above:

(xi) $$\mathcal{I}(N) = \mathcal{I}\mathcal{N}\mathcal{I}(N).$$

If C is an algebraically closed field, however, the condition is sufficient:

Theorem 1 (Hilbert's Nullstellensatz, algebraic form). *Let C/K be a field extension where C is **algebraically closed**. With the preceding notations, every ideal $\mathfrak{a}$ of $K[X_1, \ldots, X_n]$ satisfies*

(2) $$\mathcal{I}\mathcal{N}(\mathfrak{a}) = \sqrt{\mathfrak{a}}.$$

*Thus $\mathcal{N}$ constitutes a bijection between the set of **reduced** ideals of $K[X_1, \ldots, X_n]$ and the set of algebraic K-sets of C^n. The map inverse to $\mathcal{N}$ is $\mathcal{I}$.*

"Nullstellensatz" is German for "Theorem on zeros". We will trace Theorem 1 back to the following fact, which has intrinsic interest as well:

Theorem 2 (Hilbert's Nullstellensatz, geometric form). *Let K be a field and C an **algebraically closed** extension of K (for instance, an algebraic closure of K). Let an ideal $\mathfrak{a}$ of the polynomial ring $K[X_1, \ldots, X_n]$ be given. Provided that $\mathfrak{a}$ is not all of $K[X_1, \ldots, X_n]$, there exists in C^n a common zero $(z_1, \ldots, z_n)$ of all the polynomials $f \in \mathfrak{a}$: in symbols,*

$$\mathcal{N}_C(\mathfrak{a}) \neq \varnothing.$$

Proof. The assumption $\mathfrak{a} \neq K[X_1, \ldots, X_n]$ implies (see F12 in Chapter 6) that $K[X_1, \ldots, X_n]$ has a maximal ideal $\mathfrak{m}$ that contains $\mathfrak{a}$. We look at the quotient homomorphism

(3) $$K[X_1, \ldots, X_n] \to K[X_1, \ldots, X_n]/\mathfrak{m}.$$

Denote the images of $X_1, \ldots, X_n$ by $x_1, \ldots, x_n$, respectively. Then

(4) $$K[X_1, \ldots, X_n]/\mathfrak{m} = K[x_1, \ldots, x_n],$$

and because $\mathfrak{m}$ is maximal, $K[x_1, \ldots, x_n]$ is a *field*. This implies, by Theorem 5 in Chapter 18, that the extension $K[x_1, \ldots, x_n]/K$ is *algebraic*. Thus there exists (by Theorem 3 in Chapter 6) a K-homomorphism

$$\sigma : K[x_1, \ldots, x_n] \to C.$$

Let $z_1 = \sigma x_1$, $z_2 = \sigma x_2$, ..., $z_n = \sigma x_n$ be the images of $x_1, \ldots, x_n$ under σ. In view of (3) and (4) we have

$$f(z_1, \ldots, z_n) = 0 \quad \text{for all } f \in \mathfrak{m}.$$

In particular this equality is fulfilled for all $f \in \mathfrak{a}$, since $\mathfrak{a} \subseteq \mathfrak{m}$. Thus we have found a point $(z_1, \ldots, z_n) \in C^n$ where all the polynomials in $\mathfrak{a}$ vanish. $\square$

Now that Theorem 2 has been proved, we derive Theorem 1 from it using the so-called "Rabinovich trick", which is no more than an adaptation of the elementary technique of *clearing denominators*:

Let $\mathfrak{a}$ be any ideal of $K[X_1, \ldots, X_n]$, and let $f \in \mathcal{IN}(\mathfrak{a})$ be a polynomial of $K[X_1, \ldots, X_n]$ that vanishes on the zero set $\mathcal{N}(\mathfrak{a})$. We must show that there exists a natural number m such that

$$f^m \in \mathfrak{a}.$$

To this effect we will regard $K[X_1, \ldots, X_n]$ as a subring of the polynomial ring $K[X_1, \ldots, X_n, X_{n+1}]$ in the $n + 1$ variables $X_1, \ldots, X_n, X_{n+1}$, and consider the ideal $\mathfrak{A}$ of $K[X_1, \ldots, X_n, X_{n+1}]$ generated by $\mathfrak{a}$ and the element $1 - X_{n+1} f$:

$$(5) \qquad \mathfrak{A} = (\mathfrak{a}, 1 - X_{n+1} f).$$

First assume that $\mathfrak{A}$ is not all of $K[X_1, \ldots, X_{n+1}]$. By Theorem 2, then, $\mathfrak{A}$ has a zero $(z_1, \ldots, z_{n+1}) \in C^{n+1}$. By definition the point $(z_1, \ldots, z_{n+1}) \in \mathcal{N}(\mathfrak{A})$ satisfies

$$(6) \qquad g(z_1, \ldots, z_n) = 0 \quad \text{for all } g \in \mathfrak{a},$$

$$(7) \qquad f(z_1, \ldots, z_n) z_{n+1} = 1.$$

From (6) it follows that $(z_1, \ldots, z_n) \in \mathcal{N}(\mathfrak{a})$; thus our $f \in \mathcal{IN}(\mathfrak{a})$ must satisfy $f(z_1, \ldots, z_n) = 0$. But this contradicts (7), and we conclude that our assumption is untenable; that is, the ideal $\mathfrak{A}$ in (5) is all of $K[X_1, \ldots, X_{n+1}]$, so $1 \in \mathfrak{A}$. Thus by looking at (5) we see there is a relation of the form

$$(8) \qquad 1 = \sum_i h_i g_i + h(1 - X_{n+1} f),$$

for certain polynomials $g_i \in \mathfrak{a}$ and $h_i, h \in K[X_1, \ldots, X_{n+1}]$. We can assume from the beginning that $f \neq 0$. Applying to (8) the homomorphism $K[X_1, \ldots, X_{n+1}] \to K(X_1, \ldots, X_n)$ defined by the substitutions

$$X_i \mapsto X_i \text{ for } 1 \leq i \leq n \quad \text{and} \quad X_{n+1} \mapsto 1/f,$$

we get a relation

$$1 = \sum_i h_i(X_1, \ldots, X_n, 1/f) g_i(X_1, \ldots, X_n).$$

Multiplying by an appropriate power f^m of f, then, we obtain

$$(9) \qquad f^m = \sum_i \tilde{h}_i(X_1, \ldots, X_n) g_i(X_1, \ldots, X_n)$$

for certain polynomials $\tilde{h}_i \in K[X_1, \ldots, X_n]$. Since $g_i \in \mathfrak{a}$ this implies $f^m \in \mathfrak{a}$, and Theorem 1 is proved. $\square$

2. We now complement the preceding discussion of *algebraic sets* with the following fundamental theorem:

Theorem 3 (Hilbert Basis Theorem). *Let K be any field. Every ideal $\mathfrak{a}$ of the polynomial ring $K[X_1, \ldots, X_n]$ is finitely generated, that is, there exist finitely many polynomials $f_1, \ldots, f_m$ in $K[X_1, \ldots, X_n]$ such that*

$$\mathfrak{a} = (f_1, \ldots, f_m).$$

*In particular, if C is any extension of K, any algebraic K-set V of C^n is the zero set of a **finite** family of polynomials $f_1, \ldots, f_m$ in $K[X_1, \ldots, X_n]$:*

$$V = \{(x_1, \ldots, x_n) \in C^n \mid f_i(x_1, \ldots, x_n) = 0 \text{ for } 1 \leq i \leq m\}.$$

In this situation we write $V = \mathcal{N}(f_1, \ldots, f_m)$.

Theorem 3 arises directly from the following general statement, which is also often called the Hilbert Basis Theorem, although Hilbert himself never stated it explicitly:

Theorem 4. *If R is a commutative ring with unity in which every ideal is generated by finitely many elements* (such a ring is called **Noetherian**), *then every ideal of the ring $R[X]$ of polynomials in one variable X over R is also finitely generated.*

Proof. Let $\mathfrak{a}$ be an ideal of $R[X]$. For each integer $m \geq 0$, consider the set

$$\mathfrak{c}_m = \{a \mid a \text{ is an } m\text{-th coefficient of a polynomial } f \in \mathfrak{a} \text{ of degree at most } m\}.$$

Apart from 0, then, the elements of $\mathfrak{c}_m$ are the highest coefficients of polynomials of degree m contained in $\mathfrak{a}$. Clearly $\mathfrak{c}_m$ is an ideal in R. If $f(X)$ is a polynomial of degree m in $\mathfrak{a}$, then $Xf(X)$ is a polynomial of degree $m + 1$ in $\mathfrak{a}$; we thus obtain a chain

$$(10) \qquad \mathfrak{c}_0 \subseteq \mathfrak{c}_1 \subseteq \cdots \subseteq \mathfrak{c}_m \subseteq \mathfrak{c}_{m+1} \subseteq \cdots$$

of ideals in R. Therefore the union $\mathfrak{c}$ of all the $\mathfrak{c}_i$ is also an ideal in R. Because R is assumed Noetherian, $\mathfrak{c}$ is finitely generated, so the chain (10) terminates, meaning that there is some n for which

$$(11) \qquad \mathfrak{c}_0 \subseteq \mathfrak{c}_1 \subseteq \cdots \subseteq \mathfrak{c}_n = \mathfrak{c}_{n+1} = \cdots = \mathfrak{c}.$$

By assumption, all the ideals $\mathfrak{c}_0, \mathfrak{c}_1, \ldots, \mathfrak{c}_n$ are finitely generated; suppose, say, that

$$(12) \qquad \mathfrak{c}_i = (c_{i1}, \ldots, c_{ir}) \quad \text{for } 0 \leq i \leq n$$

(where we have uniformized the size r of the generating sets for $0 \leq i \leq n$ by, say, repeating generators). By the definition of the $\mathfrak{c}_i$ there is, for all $0 \leq i \leq n$ and $1 \leq j \leq r$, a polynomial $f_{ij} \in \mathfrak{a}$ of the form

$$(13) \qquad f_{ij} = c_{ij} X^i + \text{polynomial of degree less than } i,$$

with the same c_{ij} as in (12). Now let $\tilde{\mathfrak{a}}$ be the ideal of $R[X]$ generated by all the f_{ij}. We wish to show that $\mathfrak{a} = \tilde{\mathfrak{a}}$, which will prove Theorem 4.

It is clear that $\tilde{\mathfrak{a}}$ is contained in $\mathfrak{a}$. Let f be a polynomial of degree m in $\mathfrak{a}$, and let a be its leading coefficient:

$$f(X) = aX^m + \cdots, \quad \text{with } a \neq 0.$$

By definition, $a \in \mathfrak{c}_m$. We claim that if n is chosen as in (11), then

$$a \in \mathfrak{c}_i \quad \text{for some } i \leq m, n.$$

If $m \leq n$ this is clear: just take $i = m$. For $m > n$, on the other hand, the claim follows from (11). Now a, being an element of $\mathfrak{c}_i$ with $0 \leq i \leq n$, can be written as

$$a = a_1 c_{i1} + a_2 c_{i2} + \cdots + a_r c_{ir}, \quad \text{with } a_j \in R.$$

Then the polynomial $\tilde{f}$ defined by

(14)
$$\tilde{f} = f - \sum_{j=1}^{r} a_j f_{ij} X^{m-i}$$

had degree less than m, because of (13). On the other hand, the same polynomial lies in $\mathfrak{a}$, since $f_{ij} \in \tilde{\mathfrak{a}} \subseteq \mathfrak{a}$. By induction we can therefore assume that $\tilde{f} \in \tilde{\mathfrak{a}}$. By (14), then, f itself lies in $\tilde{\mathfrak{a}}$.   □

3. The Hilbert Basis Theorem has some fundamental consequences for algebraic sets, which we briefly discuss. As before, C/K will be an arbitrary but fixed field extension.

F2. *Every nonempty set of algebraic K-subsets of C^n has a minimal element.*

Proof. Clearly this is equivalent to saying that any descending chain

(15)
$$V_1 \supseteq V_2 \supseteq \cdots \supseteq V_m \supseteq V_{m+1} \supseteq \cdots$$

of algebraic K-sets V_i in C^n terminates, that is, satisfies $V_n = V_{n+1} = \cdots$ for some n. An application of $\mathscr{I}$ to (15) yields the ascending chain of ideals

(16)
$$\mathscr{I}(V_1) \subseteq \mathscr{I}(V_2) \subseteq \cdots \subseteq \mathscr{I}(V_m) \subseteq \mathscr{I}(V_{m+1}) \subseteq \cdots.$$

This chain we already know to be terminating, since the union of the $\mathscr{I}(V_i)$ is itself an ideal of $K[X_1, \ldots, X_n]$, and so must be finitely generated by Theorem 3. Applying $\mathscr{N}$ to (16) and keeping in mind F1(ix) we recover the chain (15), which is thus seen to terminate.   □

Definition 3. An algebraic K-set $V \neq \varnothing$ in C^n is called *irreducible* if it *cannot* be expressed as a union $V = V_1 \cup V_2$ of algebraic K-sets V_1, V_2 of C^n distinct from V. An irreducible (affine) algebraic K-set in C^n is also called an (affine) *K-variety* of C^n.

F3. *An algebraic K-set V of C^n is irreducible if and only if its ideal $\mathcal{I}_K(V)$ is a prime ideal of $K[X_1, \ldots, X_n]$.*

Proof. (1) Let V be irreducible and suppose $fg \in \mathcal{I}(V)$. Then

$$V = \mathcal{N}(\mathcal{I}(V)) \subseteq \mathcal{N}(fg) = \mathcal{N}(f) \cup \mathcal{N}(g),$$

so $V = V_1 \cup V_2$ where $V_1 = \mathcal{N}(f) \cap V$ and $V_2 = \mathcal{N}(g) \cap V$ are algebraic K-sets. Because V is assumed irreducible we have (say) $V_1 = V$, so $V \subseteq \mathcal{N}(f)$. But then $f \in \mathcal{I}(\mathcal{N}(f)) \subseteq \mathcal{I}(V)$. Since V is nonempty, moreover, we have $1 \notin \mathcal{I}(V)$.

(2) Let $\mathcal{I}(V)$ be a prime ideal, and suppose $V = V_1 \cup V_2$ with V_1, V_2 algebraic K-sets in K^n; suppose moreover that $V_1 \neq V$. First note that

(17) $\mathcal{I}(V) = \mathcal{I}(V_1 \cup V_2) = \mathcal{I}(V_1) \cap \mathcal{I}(V_2) \supseteq \mathcal{I}(V_1)\mathcal{I}(V_2).$

Now $\mathcal{I}(V_1) \neq \mathcal{I}(V)$, otherwise we would have $V_1 = \mathcal{N}(\mathcal{I}(V_1)) = \mathcal{N}(\mathcal{I}(V)) = V$, contrary to the assumption $V_1 \neq V$. Thus there exists $f \in \mathcal{I}(V_1)$ such that $f \notin \mathcal{I}(V)$. But $\mathcal{I}(V)$ is a prime ideal; therefore $\mathcal{I}(V_2) \subseteq \mathcal{I}(V)$, by (17). It follows that $V \subseteq V_2$ and hence that V is irreducible.   □

Remark. *Affine space K^n* (being the zero set of the zero polynomial) is itself an algebraic K-set of K^n. If the field K is *infinite*, moreover, $\mathcal{I}(K^n)$ is the zero ideal and hence K^n is *irreducible*, by F3. In particular, K^n cannot, in the case of K infinite, be expressed as a union of finitely many proper subspaces (this a favorite linear algebra exercise). If K is finite, of course, K^n is not irreducible.

Note also that every algebraic K-set V of C^n is also an algebraic C-set of C^n. But clearly being irreducible as an algebraic K-set of C^n does not imply being irreducible as an algebraic C-set of C^n.

F4. *Any algebraic K-set V of C^n can be written as a finite union*

(18) $V = V_1 \cup V_2 \cup \cdots \cup V_r$

of irreducible algebraic K-sets V_i. If we demand that $V_i \nsubseteq V_k$ for $i \neq k$ in (18), this representation is unique up to order; the V_i are called the **irreducible K-components** *of V. Every K-variety W contained in V is contained in one of the V_i.*

Proof. (1) Let $\mathfrak{U}$ be the set of all algebraic K-sets of C^n that *cannot* be expressed as in (18). If $\mathfrak{U}$ is nonempty, it has a minimal element V by F2. This V is not irreducible by assumption, and so it is of the form $V = V_1 \cup V_2$, where V_1, V_2 are algebraic sets strictly contained in V. But by definition V_1 and V_2 both have representations of the form (18), and therefore so does V, contradicting the assumption that it belongs to $\mathfrak{U}$.

(2) We next show the last assertion of F4. From (18) we have

$$W = (W \cap V_1) \cup \cdots \cup (W \cap V_r).$$

Since each $W \cap V_i$ is algebraic and W is irreducible, there exists V_i such that $W = W \cap V_i$, and thus $W \subseteq V_i$.

(3) To prove uniqueness, let $V = V_1 \cup \cdots \cup V_r$ and $V = W_1 \cup \cdots \cup W_s$ be two ways to write V as a finite union of irreducible algebraic K-sets. By (2) there exists for each W_j some V_i such that $W_j \subseteq V_i$. Likewise there exists for this V_i some W_k such that $V_i \subseteq W_k$; thus $W_j \subseteq V_i \subseteq W_k$. The noninclusion assumption then implies that $j = k$ and hence $W_j = V_i$; that is, each W_j coincides with some V_i. Likewise for each V_k there is some W_l such that $V_k = W_l$. The assertion follows.   □

Definition 4. Let V be an algebraic K-set of C^n. The K-algebra

$$(19) \qquad K[V] := K[X_1, \ldots, X_n]/\mathscr{I}(V)$$

is called the *affine coordinate ring of* V. Clearly $K[V]$ can be identified with the ring of all functions $V \to C$ that arise from polynomials in $K[X_1, \ldots, X_n]$. Being a homomorphic image of $K[X_1, \ldots, X_n]$, this K-algebra has the form

$$(20) \qquad K[V] = K[x_1, \ldots, x_n],$$

and so is finitely generated as a K-algebra. In the sequel we will call any finitely generated commutative K-algebra A an *affine K-algebra*. If the algebraic K-set V of C^n is irreducible — in which case we speak of an *(affine) K-variety of C^n* — then $K[V]$ is an integral domain (see F3); the fraction field

$$(21) \qquad K(V) := \mathrm{Frac}\, K[V]$$

is then called the *field of rational functions of the K-variety V*.

Remarks.   (i) If C is algebraically closed, $V = C^n$ is a K-variety, and $K(V)$ is the field $K(X_1, \ldots, X_n)$ of rational functions in n variables over K.

(ii) Suppose that the affine K-algebra $A = K[x_1, \ldots, x_n]$ is an integral domain. Then A, being a homomorphic image of $K[X_1, \ldots, X_n]$, is isomorphic to $K[X_1, \ldots, X_n]/\mathfrak{p}$, where $\mathfrak{p}$ is a prime ideal. Thus, if C is algebraically closed, A is isomorphic to the affine coordinate algebra of the K-variety $V = \mathscr{N}(\mathfrak{p})$ of C^n (see Theorem 1).

(iii) Again let $A = K[x_1, \ldots, x_n]$ be an integral domain. The *transcendence degree* of A is of course defined as that of the field extension $\mathrm{Frac}(A)/K$:

$$\mathrm{TrDeg}(A/K) := \mathrm{TrDeg}(\mathrm{Frac}(A)/K).$$

Since $\mathrm{Frac}\, A = K(x_1, \ldots, x_n)$, we have $\mathrm{TrDeg}(A/K) \leq n$.

(iv) Any affine K-algebra is *Noetherian*, since it is a homomorphic image of a polynomial ring $K[X_1, \ldots, X_n]$ (see Theorem 4).

4.   In this last section we examine the notion of *dimension* for algebraic sets. We continue to work with a fixed field extension C/K, and make the additional assumption throughout the section that C is *algebraically closed*.

Definition 5. The *dimension* $\dim V$ of an algebraic K-set V of C^n is the supremum of all integers m for which there is a strict chain $V_0 \subset V_1 \subset \cdots \subset V_m$ of K-varieties of C^n, all contained in V.

Natural as this definition appears, it is not self-evident that dim V is finite. In view of Theorem 1 and F3, however, Definition 5 suggests two related notions:

Definition 6. Let A be any commutative ring with unity. The *height* $h(\mathfrak{p})$ of a prime ideal $\mathfrak{p}$ of A is the supremum of all integers m for which there is a strict chain $\mathfrak{p}_0 \subset \mathfrak{p}_1 \subset \cdots \subset \mathfrak{p}_m = \mathfrak{p}$ of prime ideals $\mathfrak{p}_i$ in A. The *dimension* (more precisely, *Krull dimension*) of the ring A is the supremum of the heights of all prime ideals of A, and is denoted by dim A.

Even if A is assumed to be *Noetherian*, it is not *a priori* clear that every prime ideal $\mathfrak{p}$ of A satisfies $h(\mathfrak{p}) < \infty$. This nonetheless turns out to be the case; on the other hand, there exist Noetherian rings A such that dim $A = \infty$ (see Matsumura, *Commutative algebra*).

Regardless of the finiteness of the numbers in question, we have:

F5. *For every algebraic K-set V of C^n,*

$$(22) \qquad \dim V = \dim K[V].$$

Proof. By F3 and Theorem 1, K-varieties W contained in V are in one-to-one (and inclusion-reversing) correspondence with prime ideals of the polynomial ring $K[X_1, \ldots, X_n]$ that contain $\mathcal{I}(V)$. But the latter correspond exactly to the prime ideals of $K[X_1, \ldots, X_n]/\mathcal{I}(V) = K[V]$. The assertion follows.   □

Remark. It is easy to see that

$$\dim V = \max(\dim V_1, \ldots, \dim V_r)$$

if $V = V_1 \cup \cdots \cup V_r$ is the decomposition of V into *irreducible components* given by F4. Therefore we will restrict our attention from now on to K-*varieties V*. Take the conceivably simplest case of the K-variety $V = C^1$. Then $K[V] = K[X]$ is the polynomial ring in one variable over K. But $K[X]$ is a principal ideal domain, so every prime ideal of $K[X]$ is either a maximal ideal or the zero ideal of $K[X]$. Thus the only K-varieties contained in $V = C^1$ are (besides V itself) the root sets of irreducible polynomials of $K[X]$. Thus for $V = C^1$ we have

$$\dim V = \dim K[V] = 1 = \mathrm{TrDeg}(K[V]/K).$$

In general the dimension of an arbitrary K-variety V can be given the following description, which also makes the finiteness of dim V manifest:

Theorem 5. *For any K-variety V of C^n,*

$$(23) \qquad \dim V = \mathrm{TrDeg}(K[V]/K).$$

In particular, dim V *is at most n (and for instance* dim $C^n = n$*).*

We refer the proof of Theorem 5 to an auxiliary result:

Lemma 1. *Let the K-algebra A be an integral domain and assume that $\mathrm{TrDeg}_K(A)$ is finite. For any nonzero prime ideal $\mathfrak{p}$ of A we have*

$$(24) \qquad \mathrm{TrDeg}_K(A/\mathfrak{p}) < \mathrm{TrDeg}_K(A).$$

Proof. For notational simplicity set $\bar{A} = A/\mathfrak{p}$ and denote by

$$\varphi : A \to \bar{A},$$
$$x \mapsto \bar{x}$$

the quotient map. If $\bar{x}_1, \ldots, \bar{x}_s$ are algebraically independent over K, so are $x_1, \ldots, x_s$. Thus we already know that $\mathrm{TrDeg}_K(\bar{A}) \leq \mathrm{TrDeg}_K(A)$. Now assume, in contradiction with (24), that

$$\mathrm{TrDeg}_K(\bar{A}) = \mathrm{TrDeg}_K(A) =: r.$$

Then there exist r algebraically independent elements $\bar{x}_1, \ldots, \bar{x}_r$ of $\bar{A}$. The map φ is injective on $K[x_1, \ldots, x_r]$, so φ can be extended to a homomorphism

$$\tilde{\varphi} : K(x_1, \ldots, x_r)[A] \to K(\bar{x}_1, \ldots, \bar{x}_r)[\bar{A}],$$

where the K-algebras in question are regarded as subalgebras of the fraction field of A or $\bar{A}$ as the case may be. Let $\tilde{A}$ be the K-algebra on the left. Since $r = \mathrm{TrDeg}_K(A)$, every element of $\tilde{A}$ is algebraic over $K(x_1, \ldots, x_n)$. Thus $\tilde{A}$ is a field. But then $\tilde{\varphi}$ is injective, and hence so is φ. Because $\mathfrak{p} = \ker \varphi$ we obtain a contradiction with $\mathfrak{p} \neq 0$; this proves the lemma. $\qquad\square$

With the help of this result we now show a result which is equivalent, thanks to (22), to Theorem 5:

Theorem 5′. *Let A be an affine K-algebra without zero-divisors. Then*

$$(25) \qquad \dim A = \mathrm{TrDeg}(A/K).$$

Proof. Clearly $\mathrm{TrDeg}(A/K)$ is finite, since A is finitely generated as a K-algebra. We now use induction on $r := \mathrm{TrDeg}(A/K)$.

(i) We first want to show that $\dim A \leq r$. Let

$$0 = \mathfrak{p}_0 \subset \mathfrak{p}_1 \subset \mathfrak{p}_2 \subset \cdots \subset \mathfrak{p}_m$$

be a strict chain of prime ideals of A; we mush show that $m \leq r$. If $r = 0$ then A is a field, so $m = 0$. Now let $r \geq 1$ and (avoiding triviality) $m \geq 1$. Applying the natural map $A \to \bar{A} = A/\mathfrak{p}_1$ we get a strict chain of prime ideals

$$0 = \bar{\mathfrak{p}}_1 \subset \bar{\mathfrak{p}}_2 \subset \cdots \subset \bar{\mathfrak{p}}_m$$

of $\bar{A}$. By the induction assumption and Lemma 1 it follows that $m-1 \leq \dim \bar{A} \leq r-1$, and hence that $m \leq r$.

(ii) Now we wish to show that $r \leq \dim A$. For $r = 0$ there is nothing to show, so let $r \geq 1$. By *Noether's Normalization Theorem* (Chapter 18, Theorem 4), A contains a polynomial ring $R = K[y_1, \ldots, y_r]$ in r variables as a subalgebra such that A/R is *integral*. For the prime ideal $\mathfrak{p} = (y_r)$ of R there is a prime ideal $\mathfrak{P}$ of A such that $\mathfrak{P} \cap R = \mathfrak{p}$, by §16.12 in the Appendix. Setting

$$\bar{A} = A/\mathfrak{P}, \quad \bar{R} = R/\mathfrak{p},$$

so that $\bar{R} \simeq K[y_1, \ldots, y_{r-1}]$, we obtain in this way a ring extension $\bar{A}/\bar{R}$. But since A/R is integral, so is $\bar{A}/\bar{R}$; it follows that $\mathrm{TrDeg}(\bar{A}) = \mathrm{TrDeg}(\bar{R}) = r - 1$. By the induction assumption this implies the existence of a strict chain of prime ideals

$$0 = \bar{\mathfrak{P}}_1 \subset \bar{\mathfrak{P}}_2 \subset \cdots \subset \bar{\mathfrak{P}}_r$$

in $\bar{A}$. Denoting by $\mathfrak{P}_1, \ldots, \mathfrak{P}_r$ the inverse images of these ideals in A, we obtain a strict chain of prime ideals $0 = \mathfrak{P}_0 \subset \mathfrak{P}_1 \subset \cdots \subset \mathfrak{P}_r$. This shows that indeed $r \leq \dim A$. $\qquad\square$

Remark. Let the situation be as above. In part (ii) of the proof we showed that $r \leq \dim A$. If A is a *field*, so that $\dim A = 0$, it follows that $r = 0$, that is, A/K is *algebraic*. Thus the considerations in (ii) amount to a generalization of the argument used in Chapter 18 to prove Theorem 5 (page 215).

We will now supplement Theorem 5 with something sharper. But we again need preparatory results:

Lemma 2. *If R is a UFD, a prime ideal $\mathfrak{p}$ has height 1 if and only if it is a principal ideal.*

Proof. Suppose $h(\mathfrak{p}) = 1$. Then $\mathfrak{p} \neq 0$, and we may take a nonzero $f \in \mathfrak{p}$. Since $\mathfrak{p}$ is a prime ideal, at least one prime factor f_1 of f lies in $\mathfrak{p}$. Now (f_1) is likewise a prime ideal, and we have $0 \neq (f_1) \subseteq \mathfrak{p}$. Since $h(\mathfrak{p}) = 1$, we get $\mathfrak{p} = (f_1)$.

Conversely, suppose $\mathfrak{p} = (f)$, and assume for a contradiction that $h(\mathfrak{p}) \neq 1$. Then there is a prime ideal $\mathfrak{q}$ such that $0 \neq \mathfrak{q} \subset \mathfrak{p} = (f)$. If g is a nonzero element of $\mathfrak{q}$ there must be a prime factor g_1 of g in $\mathfrak{q}$. There follows $(g_1) \subseteq (f)$, and because g_1 is irreducible we must have $(f) = (g_1) \subseteq \mathfrak{q}$. Contradiction. $\qquad\square$

Lemma 3. *Let $R = K[y_1, \ldots, y_r]$ be a polynomial ring in the r indeterminates $y_1, \ldots, y_r$ over the field K. For every prime ideal $\mathfrak{p}$ of height $h(\mathfrak{p}) = 1$ in R, the quotient ring $\bar{R} = R/\mathfrak{p}$ has transcendence degree $r - 1$ over K.*

Proof. By Lemma 2, $\mathfrak{p}$ has the form $\mathfrak{p} = (p)$, where $p \in k[y_1, \ldots, y_r]$ is a nonconstant polynomial. If, say, the variable $y := y_r$ really does appear in p, we also have $\deg_y p \geq 1$. It follows that

$$\mathfrak{p} \cap K[y_1, \ldots, y_{r-1}] = 0.$$

Thus the quotient map $R \to \bar{R} = K[\bar{y}_1, \ldots, \bar{y}_r]$ gives rise to an *isomorphism* on the subring $K[y_1, \ldots, y_{r-1}]$; that is, $\bar{y}_1, \ldots, \bar{y}_{r-1}$ are algebraically independent over

K. At the same time, $p(\bar{y}_1, \ldots, \bar{y}_{r-1}, \bar{y}_r) = 0$ is a (nontrivial) algebraic equation for $\bar{y}_r$ over $K(\bar{y}_1, \ldots, \bar{y}_{r-1})$. Putting the two facts together we conclude that indeed

$$\operatorname{TrDeg}(\bar{R}/K) = \operatorname{TrDeg}(K(\bar{y}_1, \ldots, \bar{y}_r)/K) = r - 1. \qquad \square$$

For the announced sharpening of Theorem 5 we also need a basic fact from *commutative algebra*, whose justification we leave for the Appendix (§19.6):

F6* (Krull's Descent Lemma). *Let A/R be an **integral extension** of **integral domains**, and assume that R is **integrally closed** (in its fraction field). Suppose given prime ideals $\mathfrak{p}$ and $\mathfrak{q}$ of R with $\mathfrak{q} \subseteq \mathfrak{p}$, and also a prime ideal $\mathfrak{P}$ of A such that $\mathfrak{P} \cap R = \mathfrak{p}$. Then there is a prime ideal $\mathfrak{Q}$ of A such that $\mathfrak{Q} \cap R = \mathfrak{q}$ and $\mathfrak{Q} \subseteq \mathfrak{P}$.*

Theorem 6. *Let $A = K[x_1, \ldots, x_n]$ be an affine K-algebra, and assume that A is an integral domain. Let*

$$(26) \qquad 0 = \mathfrak{P}_0 \subset \mathfrak{P}_1 \subset \cdots \subset \mathfrak{P}_m$$

be a nonrefinable strict chain of prime ideals in A (such a chain must exist because $\dim A < \infty$; see Theorem 5'). Then

$$(27) \qquad m = \operatorname{TrDeg}(A/K);$$

in particular all maximal ideals of A have the same height, namely the transcendence degree of A over K. (We thus recover the result that $\dim A = \operatorname{TrDeg}(A/K)$.)

Proof. We proceed inductively. Let $m = 0$. Then $\mathfrak{P}_m = 0$ is a maximal ideal of A, so A is a field. Using Chapter 18, Theorem 5 or equation (25), we conclude that $\operatorname{TrDeg}(A/K) = 0$.

Now suppose $m \geq 1$. By the *Noether Normalization Theorem*, A contains a polynomial algebra $R = K[y_1, \ldots, y_r]$ as a subalgebra such that A/R is *integral*. Since (26) cannot be refined, we have $h(\mathfrak{P}_1) = 1$. We claim that the prime ideal $\mathfrak{p}_1 = \mathfrak{P}_1 \cap R$ also has height 1. Otherwise there is a prime ideal $\mathfrak{q} \subseteq \mathfrak{p}_1$ such that $0 \neq \mathfrak{q} \neq \mathfrak{p}_1$. *Krull's Descent Lemma* then provides a prime ideal $\mathfrak{Q}$ of A such that $\mathfrak{Q} \subseteq \mathfrak{P}_1$ and $\mathfrak{Q} \cap R = \mathfrak{q}$. Since $\mathfrak{q}$ is not the zero ideal, neither is $\mathfrak{Q}$. Thus $h(\mathfrak{P}_1) = 1$ implies $\mathfrak{P}_1 = \mathfrak{Q}$, and we get $\mathfrak{q} = \mathfrak{Q} \cap R = \mathfrak{P}_1 \cap R = \mathfrak{p}_1$, contradicting $\mathfrak{q} \neq \mathfrak{p}_1$. Now setting

$$\bar{A} = A/\mathfrak{P}_1 \quad \text{and} \quad \bar{R} = R/\mathfrak{p}_1,$$

we obtain a ring extension $\bar{A}/\bar{R}$. Since A/R is *integral*, so is $\bar{A}/\bar{R}$. There follows $\operatorname{TrDeg}(\bar{A}/K) = \operatorname{TrDeg}(\bar{R}/K)$. But by Lemma 3 we have $\operatorname{TrDeg}(\bar{R}/K) = r - 1$, so

$$\operatorname{TrDeg}(\bar{A}/K) = r - 1.$$

Passing from A to $\bar{A}$, we get from (26) a strict and *nonrefinable* chain of prime ideals

$$0 = \bar{\mathfrak{P}}_1 \subset \bar{\mathfrak{P}}_2 \subset \cdots \subset \bar{\mathfrak{P}}_m$$

of length $m-1$ in $\bar{A}$. By induction (on m or the transcendence degree of A) we then get $m-1 = \operatorname{TrDeg}(\bar{A}/K) = r - 1$. Therefore $m = r = \operatorname{TrDeg}(A/K)$, as claimed. $\square$

We prove an application of Theorem 6:

F6. *A K-variety V of C^n has dimension $n-1$ if and only if it is of the form $V = \mathcal{N}(f)$, where f is an irreducible polynomial in $K[X_1, \ldots, X_n]$.*

Proof. (i) Since V is irreducible, its ideal $\mathfrak{p}_1$ is a prime ideal of $K[X_1, \ldots, X_n] =: A$. Now suppose $\dim V = n - 1$. Because $\bar{A} = A/\mathfrak{p}_1$ we have $n - 1 = \dim V = \dim K[V] = \dim \bar{A}$ (see F5). Thus there exists a strict chain of prime ideals $0 = \bar{\mathfrak{p}}_1 \subset \cdots \subset \bar{\mathfrak{p}}_n$ in $\bar{A}$. Taking inverse images we then get a strict chain of prime ideals $0 = \mathfrak{p}_0 \subset \mathfrak{p}_1 \subset \cdots \subset \mathfrak{p}_n$ in A. Since $\dim A = n$, we must have $h(\mathfrak{p}_1) = 1$. Therefore, by Lemma 2, $\mathfrak{p}_1 = (f)$ is a principal ideal. Thus $V = \mathcal{N}(f)$, with $f \in K[X_1, \ldots, X_n]$ irreducible.

(ii) Suppose $V = \mathcal{N}(f)$, where $f \in K[X_1, \ldots, X_n] =: A$ is irreducible. Set $\mathfrak{p}_1 = (f)$. By Lemma 2, $h(\mathfrak{p}_1) = 1$. Since $\dim A < \infty$ there is then a nonrefinable strict chain of prime ideals $0 = \mathfrak{p}_0 \subseteq \mathfrak{p}_1 \subseteq \cdots \subseteq \mathfrak{p}_m$ (containing the given prime ideal!). Now we resort to Theorem 6, according to which $m = n$. Passing to $\bar{A} = A/\mathfrak{p}_1$ we get a strict and nonrefinable chain of prime ideals of length $n - 1$ in $\bar{A}$. Using Theorem 6 again we conclude that $n - 1 = \dim \bar{A} = \dim K[V] = \dim V$, as desired. $\square$

From Theorem 6 some consequences of a general sort can be drawn. We leave the simple demonstrations to the reader:

F7. *Let A be an affine K-algebra and an integral domain. For every prime ideal $\mathfrak{p}$ of A,*

$$(28) \qquad h(\mathfrak{p}) + \dim A/\mathfrak{p} = \dim A.$$

Moreover, if $0 = \mathfrak{p}_0 \subseteq \mathfrak{p}_1 \subseteq \cdots \subseteq \mathfrak{p}_r = \mathfrak{p}$ is a strict chain of prime ideals in A, with last element $\mathfrak{p}$, and if there is no finer chain of the same description, then $r = h(\mathfrak{p})$.

In the context of K-varieties, equality (28) becomes: *For every prime ideal $\mathfrak{p}$ of $K[X_1, \ldots, X_n]$, the algebraic set $V = \mathcal{N}_C(\mathfrak{p})$ — where, as we recall, C is an algebraically closed extension of K — has dimension*

$$(29) \qquad \dim V = n - h(\mathfrak{p}).$$

One immediate consequence of the first statement in F8 is that, for $1 \leq r \leq n$, the prime ideal $(X_1, \ldots, X_r)$ of $K[X_1, \ldots, X_n]$ has height r; the strict chain of prime ideals $0 \subseteq (X_1) \subseteq (X_1, X_2) \subseteq \cdots \subseteq (X_1, \ldots, X_r)$, therefore, admits no refinement.

Here is an addendum to F7:

F7′. *For every nonzero f in $K[X_1, \ldots, X_n]$, all the irreducible K-components of $\mathcal{N}(f) = \mathcal{N}_C(f)$ have dimension $n - 1$. Conversely, if V is an algebraic K-set of C^n whose irreducible K-components all have dimension $n-1$, then V is a hypersurface, that is, it can be expressed as $V = \mathcal{N}(f)$ for some nonzero f in $K[X_1, \ldots, X_n]$.*

Proof. If $V = \mathcal{N}(f)$ and $f = f_1^{e_1} \ldots f_r^{e_r}$ is the prime factorization of f, we have $V = V_1 \cup \cdots \cup V_r$, where the $V_i = \mathcal{N}(f_i)$ are K-varieties. By F7, each V_i has dimension $n - 1$.

Conversely, if the irreducible K-components $V_1, \ldots, V_r$ of an algebraic K-set V of C^n all have dimension $n - 1$, then by F7 each V_i is of the form $V_i = \mathcal{N}(f_i)$. Setting $V = V_1 \cup \cdots \cup V_r = \mathcal{N}(f_1) \cup \cdots \cup \mathcal{N}(f_2) = \mathcal{N}(f_1 \ldots f_r)$ we get the desired result.   □

Remark. Let V be any algebraic K-set of C^n, and suppose that $f \in K[X_1, \ldots, X_n]$ does not vanish on any of the irreducible K-components of V. Then

$$(30) \qquad \dim(V \cap \mathcal{N}(f)) \leq \dim V - 1.$$

To see this, we start by assuming without loss of generality (see F4) that V is a K-variety. Every K-component W of $V \cap \mathcal{N}(f)$ is a proper subset of V. Thus the assertion follows directly from Definition 5. In Section 27.5 in volume II we will show, for the case $K = C$, that in fact $\dim W = \dim V - 1$; this represents an important generalization of F7' in the case $K = C$.

Finally, using (30) one can prove the following fact (see §19.10 in the Appendix):

F9 (Kronecker). *An algebraic K-set V of C^n can always be represented as the zero set of at most $n + 1$ polynomials.*

As Kronecker presumably already knew, and as proved for example by U. Storch ("Bemerkung zu einem Satz von M. Kneser", *Archiv der Math.* **23**, 1972), one can actually replace $n + 1$ by n in this statement. It is an interesting — and difficult — question to determine under what conditions V has a representation involving exactly $n - \dim V$ polynomials.

Appendix:

Problems and Remarks

References preceded by § are to this appendix.

Chapter 1: Constructibility with Ruler and Compass

1.1 Let K be a subfield of $\mathbb{C}$ and a, b elements in $K^\times$. Show equivalence between:

(i) $K(\sqrt{a}) = K(\sqrt{b})$.

(ii) There exists $c \in K^\times$ with $a = bc^2$.

(The assumption $K \subseteq \mathbb{C}$ is not essential; the statement holds for any field K where $1 + 1 \neq 0$.)

1.2 Let K be a subfield of $\mathbb{C}$ with $\overline{K} = K$ and let w be a complex number such that $w^2 \in K$. It is then always that case that $\overline{K(w)} = K(w)$? The answer is no (that is why in the proof of Theorem 1 we had to take a certain precaution). *Hint:* Consider the example $K = \mathbb{Q}(i)$ with $w^2 = 1 + i$ (and use §1.1 with $a = 1 + i$, $b^{-1} = 1 - i$).

1.3 In F9, the assumption $K = \overline{K}$ cannot be dispensed with — nor can the field $\mathbb{Q}(M \cup \overline{M})$ in Theorem 1 by replaced by, say, $\mathbb{Q}(M)$. One can see this most spectacularly as follows (while peeking at some concepts that will only be treated later, in Chapters 2 and 18): choose algebraically independent real numbers x, y and set $K = \mathbb{Q}(x + iy)$, $E = \mathbb{Q}(i, x, y)$. Then $E \subseteq_{\mathbb{A}} K$, but E/K is *not* algebraic, since otherwise by looking at transcendence degrees we would get the contradiction $2 = \mathrm{TrDeg}(E/\mathbb{Q}) \leq \mathrm{TrDeg}(K/\mathbb{Q}) \leq 1$.

1.4 Let E/K be a finite field extension. Prove that if $E : K$ is a prime number, E/K has no proper intermediate fields, and for each $\alpha \in E$ such that $\alpha \notin K$ we therefore have $E = K(\alpha)$.

1.5 Prove that $\mathbb{Q}(\sqrt{2}, \sqrt{3}) : \mathbb{Q} = 4$ and $\mathbb{Q}(\sqrt{2} + \sqrt{3}) = \mathbb{Q}(\sqrt{2}, \sqrt{3})$.

1.6 Let K be an infinite field and E an extension of degree $n > 1$ over K. Show that the quotient group $E^\times / K^\times$ of the multiplicative groups of E and K is infinite. (*Hint:* Otherwise the K-vector space $E \simeq K^n$ would be a union of finitely many one-dimensional subspaces; in other words, the projective space $P_{n-1}(K)$ would be finite.)   Remark: $E^\times / K^\times$ is not even finitely generated; but this is a much deeper result, for which see A. Brandis, "Über die multiplikative Struktur von Körpererweiterungen", *Math. Zeitschrift* **87** (1985).

1.7 A Danish schoolbook of 1854, published in Flensburg, contains the following recipe for constructing the heptagon inscribed in a circle S of radius 1 around the origin: Let the circle of radius 1 centered at $z_1 = 1$ intersect S at z_2 and z_3. Let z_4 be the intersection of the line through z_2 and z_3 with the line through 0 and z_1. Beginning at z_1, mark off the distance $|z_4 - z_2|$ against the circle S, seven times in succession.

Does this mean that Gauss's statement (see F12 in chapter 5) that $e^{2\pi i/7} \notin \triangle\{0, 1\}$ is in error? Show that the points obtained according to the procedure above are the powers $z, z^2, \ldots, z^7$ of the complex number $z = \frac{5}{8} + \frac{1}{8}\sqrt{39}\, i$. It follows that $65536 z^7 = 65530 - 142\sqrt{39}\, i$.

Chapter 2: Algebraic Extensions

2.1 Let E/K be a field extension and let L_1, L_2 be intermediate fields of E/K with $L_i : K < \infty$. Prove that $L_1 L_2 : K = [L_1 : K] \cdot [L_2 : K]$ implies $L_1 \cap L_2 = K$. (The converse does not hold; see the Example in Section 3.1.)

2.2 Show that $\mathbb{Q}(\sqrt{2}, \sqrt{1+i}) : \mathbb{Q} = 8$. *Hint:* For $w = \sqrt{1+i}$ we have $w\bar{w} = \sqrt{2}$, so $\mathbb{Q}(\sqrt{2}, w) = \mathbb{Q}(i, w, \bar{w})$. Now see §1.2.

2.3 Let E/K be an extension. Prove that E/K is algebraic if and only if every subring R of E containing K is a field.

2.4 Let R be a commutative ring with unity and K a subring of R. Prove: If K is a field and R has no zero-divisors, then $1_R = 1_K$ (in particular $1_R \neq 0$, so R is an integral domain and a K-vector space). Show by example that 1_R and 1_K can be distinct if R has zero-divisors.

2.5 Define a sequence $(\alpha_n)_n$ of real numbers $\alpha_n > 0$ through the recursion $\alpha_1 = 2$, $\alpha_{n+1} = \sqrt{\alpha_n}$. Then $\alpha_n \in \triangle\mathbb{Q}$ for every n, and

$$(1) \qquad\qquad \mathbb{Q}(\alpha_{n+1}) : \mathbb{Q} = 2^n,$$

so the algebraic extension $\triangle\mathbb{Q}/\mathbb{Q}$ *cannot be finite*. For the proof, we will show by induction over n that

$$(2) \qquad\qquad \mathbb{Q}(\alpha_{n+1}) : \mathbb{Q}(\alpha_n) = 2$$

for every n. The initial case $n = 1$ follows from $\sqrt{2} \notin \mathbb{Q}$. Assume that for some $n > 1$ equation (2) is false; then $\alpha_{n+1} \in \mathbb{Q}(\alpha_n) = \{a + b\alpha_n \mid a, b \in \mathbb{Q}(\alpha_{n-1})\}$. But

$\alpha_{n+1} = a + b\alpha_n$ implies $\alpha_n = \alpha_{n+1}^2 = a^2 + b^2\alpha_n^2 + 2ab\alpha_n = a^2 + b^2\alpha_{n-1} + 2ab\alpha_n$. From the induction hypothesis we get $a^2 + b^2\alpha_{n-1} = 0$, and since $\alpha_{n-1} > 0$ we obtain $a = b = 0$.

It's worth mentioning that the equality in (1) can be read off in a trice from certain later results from Chapters 3 and 5: α_{n+1} is a root of the polynomial $X^{2^n} - 2$, which is *irreducible* over $\mathbb{Q}$.

2.6 Let $(p_n)_n$ be the sequence of prime natural numbers. Working just as in §2.5, prove by induction that

$$\mathbb{Q}(\sqrt{p_1}, \ldots, \sqrt{p_n}) : \mathbb{Q} = 2^n.$$

In Chapter 14 we will see that this equality, and also that of §1.1, follow directly from *Kummer theory*.

2.7 Prove that the numbers $a = \sin 45°$ and $b = \cos 72°$ are *algebraic* and *irrational*. (*Hint:* For b see Chapter 1, Example 3 after Definition 2.)

2.8 What is the minimal polynomial of $\sqrt{2} + \sqrt{3}$ over $\mathbb{Q}$? (See §1.5.) Consider the real number

$$\alpha = \sqrt{5 + 2\sqrt{6}}$$

and show that $1, \alpha, \alpha^2, \alpha^3$ form a $\mathbb{Q}$-basis of $\mathbb{Q}(\alpha)$, as do $1, \sqrt{2}, \sqrt{3}, \sqrt{6}$. Is the polynomial $X^4 - 10X^2 + 1$ irreducible in $\mathbb{Q}[X]$?

Chapter 3: Simple Extensions

3.1 Let E/K be an extension and L_1, L_2 intermediate fields of E/K with $L_i : K$ finite. Then necessarily $L_1L_2 : L_2 \leq L_1 : K$ (Chapter 2, F11), but $L_1L_2 : L_2$ is *not necessarily a factor* of $L_1 : K$. *Hint:* Consider $L_1 = \mathbb{Q}(\sqrt[3]{2})$ and $L_2 = \mathbb{Q}(\zeta_3 \sqrt[3]{2})$ with $\zeta_3 = e^{2\pi i/3}$.

3.2 Let E/K be a field extension and suppose $\alpha, \beta \in E$ are algebraic over K. Set $f = \mathrm{MiPo}_K(\alpha)$ and $g = \mathrm{MiPo}_K(\beta)$. Prove that f is irreducible over $K(\beta)$ if and only if g is irreducible over $K(\alpha)$.

3.3 Let K be a field and $f \in K[X]$ a polynomial of degree $n > 0$. Using induction on n and Theorem 4 (Kronecker), show that there exists an extension E of K such that f can be expressed as a product of linear factors over E:

$$f(X) = \gamma(X - \alpha_1)(X - \alpha_2)\ldots(X - \alpha_n).$$

Therefore the subfield $K(\alpha_1, \ldots, \alpha_n)$ of E satisfies $K(\alpha_1, \ldots, \alpha_n) : K \leq n!$.

3.4 Keeping the assumptions and notations of §3.3, show that $K(\alpha_1, \ldots, \alpha_n) : K$ is in fact a *divisor* of $n!$. (*Hint:* Use induction on $n = \deg f$. If $f = gh$ in $K[X]$ with $\deg g = r$ and $\deg h = n - r$, then $r!(n-r)!$ is a divisor of $n!$). Show furthermore that if $K(\alpha_1, \ldots, \alpha_n) : K \geq (n-1)!$, then f is irreducible over K or f already has a root in K.

3.5 How does it follow from Kronecker's Theorem that there must exist a field with say exactly 9 elements?

3.6 Let $K(X)$ be the field of rational functions in one variable over the field K. Prove:

(a) If $T = T(X)$ is an element of $K(X)$ not contained in K, the extension $K(X)/K(T)$ is *finite*. *Hint:* Let Y be an indeterminate over $K(X)$ and let T be of the form $T = f/g$, with $f, g \in K[X]$. Then X is a root of the polynomial $f(Y) - Tg(Y) \in K(T)[Y]$, and since $T \notin K$, this polynomial is nonzero.

(b) K is *algebraically closed* in $K(X)$, that is, the algebraic closure of K in $K(X)$ is K. (*Hint:* Using (a), this follows from the transitivity of algebraicness.)

(c) $K(X)/K$ is a *simple* field extension possessing infinitely many intermediate fields. (*Hint:* If $K(\alpha)/K$ is not algebraic, α^2 is transcendental over K and $K(\alpha^2) \neq K(\alpha)$.)

(d) If K/k is a finite field extension of degree n, so is $K(X)/k(X)$.

3.7 Let E/K be a field extension and take $\alpha, \beta \in E^{\times}$. Suppose that $\alpha^m \in K$ and $\beta^n \in K$ for certain relatively prime natural numbers m, n. Prove that $\alpha\beta$ is a *primitive element* of the extension $K(\alpha, \beta)/K$. (*Hint:* there exist $x, y \in \mathbb{Z}$ such that $1 = xm + yn$.)

3.8 Let E be a subfield of $\mathbb{C}$ and $E_0 = E \cap \mathbb{R}$. Prove:

(a) It is not always the case that $E : E_0 \leq 2$.

(b) If $E = \mathbb{Q}(\zeta_n)$ with $\zeta_n = e^{2\pi i/n}$, then $E_0 = \mathbb{Q}(\eta_n)$ with $\eta_n = \zeta_n + \zeta_n^{-1} = \zeta_n + \bar{\zeta}_n = 2\cos(2\pi/n)$, and $E : E_0 = 2$ for $n > 2$.

(c) For $n \neq 1, 2, 3, 4, 6$, the number $\zeta_n + \zeta_n^{-1}$ is irrational. (*Hint:* If $\alpha \in \mathbb{Q}$ is a zero of a normalized polynomial $f \in \mathbb{Z}[X]$, then $\alpha \in \mathbb{Z}$.)

(d) $\mathbb{Q}(\zeta_7) : \mathbb{Q} = 6$, so the regular heptagon cannot be constructed with ruler and compass. (Show that η_7 is a zero of a cubic polynomial over $\mathbb{Q}$.)

3.9 Let G be an *abelian* group of order n (written multiplicatively), and denote by $\mu(G)$ the product of all elements of G. Consider the subgroups $G_2 = \{x \in G \mid x^2 = 1\}$ and $G^2 = \{x^2 \mid x \in G\}$ of G. Prove:

(a) $\mu(G) = \mu(G_2)$.

(b) If G is a subgroup of the multiplicative group of a field K and $-1 \in G$, then $\mu(G) = -1 = (-1)^{n/2}\mu(G^2)$.

Deduce that:

(α) In the field $\mathbb{F}_p$ we have $(p-1)! = -1$.

(β) For $p \neq 2$ the element -1 is a square in $\mathbb{F}_p$ if and only if $p \equiv 1 \bmod 4$.

3.10 Let the extension $E = K(\alpha, \beta)$ of K satisfy $K(\alpha) : K = p$ and $K(\beta) : K = q$, with $p > q$ both prime. Assume moreover that $\operatorname{char} K \neq p$. Prove that $E = K(\alpha + \beta)$.
    Hint: If this were not so and we set $h(X) = \operatorname{MiPo}_K(\alpha + \beta)$, we would have $h(X + \beta) = \operatorname{MiPo}_K(\alpha)$.

3.11 Let K be a field and, for elements a, b in $K^\times$, write $a \sim b$ if ab is a sum of two squares in K.

(a) Why is $\sim$ an *equivalence relation*?

(b) Does the analogous statement hold if one replaces 2 by some arbitrary power 2^n? (*Hint:* See LA II, p. 187, Problem 87.)

(c) In the case $K = \mathbb{R}(X)$, prove: $f \sim 1$ if and only if $f(x) \geq 0$ for all $x \in \mathbb{R}$ where $f(x)$ is defined. (*Hint:* Look first into the case $f \in \mathbb{R}[X]$, and take into account Section 4.3, Remark 2 after Definition 5.)

3.12 The construction of the fraction field of an integral domain given in the text can be generalized. Let R be a commutative ring with unity, and let S be a *multiplicative subset* of R, that is, a set S containing 1 and for which $s, t \in S$ implies $st \in S$. On the set $M = R \times S$, form the relation $\sim$ given by

$$(x, s) \sim (y, t) \quad \text{means} \quad \exists u \in S \text{ such that } (xt - ys)u = 0.$$

It is easy to prove that $\sim$ is an equivalence relation.[1] Denote by x/s the equivalence class of (x, s), and let $S^{-1}R$ be the set of all such equivalence classes. If the classes x/s are added and multiplied in the usual way (prove that this is well-defined), the set $S^{-1}R$ becomes a commutative ring with unity; it is called the *ring of formal fractions of R with denominators in S*, or the *localization of R relative to S*. For an integral domain R and $S = R \smallsetminus \{0\}$, we have $S^{-1}R = \operatorname{Frac} R$. Now, if we denote by $\iota : R \to S^{-1}R$ the homomorphism defined by $\iota(x) = x/1$, we have a *universal property* analogous to F7: *Let $\kappa : R \to B$ be a ring homomorphism (of commutative rings with unity) such that $\kappa(s)$ is invertible for every $s \in S$. There exists exactly one ring homomorphism $\lambda : S^{-1}R \to B$ such that $\lambda \circ \iota = \kappa$.*

The ring $A := S^{-1}R$ and the homomorphism $\iota : R \to A$ have the following properties:

(i) $\iota(s)$ is a unit in A for every $s \in S$.

(ii) Every element of A is of the form $\iota(x)\iota(s)^{-1}$, with $x \in R$ and $s \in S$.

(iii) $\iota(x) = 0$ if and only if there exists $s \in S$ with $xs = 0$.

Because of this last property, ι is not always injective, so R in general cannot be regarded as a subring of $S^{-1}R$. But, analogously with F8, we have: If $\iota : R \to A$ is a ring homomorphism with properties (i)–(iii), then A is isomorphic to the localization of R relative to S. Incidentally, $S^{-1}R$ is the zero ring if and only if $0 \in S$.

3.13 Let R be an integral domain with fraction field K. Assume R is a *Bézout ring*, that is, given $a, b \in R$ there is always some $d \in R$ with that $aR + bR = dR$. Prove that every subring A of K containing R is of the form $A = S^{-1}R$, for some multiplicative subset S of R. *Hint:* Consider $S = A^\times \cap R$.

3.14 (a) Find a subfield K of $\mathbb{C}$ and $z \in \mathbb{A}K$ such that $K(z):K$ is finite but not a power of 2. How is this to agree with Chapter 1, F9? (*Hint:* Choose $K = \mathbb{Q}(x)$ with $x = \sqrt[3]{2} + it$, where t is a transcendental real number.)

---

[1] The proof will make it clear why we can't just demand $xt = ys$ in the defining condition.

(b) Find a subfield K of $\mathbb{C}$ and $z \in \triangle K$ such that $K(z)/K$ is not algebraic. How is this to agree with Chapter 2, Theorem 1? (*Hint:* Choose $K = \mathbb{Q}(s + it)$ with $s, t \in \mathbb{R}$, where t is transcendental and s is not in the algebraic closure of $\mathbb{Q}(t)$ in $\mathbb{C}$. Then $x = s + it$ is not algebraic over $\mathbb{Q}(t)$. Prove that if t is algebraic over $\mathbb{Q}(x)$, then x is algebraic over $\mathbb{Q}(t)$.)

Chapter 4: Fundamentals of Divisibility

4.1 Find the gcd of $a = 17017$ and $b = 1114129$ and write it as an integer linear combination of a and b. Note how much faster it is, already in this simple example, to use the *Euclidean algorithm* than to find prime factorizations. (The factorizations, by the way, are $a = 17 \cdot 13 \cdot 11 \cdot 7$ and $b = 17 \cdot 65537$.)

4.2 Let R be a UFD. Using F12, prove:

(a) If $\gamma \in R$ and δ is a gcd of $\alpha_1, \ldots, \alpha_n$ in R, then $\gamma\delta$ is a gcd of $\gamma\alpha_1, \ldots, \gamma\alpha_n$.

(b) If $\alpha_1, \ldots, \alpha_n$ are pairwise relatively prime elements of R and their product $\alpha_1 \alpha_2 \ldots \alpha_n$ is an m-th power in R, each α_i is associated to an m-th power in R.

(c) If $\mathcal{P}$ is a directory of primes of R and $K = \text{Frac } R$, the multiplicative group of K satisfies $K^\times \simeq R^\times \times \mathbb{Z}^{(\mathcal{P})}$.

4.3 Find the prime factorization of $X^4 + 1$ in $\mathbb{C}[X]$, in $\mathbb{R}[X]$ and in $\mathbb{Q}[X]$. (*Hint:* In $\mathbb{C}[X]$ the answer is clear; the rest follows.) How about the prime factorization of $X^5 + X + 1$ in $\mathbb{Q}[X]$, in $\mathbb{F}_2[X]$ and in $\mathbb{F}_{19}[X]$? Show that $X^4 + 4$ is not irreducible in $\mathbb{Q}[X]$.

4.4 Show that:

(a) The number 2 is irreducible in $R = \mathbb{Z}[\sqrt{-5}]$, but not prime. (*Hint:* Use the fact that the function $N\alpha = \alpha\bar\alpha$ is multiplicative; why is $\alpha \in R^\times$ equivalent to $N\alpha = 1$?)

(b) $\mathbb{Z}[\sqrt{-1}]$, $\mathbb{Z}[\sqrt{-2}]$, $\mathbb{Z}[\frac{1}{2}(-1+\sqrt{-3})]$ and $\mathbb{Z}[\sqrt{2}]$ are all Euclidean domains. (*Hint:* For $\mathbb{Z}[\sqrt{2}]$ consider $\tilde{N}\alpha := |\alpha\alpha'|$, where $\alpha' = a - b\sqrt{2}$ for $\alpha = a + b\sqrt{2}$.)

4.5 The statement that $Y^2 = X^3 - 2$ has exactly one solution (x, y) in natural numbers goes back as far as Fermat. Prove its truthfulness, by working in the ring $\mathbb{Z}[\sqrt{-2}]$ and making use of §4.4b and §4.2b. Remark: However, the same equation has a whole series of *rational solutions* (x, y), such as $(129/100, 383/1000)$, $(164323/171^2, 66234835/171^3)$, and so on; this is connected with the operation of *addition on elliptic curves.*

4.6 Let R be a principal ideal domain and consider an infinite strictly decreasing chain of ideals in R, say $I_1 \supset I_2 \supset I_3 \supset \cdots$. Show that $\bigcap_{i=1}^{\infty} I_i = (0)$.

4.7 (a) Let R be a subring of an integral domain R'. Assume that R is a *principal ideal domain* and that $a, b \in R$. Show that a gcd of a, b in R is also a gcd of a, b in R'.

(b) Let E/K be a field extension and let $f, g \in K[X]$ be polynomials over K. Show that if a normalized polynomial $h \in E[X]$ is the gcd of f, g in $E[X]$, all the coefficients of h already lie in K.

4.8 Let R be an integral domain in which any two elements $x, y \in R$ have an lcm. Show that every irreducible element of R is prime. (*Hint:* If $\pi \nmid a$, then $a\pi$ is an lcm of a, π.)

4.9 Let R be an integral domain in which any two elements have a gcd. Show that any two elements a, b also have an lcm (namely $ab/\gcd(a, b)$ if $a, b \neq 0$).

4.10 Here is an elementary proof that $\mathbb{Z}$ is a UFD: by induction, it is clear that any $n > 1$ is a product of prime numbers. Now use §4.8. Prove that if m is the smallest of all common natural multiples of two given integers $x, y \neq 0$, then m is an lcm of x, y. (*Hint:* A one-time application of division with remainder.)

4.11 Set $R = \mathbb{Z}[\sqrt{10}]$. Show that in R every element $\alpha \neq 0$ is the product of irreducible elements, but R is not a unique factorization domain. (*Hint:* Consider the *multiplicative* function $\tilde{N}$ defined as in §4.4b, and note that $\alpha \in R^\times$ if and only if $\tilde{N}\alpha = 1$.)

4.12 Let R be a commutative ring with unity and S a multiplicative subset of R (see §3.12). Form the localization $S^{-1}R$ of R relative to S, with canonical map $\iota : R \to S^{-1}R$. If $\mathfrak{a}$ is an ideal of R, denote by $S^{-1}\mathfrak{a}$ the ideal of $S^{-1}R$ generated by $\iota(\mathfrak{a})$. It is easy to check that $S^{-1}\mathfrak{a}$ consists of all elements of the form a/s with $a \in \mathfrak{a}$ and $s \in S$; moreover $S^{-1}\mathfrak{a} = (1)$ if and only if $\mathfrak{a} \cap S \neq \varnothing$. Conversely, if $\mathfrak{A}$ is an ideal of $S^{-1}R$, denote the ideal $\iota^{-1}(\mathfrak{A})$ of R by $\mathfrak{A} \cap R$. Then $\mathfrak{a}$ is of the form $\mathfrak{a} = \iota^{-1}(\mathfrak{A})$ if and only if no element of S gives rise to a zero-divisor of $R/\mathfrak{a}$. Prove that *the maps $\mathfrak{P} \mapsto \mathfrak{P} \cap R$ and $\mathfrak{p} \mapsto S^{-1}\mathfrak{p}$ establish a one-to-one correspondence between prime ideals of $S^{-1}R$ and prime ideals of R that are disjoint from S.*

4.13 Let R be a commutative ring with unity and $\mathfrak{p}$ a prime ideal of R. Then $S := R \setminus \mathfrak{p}$ is *multiplicative*, in the sense of §3.12. In this case we denote the ring $S^{-1}R$ by $R_\mathfrak{p}$, and call it the *localization of R at $\mathfrak{p}$*. Set $\mathfrak{M} = \mathfrak{p}R_\mathfrak{p} = S^{-1}\mathfrak{p}$; then $1 \notin \mathfrak{M}$. *Every element of $R_\mathfrak{p}$ not belonging to $\mathfrak{M}$ is a unit of $R_\mathfrak{p}$, and conversely*. In other words: each ideal $\mathfrak{A} \neq (1)$ of $R_\mathfrak{p}$ is contained in $\mathfrak{M}$. In yet different words: $\mathfrak{M}$ is the unique maximal ideal of $R_\mathfrak{p}$. (A commutative ring with unity that has a unique maximal ideal is called a *local ring*; in this connection see Chapter 6, F12.) From §4.12 it follows that *prime ideals in the local ring $R_\mathfrak{p}$ are in one-to-one correspondence with prime ideals of R contained in $\mathfrak{p}$.*

4.14 Let R be a commutative ring with unity. An element f of R is called *nilpotent* if there is a natural number n such that $f^n = 0$. Denote by $\mathfrak{N}$ the set of nilpotent elements of R, called the *nilradical* of R. Prove that *the nilradical of R is the intersection of all prime ideals of R*. (*Hint:* Given $f \in R$, consider $S = \{f^n \mid n \in \mathbb{N}_0\}$. If f is not nilpotent, $S^{-1}R$ is not the zero ring and thus has a maximal ideal $\mathfrak{P}$; see Chapter 6, F12. Now apply §4.12.)

4.15 Let R be a commutative ring with unity $1 \neq 0$. Show that if every principal ideal of R distinct from R is prime, R is a field. (*Hint:* Consider the principal ideals (0) and (a^2) for all $a \neq 0$.)

4.16 Find an example of a *noncommutative simple ring*. (*Hint:* Consider the matrix group $M_n(K)$; see LA II, p. 179, Problem 46.)

4.17 Let m, n be natural numbers with $m \mid n$. Show that the canonical map

$$(\mathbb{Z}/n\mathbb{Z})^{\times} \longrightarrow (\mathbb{Z}/m\mathbb{Z})^{\times}$$

is surjective. (Hint for *one* possible solution: Look first at the case of a prime power $n = p^r$ and then bring the *Chinese Remainder Theorem* to bear.)

4.18 The ring $\mathbb{Z}[i]$, where $i = \sqrt{-1}$, is called the *ring of Gaussian integers*. By §4.4b above it is a Euclidean domain, and so has the same nice divisibility properties as $\mathbb{Z}$. Thus it is pertinent to ask: (a) What primes p of $\mathbb{Z}$ remain prime in $\mathbb{Z}[i]$? (b) How does one locate all primes in $\mathbb{Z}[i]$?

Answer: For each Gaussian prime π there exists a $\mathbb{Z}$-prime p with $\pi \mid p$. Now, either $(p) = (\pi)$, or there exists $(p) = (\pi\bar{\pi})$ such that $(\pi) \neq (\bar{\pi})$, except in the case $p = 2$. The first case happens when $p \equiv 3 \bmod 4$, and the second when $p \equiv 1 \bmod 4$ or $p = 2$.

Therefrom deduce *Fermat's Theorem:* A prime number $p \neq 2$ can be written in the form $p = x^2 + y^2$ with x, y integers if and only if $p \equiv 1 \bmod 4$. (*Hint:* For $p \equiv 1 \bmod 4$, by §3.9, there is some x such that p is a divisor of $x^2 + 1 = (x + i)(x - i)$. Now, if p is prime in $\mathbb{Z}[i]$, it follows that $p \mid x + i$.)

What would one have to know in order to perform an exactly similar analysis of, say, the Euclidean domain $\mathbb{Z}[\sqrt{2}]$?

4.19 Let E be a field, and assume that $f \in E[X]$ has a decomposition $f = f_1 f_2 \ldots f_r$ into pairwise relatively prime factors. Show that

$$E[X]/f \simeq E[X]/f_1 \times \cdots \times E[X]/f_r.$$

(This is an application of the Chinese Remainder Theorem, but it also can be proved easily by a direct dimension argument.)

4.20 Let A be a ring with unity. Assume that $A = A_1 \times \cdots \times A_r$ is a direct product of (sub)rings A_i.

(a) Prove that the ideals of A are precisely those subsets of the form $I_1 \times \cdots \times I_r$, where each I_k is an ideal of A_k. This statement also holds for left ideals instead of two-sided ideals.

(b) Assume each ring A_i is *simple* and prove that the A_i coincide with the *minimal ideals* of A, and therefore are *uniquely determined*. (An ideal I of a ring A is called minimal if it is minimal in the set of all nonzero ideals of A.)

4.21 Let R be a UFD where the ideal generated by any two elements is a principal ideal. Prove that R is a principal ideal domain.

4.22 Let $R = \mathfrak{C}(X)$ be the ring of all *continuous* functions on $X = [0, 1]$ with values in $\mathbb{R}$. For $a \in X$ let I_a be the set of $f \in R$ such that $f(a) = 0$. Show that the I_a are the only maximal ideals of R. (*Hint:* X is compact.) Is the same true if X is the open interval $X = (0, 1)$? (*Hint:* This question cannot be answered if F12 in Chapter 6 is not accepted.)

4.23 Let R be a UFD, and suppose there is a prime element q of R such that no unit $e \neq 1$ of R satisfies the congruence $e \equiv 1 \bmod q$. Show that R has *infinitely many* prime principal ideals (p). In particular this is true if $R = \mathbb{Z}$ or $R = K[X]$ for K a field.

Hint (Euclid): Let $p_1, \ldots, p_n$ be primes in R with $p_1 = q$. Then $1 + p_1 \ldots p_n$ has at least one prime divisor p.

4.24 For $m, n \in \mathbb{N}$ relatively prime, show that:

(a) $\mathbb{Q}(\sqrt[m]{2}, \sqrt[n]{3}) = \mathbb{Q}(\sqrt[m]{2}\,\sqrt[n]{3})$.

(b) A nonzero element a of a field K is an mn-th power in K if and only if a is both an m-th power and an n-th power in K.

4.25 Determine all maximal ideals of the following rings:
(a) $\mathbb{Q} \times \mathbb{Q}$; (b) $\mathbb{Q}[X]/X^2 + X + 1$; (c) $\mathbb{Q}[X]/X^3$; (d) $\mathbb{Q}[X]/X^2 - 3X + 2$.

4.26 Let K be a field of characteristic 0, and denote by $E = K(X)$ the field of rational functions in the variable X over K. For given $a, b \in K$ consider the subfield $F = K(X^2 + aX + b)$. Set $F_0 = K(X^2)$ and show that the extensions E/F and E/F_0 are finite (of degree 2), but if $a \neq 0$ the extension $E/F \cap F_0$ is not algebraic. (*Hint:* For each $f \in F$ we have $f(-X - a) = f(X)$. Show that any $f \in K(X)$ satisfying $f(X + a) = f(X)$ must be constant.)

Chapter 5: Prime Factorization in Polynomial Rings. Gauss's Theorem

5.1 Show that the following polynomials are irreducible in $\mathbb{Q}[X]$:
(a) $3X^4 + 6X^2 - 12X + 10$; (b) $\frac{7}{8}X^4 + \frac{1}{2}X^3 + 5X^2 + 6X + 12$.

5.2 Why is $\sqrt[3]{2}$ not an element of $\mathbb{Q}(\sqrt[7]{5})$? Why is there no extension E of $\mathbb{R}$ such that $E : \mathbb{R} = 3$?

5.3 Prove:

(a) $X^2 + X + 1$ is the only prime polynomial of degree 2 in $\mathbb{F}_2[X]$.

(b) The polynomial $f(X) = X^4 + 3X^3 + X^2 - 2X + 1$ is irreducible in $\mathbb{Z}[X]$, therefore also in $\mathbb{Q}[X]$. (*Hint:* First, f has no zeroes in $\mathbb{Q}$ — see F8. Now work mod 2; see the Remark after F9.)

(c) $X^m + 1 \in \mathbb{Q}[X]$ is irreducible for every $m = 2^n$ (see F13).

5.4 Show that no element of the field $\mathbb{R}(X, Y)$ of rational functions in two variables over $\mathbb{R}$ is a square root of $X^4 + X^2Y^2 + XY + X$.

5.5 Show that if $a_1, \ldots, a_n$ are distinct integers, the polynomial

$$f(X) = (X - a_1)^2 (X - a_2)^2 \ldots (X - a_n)^2 + 1$$

is irreducible in $\mathbb{Q}[X]$. (*Hint:* f is a normalized polynomial of degree $2n$ in $\mathbb{Z}[X]$ taking positive values at all $\alpha \in \mathbb{R}$ and taking the value 1 at the n distinct points $a_1, \ldots, a_n$. A nontrivial factorization $f = gh$ with $g, h \in \mathbb{Q}[X]$ normalized would therefore imply $g = h = (X - a_1) \ldots (X - a_n) + 1$.)

5.6 Let $f(X) = X^n - up^m$, with p a prime and $u \in \mathbb{Z}$ relatively prime to p. Show that if m, n are relatively prime, f is irreducible. (*Hint:* At least the case $u = 1$ is clear, because then $\mathbb{Q}\left((\sqrt[n]{p})^m \right) = \mathbb{Q}(\sqrt[n]{p})$.)

5.7 Consider $f(X) = X^4 + X + 1 \in \mathbb{Q}[X]$. Show that:

(i) f is irreducible.

(ii) In $\mathbb{C}[X]$ we have $f(X) = (X - z)(X - \bar{z})(X - w)(X - \bar{w})$; the following relations are satisfied: $z + \bar{z} + w + \bar{w} = 0$, $z\bar{z} + w\bar{w} + (z + \bar{z})(w + \bar{w}) = 0$, $z\bar{z}(w + \bar{w}) + w\bar{w}(z + \bar{z}) = -1$, $z\bar{z}w\bar{w} = 1$.

(iii) $\alpha := z\bar{z} + w\bar{w} = (z + \bar{z})^2$ is a zero of the polynomial $X^3 - 4X - 1$.

(iv) *Although* $\mathbb{Q}(z) : \mathbb{Q} = 4$, *the number z does not belong to* $\mathbb{A}\mathbb{Q}$.

5.8 In the situation of §3.6, choose for $T \in K(X)$ a representation $T = f/g$, with f, g *relatively prime*. Show that

$$K(X) : K(T) = \max(\deg f, \deg g).$$

(*Hint:* The polynomial $f(Y) - Tg(Y)$ in §3.6 is irreducible in $K[T][Y] = K[Y][T]$, therefore also in ....)

5.9 Let E be a subfield of the field of rational functions $K(X)$ in one variable over K, with $E \neq K$.

(i) Why is $K(X) : E < \infty$? Consider the minimal polynomial of X over E, say $Y^n + t_{n-1}(X)Y^{n-1} + \cdots + t_0(X)$. Show that, up to multiplication by an element of $K[X]$, this polynomial coincides with a polynomial

$$F(X, Y) = c_n(X)Y^n + c_{n-1}(X)Y^{n-1} + \cdots + c_0(X) \in K[X][Y]$$

that is primitive over $K[X]$.

(ii) Let t be an element *of* E such that $t \notin K$ and $t = f/g$ with $f, g \in K[X]$ relatively prime. Show that $F(X, Y)$ divides $g(X)f(Y) - f(X)g(Y)$ in $K[X][Y]$. Then, using §5.8, deduce that $\deg_X F \leq K(X) : K(t)$.

(iii) In the same situation as (ii), assume that $\deg f, \deg g \leq \deg_X F$ and deduce that $g(X)f(Y) - f(X)g(Y) = a F(X, Y)$ with $a \in K^\times$; and hence that $K(t) = E$.

(iv) Using (iii), show that E/K is *purely transcendental*, that is, E is itself a field of rational functions in one variable over K (*Lüroth's Theorem*). *Hint:* At least one $t_i(X)$ does not lie in K, and therefore satisfies $E = K(t_i)$.

5.10 Let R be an integral domain and $f = aX^n + \cdots + a_1 X + a_0$ a *primitive* polynomial in $R[X]$. Suppose there is a prime element π of R and polynomials $\varphi, \psi \in R[X]$ such that

$$f(X) = a\varphi(X)^m + \pi\psi(X), \quad \text{with } m \in \mathbb{N}.$$

Let $x \mapsto \bar{x}$ be the quotient map from R onto R/π. Prove that, if $\bar{\varphi}$ is a prime polynomial in $\bar{R}[X]$ and $\bar{\psi} \not\equiv 0 \bmod \bar{\varphi}$, then f irreducible in $R[X]$.

This generalization of Eisenstein's irreducibility criterion goes back to T. Schöne-mann, who taught at the Havel Gymnasium in Berlin. He published his result (in *Crelle's Journal*) in 1846, a couple of years before Eisenstein.

5.11 Let $f = X^n + a_{n-1} X^{n-1} + \cdots + a_0$ be a polynomial in $\mathbb{Z}[X]$ with $a_0 \neq 0$, and suppose $f(X) = (X - \alpha_1) \ldots (X - \alpha_n)$ over $\mathbb{C}$. Show that f is irreducible over $\mathbb{Z}$ if $\alpha_1, \ldots, \alpha_{n-1}$ have absolute value less than 1.

5.12 Prove:

(a) For every prime number p and every normalized $f \in \mathbb{Z}[X]$ whose image $\bar{f}$ is irreducible in $\mathbb{F}_p[X]$, the ideal (p, f) is a *maximal ideal* of $\mathbb{Z}[X]$. (*Hint:* Why is there a homomorphism from $\mathbb{F}_p[X]/\bar{f}$ onto $\mathbb{Z}[X]/(p, f)$?)

(b) If P is a prime ideal of $\mathbb{Z}[X]$ that is not of the form given in (a), P is a *principal ideal* of $\mathbb{Z}[X]$. If P is a principal ideal of $\mathbb{Z}[X]$, then P is not a maximal ideal. (*Hint:* If a nonzero g is in P, so is some prime factor of g.)

Chapter 6: Polynomial Splitting Fields

6.1 Show that $\mathbb{Q}(\sqrt[3]{2}, \sqrt[2]{-3})$ is a splitting field of $X^3 - 2$ over $\mathbb{Q}$.

6.2 Let E/K be an algebraic field extension and F an intermediate field of E/K. Prove:

(i) If E/K is normal, so is E/F.

(ii) If $F : K = 2$, then F/K is normal.

Give an example where E/F and F/K are normal, but E/K is not. (*Hint:* See §1.2 and keep Theorem 4 in mind; see also §6.3 below.)

6.3 Prove that $E' = \mathbb{Q}(\sqrt{2}, \sqrt{1+i})$ is a *normal closure* of the degree-4 extension $\mathbb{Q}(\sqrt{1+i})/\mathbb{Q}$. Thanks to §2.2, $E' : \mathbb{Q} = 8$.

6.4 Show that if E/K is an algebraic field extension with the property that every irreducible $f \in K[X]$ over E splits into linear factors, then E is an algebraic closure of K.

6.5 (a) Let $L = K(\alpha_1, \ldots, \alpha_n)$ be a finite-degree extension over K. Show that L/K is normal if and only if L is a splitting field of $f_1 f_2 \ldots f_n$ over K, where $f_i = \text{MiPo}_K(\alpha_i)$.

(b) Let E/K and L/K be finite, normal field extensions. Show that there exists a K-homomorphism $\sigma : E \to L$ if and only if there exist $f, g \in K[X]$ satisfying the following conditions: g divides f, L is a splitting field of f over K and E is splitting field of g over K. (Consider F4.)

6.6 (a) Let E/K be an extension and take $f \in K[X]$. Prove the existence of the following isomorphism of E-algebras:

(1) $$E \otimes K[X]/f \simeq E[X]/f.$$

(b) Set $F = \mathbb{Q}(i)$. Show that the tensor product $F \otimes F$ of the $\mathbb{Q}$-algebra F with itself is not a field. (Consider that $F \simeq \mathbb{Q}[X]/(X^2 + 1)$.)

6.7 Let E_1/K, E_2/K be finite extensions and E an extension of both E_1 and E_2. Let $E_1 E_2 := E_1(E_2) = E_2(E_1)$ be the composite of E_1 and E_2 in E. Show that $E_1 E_2 \simeq E_1 \otimes E_2$ (as K-algebras) if and only if $E_1 E_2 : E_1 = E_2 : K$.

6.8 Let f, g be irreducible polynomials over the field K, both without multiple zeros (in an algebraic closure of K). Let $L = K(\alpha)$ and $E = K(\beta)$ be extensions of K with $f(\alpha) = 0$ and $g(\beta) = 0$. Show that if $f = f_1 \ldots f_r$ and $g = g_1 \ldots g_s$ are the prime factorizations of f over $K(\beta)$ and of g over $K(\alpha)$, we have $r = s$, and after reordering we also have $E[X]/f_i \simeq L[X]/g_i$ for all $1 \le i \le r$; in particular,

(2) $$[K(\beta):K] \deg f_i = [K(\alpha):K] \deg g_i.$$

(*Hint:* Use §6.6 together with §4.19 and §4.20.) This result is a significant strengthening of §3.2; it is due to Dedekind. Incidentally, it was also Dedekind who baptized what we know as fields (with the German word *Körper*, literally "body").

6.9 Let R be a commutative ring with unity, S a subset of R and $\mathfrak{a}$ an ideal such that $\mathfrak{a} \cap S = \varnothing$. Show that the set of ideals $\mathfrak{b}$ of R such that $\mathfrak{a} \subseteq \mathfrak{b}$ and $\mathfrak{b} \cap S = \varnothing$ has maximal elements. These elements are prime ideals when S is a *multiplicative* set (see §3.12). As an application, show that if $R = \mathfrak{C}(X)$ is the ring from §4.22, every maximal ideal $\mathfrak{m} = I_a$ of R contains a prime ideal $\mathfrak{p}$ distinct from $\mathfrak{m}$.

6.10 Let $A = R[X_1, \ldots, X_n]$ be the polynomial ring in n variables over an integral domain R. Denote by A_d the set of all *homogeneous polynomials of degree d* (that is, polynomials in A where only monomials $X_1^{\nu_1} \ldots X_n^{\nu_n}$ of degree $d = \nu_1 + \cdots + \nu_n$ appear). Prove that, as an R-module, A is the direct sum of the submodules A_d, $d \in \mathbb{N}_0$. We have $A_d A_e \subseteq A_{d+e}$. A polynomial $f \in A$ is homogeneous of degree d if and only if the equation $f(t X_1, \ldots, t X_n) = t^d f(X_1, \ldots, X_n)$ holds in the polynomial ring $A[t]$.

6.11 In order to prove Theorem 1, which is fundamental in our context, we had to introduce in Section 6.3 the notion of the *tensor product of K-algebras*. The notion, familiar from linear algebra, of the *tensor product* of (finitely many) K-vector spaces is not subsumed under the definition given in Section 6.3, because for K-algebras we demand the existence of a unity element (and therefore a K-vector space cannot simply be regarded as a ring with trivial multiplication). In general, suppose K is a commutative ring with unity, and let V, W be K-modules. A tensor product of V, W is a K-module $V \otimes W$ together with a *bilinear* map

$$\pi : V \times W \to V \otimes W,$$
$$(x, y) \mapsto x \otimes y,$$

satisfying the following condition: Given any bilinear map $\beta : V \times W \to Z$, there exists *a unique linear map* $f : V \otimes W \to Z$ such that $f(x \otimes y) = \beta(x, y)$ for all $x \in V$, $y \in W$. Prove:

(i) Let $(V \otimes_1 W, \pi_1)$ and $(V \otimes_2 W, \pi_2)$ be tensor products of V, W. There exists a unique isomorphism $\lambda : V \otimes_1 W \to V \otimes_2 W$ such that $\lambda(x \otimes_1 y) = x \otimes_2 y$.

(ii) Every element of $V \otimes W$ is a finite sum of elements of the form $x \otimes y$. The following relations hold:

$$(x + x') \otimes y = x \otimes y + x' \otimes y,$$
$$x \otimes (y + y') = x \otimes y + x \otimes y',$$
$$\alpha x \otimes y = x \otimes \alpha y = \alpha(x \otimes y) \quad \text{for } \alpha \in K.$$

(iii) Given K-homomorphisms $f : V \to V'$ and $g : W \to W'$, there exists a unique K-homomorphism $h : V \otimes W \to V' \otimes W'$ such that $h(x \otimes y) = fx \otimes gy$. This homomorphism is written $h = f \otimes g$.

(iv) There is a canonical isomorphism $f : K \otimes V \to V$ such that $f(\alpha \otimes x) = \alpha x$.

(v) For any family $(V_i)_{i \in I}$ of K-modules, there exists a canonical isomorphism $\left(\bigoplus_{i \in I} V_i \right) \otimes W \to \bigoplus_{i \in I} (V_i \otimes W)$ such that $\left(\sum_i x_i \right) \otimes y \mapsto \sum_i (x_i \otimes y)$.

(vi) There is a canonical isomorphism $V \otimes W \to W \otimes V$ such that $x \otimes y \mapsto y \otimes x$ and a canonical isomorphism $(V \otimes W) \otimes Z \to V \otimes (W \otimes Z)$ such that $(x \otimes y) \otimes z \mapsto x \otimes (y \otimes z)$.

6.12 Prove the *existence* of a tensor product $(V \otimes W, \pi)$ for arbitrary K-modules. *Hint:* As in Section 6.3, start with the free K-module $F = KM$ generated by the set $M = V \times W$; then form the K-submodule U generated by all elements of one of the forms

$$(x + x', y) - (x, y) - (x', y), \quad (x, y + y') - (x, y) - (x, y'), \quad (\alpha x, y) - (x, \alpha y).$$

Finally, consider the quotient module F/U.

6.13 Let W be a free K-module, with basis $(e_j)_{j \in J}$. Deduce from §6.11(v,vi) that, if V is any K-module, every $t \in V \otimes W$ has a unique representation $t = \sum_j x_j \otimes e_j$ with $x_j \in V$. Thus, for every injective homomorphism $V' \to V$, the corresponding map $V' \otimes W \to V \otimes W$ is injective as well. If V, too, is free, with basis $(d_i)_{i \in I}$, the family $(d_i \otimes e_j)_{i,j}$ is a basis of $V \otimes W$. Specializing to K-vector spaces V, W of dimensions m, n, we see that $V \otimes W$ has dimension mn.

6.14 Let A, B be K-algebras. For the moment we will write the underlying vector spaces of A, B as A_0, B_0. Show that $A_0 \otimes B_0$ can be given one and only one K-algebra structure in such a way that $(a \otimes b)(a' \otimes b') = aa' \otimes bb'$, and this algebra is then the tensor product of the K-algebras A, B in the sense of Section 6.3 (see F7).

6.15 Consider the $\mathbb{Z}$-algebras $\mathbb{Q}$ and $\mathbb{Z}/n$ $(:= \mathbb{Z}/n\mathbb{Z})$. Show that $\mathbb{Q} \otimes \mathbb{Z}/n = 0$, and that $\mathbb{Q} \otimes \mathbb{Q} \simeq \mathbb{Q}$ and $\mathbb{Z}/n \otimes \mathbb{Z}/n \simeq \mathbb{Z}/n$ (with canonical isomorphisms).

Chapter 7: Separable Extensions

7.1 Let $f \in K[X]$ be irreducible and let L/K be a *normal* field extension. Show that if g, h are normalized prime factors of f in $L[X]$, there exists σ in $G(L/K)$ such that $\sigma g = h$.

Deduce that in the prime factorization of f in $L[X]$ all prime factors have the same exponent.

7.2 Let E/K be a (not necessarily algebraic) field extension such that char $K = p > 0$, and take $\alpha \in E$. Prove:

(a) If $\alpha^p \in K$ but $\alpha \notin K$, then $K(\alpha)/K$ is purely inseparable of degree p. (*Hint:* F16, F15.)

(b) $K(\alpha^p) = K(\alpha)$ if and only if α is algebraic and separable over K. (*Hint:* Consider F12.)

7.3 Let $f \in K[X_1, \ldots, X_n]$ be a polynomial in n variables over the field K, and let M be a subset of K^n on which f vanishes. Which of the following conditions imply $f = 0$?

(i) M is infinite;

(ii) $M = K^n$;

(iii) M contains a set of the form $A_1 \times \cdots \times A_n$, where each A_i is an infinite subset of K.

Answer: In general only (iii). And the converse is false; if, for example, $K = \mathbb{R}$, $n = 2$ and f vanishes on $M = \{(m, k + \frac{1}{m}) \mid m, k \in \mathbb{N}\}$, then $f = 0$, but M does not satisfy (iii).

7.4 Let F/K be an extension and suppose K is infinite. Let $f_1, \ldots, f_k$ be polynomial in $F[X_1, \ldots, X_n]$ and set $V_i = \{x \in K^n \mid f_i(x) = 0\}$ for each i. Show that if $K^n = V_1 \cup \cdots \cup V_k$, at least one V_i coincides with K^n (that is, $f_i = 0$).

7.5 Suppose a field extension E/K satisfies $E = K(\alpha_1, \ldots, \alpha_n)$, where the α_i are algebraic over K. Suppose $\alpha_2, \ldots, \alpha_n$ *separable* over K. Show that if K has infinitely many elements, E contains an element α of the form

(1) $$\alpha = x_1\alpha_1 + \cdots + x_n\alpha_n \quad \text{with } x_i \in K,$$

for which $E = K(\alpha)$. This is the *primitive element theorem*, which is due to Abel.
    Hint: Using §7.4, show that there exists $(x_1, \ldots, x_n) \in K^n$ such that $x_1 \neq 0$ and

(2) $$\sum_{j=1}^{n} x_j \sigma_i(\alpha_j) \neq \sum_{j=1}^{n} x_j \alpha_j \quad \text{for } i = 2, \ldots, r,$$

where $\sigma_1 = 1, \sigma_2, \ldots, \sigma_r$ are the $r := [E : K]_s = E_s : K$ distinct K-homomorphisms from E into a normal closure F of E/K. Now, if (2) is satisfied and α is as in (1), it follows that $E_s = K(\alpha)_s \subseteq K(\alpha)$, and therefore $\alpha_2, \ldots, \alpha_n$ lie in $K(\alpha)$. Since $x_1 \neq 0$, we then have $\alpha_1 \in K(\alpha)$ as well. (A somewhat weaker version of the primitive element theorem will come up in the context of Chapter 8.)

7.6 Let $E = K(\alpha_1, \ldots, \alpha_n)$ be as in §7.5, but assume that $\alpha_2, \ldots \alpha_n$ are only separable *over* $K(\alpha_1)$. Prove that there exists $\alpha \in E$ with $E = K(\alpha)$.

Hint: Let F be the separable closure of K in E. Then $E = F(\alpha_1)$. Now §7.5 yields the assertion. (For the case of finite fields see Chapter 9, Theorem 2.)

7.7 Suppose an extension E/K of degree 2^n is of the form $E = K(\sqrt{a_1}, \ldots, \sqrt{a_n})$, with $a_i \in K$. Prove that, if char $K \neq 2$, then $\alpha := \sqrt{a_1} + \cdots + \sqrt{a_n}$ is a primitive element of E/K. (*Hint:* Otherwise there would exist $\sigma \in G(E/K(\alpha))$ with $\sigma \neq \text{id}$.)

7.8 Find a *finite* extension E/K that has infinitely many intermediate fields. (*Hint:* Consider the field of rational functions $E = \mathbb{F}_p(X, Y)$ in two variables, the subfield $K = \mathbb{F}_p(X^p, Y^p)$, and the intermediate fields $K(X + tY)$ for $t \in K$.)

7.9 Let E/K be a *finite* field extension. Show that if E is perfect, so is K. (Compare F18.) *Hint:* Why is $E^p : K^p = E : K$?

7.10 Let E/K be a finite field extension. Then $[E : K] = [E : K]_s [E : K]_i$; see F17 and the remark following it. Deduce that for any intermediate field F of E/K,

$$[E : K]_i = [E : F]_i [F : K]_i.$$

7.11 Let L/K be an algebraic extension and $h \in L[X]$ a normalized polynomial. Prove that a necessary and sufficient condition for all the coefficients of h to be separable over K is that for every root α of h the multiplicity $\text{ord}_\alpha h$ be divisible by $[K(\alpha) : K]_i$.

Hint: Let α be a root of h and f the normalized prime factor of h such that $f(\alpha) = 0$. Then

$$f(X) = g(X^{p^m}) = \prod_{j=1}^{r} (X^{p^m} - \alpha_j^{p^m}) = \prod_{j=1}^{r} (X - \alpha_j)^{p^m},$$

where $p^m = [L(\alpha) : L]_i$ and the $\alpha_1, \ldots, \alpha_r$ are all distinct.

Chapter 8: Galois Extensions

8.1 (a) Solve §1.1 again, this time using Galois theory. (*Hint:* Consider $\sqrt{a}/\sqrt{b}$ in $K(\sqrt{a}, \sqrt{b})$.)

(b) Take $E = \mathbb{Q}(\sqrt[2]{2}, \sqrt[3]{3})$. Using Galois theory, show that $E = \mathbb{Q}(\sqrt[2]{2} + \sqrt[3]{3})$. (*Hint:* Consider $E \subseteq \mathbb{R}$.)

8.2 For $E = \mathbb{Q}(\sqrt{2}, \sqrt{5})$, show:

(i) $E/\mathbb{Q}$ is a Galois extension.

(ii) $G(E/\mathbb{Q}) \simeq \mathbb{Z}/2 \times \mathbb{Z}/2$. List all the intermediate fields of $E/\mathbb{Q}$.

8.3 Suppose that the Galois group of an irreducible and separable polynomial $f \in K[X]$ is *abelian*. Let E be a splitting field of f over K and let $\alpha_1, \ldots, \alpha_n$ be the roots of f in E. Show that $E = K(\alpha_i)$ for any i, and hence that $E : K = \deg f$.

8.4 Let $E \subset \mathbb{C}$ be a field. Show that if $E/\mathbb{Q}$ is a *Galois extension* and $E_0 = E \cap \mathbb{R}$, we must have $E : E_0 \leq 2$. Is $E_0/\mathbb{Q}$ always Galois if $E/\mathbb{Q}$ is? Prove: If E/K is a Galois extension, $E_0/\mathbb{Q}$ is Galois if and only if the generator ρ of $G(E/E_0)$ commutes with all elements of $G(E/\mathbb{Q})$.

8.5 Let K be a subfield of $\mathbb{R}$ and let E/K be a Galois extension of degree $4n$ whose Galois group is *cyclic*. Using F6 in Chapter 9, show that no element $d < 0$ of K is a square in E. (*Hint:* Consider §8.4.)

8.6 Let $a_1, a_2, \ldots, a_n$ be pairwise relatively prime square-free integers of absolute value $\neq 1$. Show that $\mathbb{Q}(\sqrt{a_1}, \ldots, \sqrt{a_n})/\mathbb{Q}$ is a Galois extension, with Galois group isomorphic to $\{+1, -1\}^n \simeq (\mathbb{Z}/2)^n$ (compare §2.6). For each $0 \leq k \leq n$, the number of subfields of degree 2^k is equal to the number of subfields of degree 2^{n-k}; find this number.

8.7 Let E/K be a *normal* field extension, and let E_s be the separable closure of K in E. Show that there exists a (canonical) intermediate field F of E/K such that $F \cap E_s = K$ and $FE_s = E$; in particular, F/K is *purely inseparable* and E/F is separable. If E/K is finite, $F : K$ is the inseparable degree of E/K. (*Hint:* Consider the fixed field of the automorphism group $G(E/K)$ of E/K.)

8.8 Let L/K be an algebraic extension with the property that every irreducible $f \in K[X]$ in L has at least one root (compare §6.4). Show that L is an algebraic closure of K. (*Hint:* Work in a fixed algebraic closure C of L, apply the primitive element theorem and use §8.7.)

8.9 Let k be a field of characteristic $p > 0$ and let $K = k(Y, Z)$ be the field of rational functions in two variables over k. Also let α be a root of $X^{2p} + YX^p + Z \in K[X]$ and consider $E = K(\alpha)$. Prove:

(i) E/K is an inseparable but not purely inseparable extension of degree $2p$.

(ii) E/K has no proper intermediate field inseparable over K (see §8.7). *Hint:* You can use §7.1.

8.10 Let E/K be a *finite* field extension and G a group of K-automorphisms of E. Show that G is finite and its order divides $E : K$. Moreover, $|G| = E : K$ if and only if E/K is Galois and $G(E/K) = G$.

8.11 Let α be a complex number satisfying $\alpha^6 + 3 = 0$. Show that $\mathbb{Q}(\alpha)/\mathbb{Q}$ is a Galois extension, and determine its Galois group and all its intermediate fields.

8.12 Take $f_1 = X^4 - 2$ and $f_2 = X^4 - 2X^2 + 2$ and let E_1, E_2 be splitting fields of f_1, f_2, respectively, over $\mathbb{Q}$. Prove:

(i) $E_1 = \mathbb{Q}(i, \sqrt[4]{2})$, $E_2 = \mathbb{Q}(\sqrt{-2}, \sqrt{1+i})$, and both have degree 8 over $\mathbb{Q}$.

(ii) E_1 and E_2 are not isomorphic, but $E_1/\mathbb{Q}$ and $E_2/\mathbb{Q}$ have isomorphic Galois groups.

Find all intermediate fields of $E_1/\mathbb{Q}$ and $E_2/\mathbb{Q}$.

8.13 Determine the Galois group G of the splitting field E of $X^3 + X + 1$ over $\mathbb{Q}$. How many subfields does E have, and what are their degrees over $\mathbb{Q}$? (*Hint:* Exactly one of the roots $\alpha_1, \alpha_2, \alpha_3$ is real. It follows that $G \simeq S_3$. Thus there are exactly three subfields of degree 3, namely $\mathbb{Q}(\alpha_1), \mathbb{Q}(\alpha_2), \mathbb{Q}(\alpha_3)$, and exactly one of degree 2. By the way, the degree-2 field is $\mathbb{Q}(\sqrt{-31})$; see Section 15.5.)

8.14 Let E be a splitting field of $X^5 - 2$ over $\mathbb{Q}$. Prove:

(i) $E : \mathbb{Q} = 20$.

(ii) There exists precisely one intermediate field F of $E/\mathbb{Q}$ with $E : F = 5$, and $F/\mathbb{Q}$ is normal.

(iii) $\sqrt[5]{2} + \zeta_5$ is a primitive element of $E/\mathbb{Q}$.

(iv) The Galois group G of $E/\mathbb{Q}$ has elements σ, τ such that ord $\sigma = 5$, ord $\tau = 4$, and $\tau \sigma \tau^{-1} = \sigma^2$.

List all intermediate fields of $E/\mathbb{Q}$. (Note: Without knowledge of group theory some labor is involved in proving the uniqueness part of (ii), or at any rate in determining the intermediate fields; but see the results in Chapter 10, particularly Theorem 1 (Sylow's Theorem), which turns the problem into a piece of cake.)

8.15 Let $E = k(X)$ be the field of rational functions in one variable over some field k. Any *automorphism* σ of E/k is characterized by the image $\sigma(X)$ of X. Why does this (plus §5.8) immediately imply that $\sigma(X)$ has the form

$$\sigma X = \frac{aX + b}{cX + d} \quad \text{with } ad - bc \neq 0?$$

The automorphism group of $k(X)/k$ is thus canonically isomorphic to the *projective linear group* $\mathrm{PGL}(2, k) = \mathrm{GL}(2, k)/k^\times$.

8.16 Let $E = k(X)$ be as in §8.15. We make the identification $\mathrm{Aut}(k(X)/k) = \mathrm{PGL}(2, k)$. Prove that *finite* subgroups G of $\mathrm{PGL}(2, k)$ are in one-to-one correspondence with intermediate fields K such that E/K is *Galois*, and in fact the correspondence is given by $G \mapsto E^G$ and $K \mapsto G(E/K)$.

Prove that if G is a *finite* subgroup of $\mathrm{PGL}(2, k)$, there exists a rational function $f \in k(X)$ such that G contains precisely those $\sigma \in \mathrm{PGL}(2, k)$ that leave f invariant:

(1) $$G = \{\sigma \in \mathrm{PGL}(2, k) \mid \sigma f = f\}.$$

Hint: Lüroth's Theorem (§5.9).

Here is an example: Take $\sigma_1(X) = X$, $\sigma_2(X) = X^{-1}$, $\sigma_3(X) = 1 - X$, $\sigma_4(X) = (1-X)^{-1}$, $\sigma_5(X) = (X-1)X^{-1}$, $\sigma_6(X) = X(X-1)^{-1}$. These elements of $\mathrm{PGL}(2, k)$

form a group G isomorphic to S_3. Find an f satisfying (1). (*Hint:* One of the coefficients of the minimal polynomial of X over the fixed field of G is not constant; see §5.9.)

8.17 In the field of rational functions $E = k(X)$, consider the element

$$J(X) = \frac{(X^2 - X + 1)^3}{X^2(X-1)^2}.$$

Show that the extension $k(X)/k(J)$ is *Galois* of degree 6, and determine its Galois group and all its intermediate fields.

Watch out: The group defined by (1) is *finite* for any nonconstant $f \in k(X)$, but $k(X)/k(f)$ is generally not normal.

8.18 Set $G = \mathrm{Aut}(k(X)/k)$. Prove:

(a) If k is an *infinite* field, k is the fixed field of G.

(b) In the case of a *finite* field k with q elements, however, the fixed field K of G is not the same as k; the extension $k(X)/K$ is Galois of order $q^3 - q$. Hence there exists a nonconstant $f \in k(X)$ with denominator of degree $q^3 - q$ and such that $K = k(f)$. Using results from Section 9.1 one can show that

$$f = \frac{(X^q - X)^{q^2+1}}{(X^{q^2} - X)^{q+1}}$$

satisfies this property.

Hint: Note that G is generated by the elements aX, $X + b$ and $1/X$, with $a \in k^\times$ and $b \in k$.

8.19 Let K be a subfield of $\mathbb{R}$ and assume $f \in K[X]$ is separable and normalized. Show that the discriminant $D(f)$ has sign $(-1)^{r_2}$, where r_2 is half the number of nonreal roots of f in $\mathbb{C}$.

8.20 Prove that the only homomorphism of the field $\mathbb{R}$ into itself is the identity.

8.21 Let E/K be a finite Galois extension with group G, and let V be an n-dimensional vector space over E on which G operates *semilinearly*, meaning that $\sigma(x + y) = \sigma(x) + \sigma(y)$ and $\sigma(\lambda x) = \sigma(\lambda)\sigma(x)$ for any $x, y \in V$, $\lambda \in E$ and $\sigma \in G$. Clearly the set V^G of elements fixed by G is a K-vector space. Prove that there exists a K-basis of V^G that is also an E-basis of V.

(*Hint:* Show first that any $v_1, \dots, v_m \in V^G$ that are linearly independent over K are also linearly independent over E. Then show that any linear functional on the E-vector space V that vanishes on V^G is trivial.)

8.22 Let F be an intermediate field of the Galois extension L/K. Set $G = G(L/K)$ and $H = G(L/F)$. Why is it that $[F:K] = G:H$, if L/K is assumed finite? Prove, more generally, that if $[F:K]$ is finite, so is $G:H$, and then $[F:K] = G:H$. (*Hint:* There is a well defined from G/H into the set $G(F/K, L/K)$ of all K-homomorphisms of F in L.)

Prove that $G:H < \infty$ implies $[F:K] < \infty$.

8.23 Let $\alpha \in \mathbb{C}$ be an algebraic number and $K \subset \mathbb{C}$ a field. Why is $K(\alpha):K$ no greater than $\mathbb{Q}(\alpha):\mathbb{Q}$? Prove that if $\mathbb{Q}(\alpha)/\mathbb{Q}$ is *normal*, $K(\alpha):K$ divides $\mathbb{Q}(\alpha):\mathbb{Q}$. (*Hint:* Why is $K(\alpha)/K$ Galois and why does every $\sigma \in G(K(\alpha)/K)$ give rise to a $\sigma_0 \in G(\mathbb{Q}(\alpha)/\mathbb{Q})$? Considerations of this sort lead to the so-called *Translation Theorem* of Galois theory; see Chapter 12, Theorem 1.)

Chapter 9: Finite Fields, Cyclic groups and Roots of Unity

9.1 Keeping the notation of Theorem 1, prove that $\mathbb{F}_{p^m}$ is a subfield of $\mathbb{F}_{p^n}$ if and only if n is divisible by m.

9.2 Let K be a finite field with q elements and let $f \in K[X]$ be irreducible. Prove that f divides $X^{q^n} - X$ in $K[X]$ if and only if n is divisible by $\deg f$.

9.3 Let $p \neq 2$ be prime.

(a) Using Theorem 2, prove again (compare §3.9) that -1 is a square in $\mathbb{F}_p$ if and only if $p \equiv 1 \bmod 4$.

(b) More generally, $a \in \mathbb{F}_p^\times$ is a square if and only if $a^{(p-1)/2} = 1$.

(c) Characterize finite fields $\mathbb{F}_{p^n}$ where -1 is a square.

9.4 Take $G = (\mathbb{Z}/23)^\times$.

(a) Find all elements of order 11 in G. (*Hint:* ord $\bar{2} = 11$.)

(b) Find all generators of G. (*Hint:* ord $\bar{5} = 22$.)

(c) Show that the splitting field of $X^{23} - 1$ over $\mathbb{F}_2$ has degree 11 (see F11).

9.5 In his memoir "Sur la théorie of the nombres", Galois (1811–1832) was the first to consider finite fields that are proper extensions of their prime fields. Actually he says nothing about their existence. He simply performs computations in them, calling the quantities that he manipulates "les imaginaires". Although Galois himself does not offer any specifically number-theoretical applications of his investigations, he is visibly convinced of their usefulness.

In his memoir Galois discusses the following example: The polynomial $X^3 - 2$ is irreducible over $\mathbb{F}_7$; thus, if we fix one of its roots, calling it $\alpha = \sqrt[3]{2}$, we have found a *primitive element* of $\mathbb{F}_{7^3}/\mathbb{F}_7$, since $\mathbb{F}_{7^3} = \mathbb{F}_7(\alpha)$. Now try to find a *primitive root* $\zeta = a\alpha^2 + b\alpha + c$ of $\mathbb{F}_{7^3}$, that is, an element ζ such that ord $\zeta = 7^3 - 1 = 342 = 19 \cdot 3^2 \cdot 2$. Clearly α itself is only a primitive 9th root of unity. Next try $\zeta = \alpha + 1$. Galois chooses $\alpha^2 + \alpha$ as a primitive root and states that $(\alpha + 1)^{10} = -1$. Has he made a mistake?

9.6 Determine the prime factorization of $f(X) = X^5 - X^4 - 6X^3 + 6X^2 - 3X + 3$ as a polynomial over $\mathbb{Q}$, $\mathbb{F}_3$, and $\mathbb{F}_5$ (one at a time). Show that the respective Galois groups all have distinct orders.

9.7 For any $k \in \mathbb{N}$ let ζ_k denote a primitive k-th root of unity in $\mathbb{C}$. Show that, for any natural numbers m, n with d as their gcd and v as their lcm, we have:

(i) $\varphi(mn)\varphi(d) = \varphi(m)\varphi(n)d$.

(ii) $\varphi(m)\varphi(n) = \varphi(v)\varphi(d)$.

(iii) $\mathbb{Q}(\zeta_m, \zeta_n) = \mathbb{Q}(\zeta_v)$.

(iv) $\mathbb{Q}(\zeta_m) \cap \mathbb{Q}(\zeta_n) = \mathbb{Q}(\zeta_d)$.

(*Hint:* See Theorem 3 and use §2.1.)

9.8 Prove that cyclotomic polynomials F_n for $n > 1$ have the following properties:

(i) $F_{2m}(X) = F_m(-X)$ if m is odd.

(ii) $F_m(X^p) = F_m(X)F_{mp}(X)$ for all primes p not dividing m.

(iii) $F_n(X) = F_m(X^{n/m})$ if m is the product of all primes dividing n.

(iv) $F_n(0) = 1$ and hence $F_n(X^{-1})X^{\varphi(n)} = F_n(X)$.

(v) $F_n(1) = p$ if n is the power of a prime p.

(vi) $F_n(1) = 1$ if n is not a prime power.

The last two properties are particularly interesting in connection with arithmetic: In case (vi), $1-\zeta_n$ is a *unit* in the ring $R = \mathbb{Z}[\zeta_n]$, whereas in (v) the element $\pi = 1-\zeta_n$ is *prime* in R and p has in R the prime factorization $p = \varepsilon \pi^{\varphi(n)}$, with ε a unit (this despite the fact that $\mathbb{Z}[\zeta_p]$ is generally *not* a UFD; more precisely, $\mathbb{Z}[\zeta_p]$ for p prime is not a UFD exactly when $p \geq 23$).

9.9 Prove, with the greatest possible economy:

(a) $F_{15}(X) = X^8 - X^7 + X^5 - X^4 + X^3 - X + 1$.

(b) For primes $p \neq q$, all coefficients of $F_{pq}(X)$ have absolute value 1. *Hint:* Look at

$$F_{pq}(x) = (1-x)\left(F_q(x^p)(1-x^q)^{-1}\right)$$

as an identity between power series.

(c) The smallest n for which not all coefficients of F_n have absolute value 1 is $n = 105$. In fact,

$$F_{105}(X) = X^{48} + X^{47} + X^{46} - X^{43} - X^{42} - 2X^{41} - X^{40} + \cdots + 1.$$

Note: I. Schur showed that the coefficients of the F_n can be arbitrarily large.

9.10 Let p be a prime and suppose that for some $a \in \mathbb{Z}$ and $v \in \mathbb{N}$ we have $a \equiv 1 \bmod p^v$ but $a \not\equiv 1 \bmod p^{v+1}$. Show that, apart from the case $p = 2$ and $v = 1$, there follows

$$a^p \equiv 1 \bmod p^{v+1}, \quad a^p \not\equiv 1 \bmod p^{v+2}.$$

Thus, for $p \neq 2$ and any $n \in \mathbb{N}$, the residue class of $1 + p$ in $(\mathbb{Z}/p^n\mathbb{Z})^\times$ has order p^{n-1}; for $p = 2$ the equality $1 + 2^2 = 5$ determines an element of order 2^{n-2} in $(\mathbb{Z}/2\mathbb{Z})^\times$.

9.11 Prove:

(a) *If p is an odd prime and $n \in \mathbb{N}$, the group $(\mathbb{Z}/p^n\mathbb{Z})^\times$ is **cyclic** of order $\varphi(p^n) = (p-1)p^{n-1}$.* (*Hint:* Let a represent a primitive root in $(\mathbb{Z}/p\mathbb{Z})^\times$. We know that $a^{p-1} \equiv 1 \bmod p$. By taking p-th powers, a can (thanks to §9.10) be modified so that $a^{p-1} \equiv 1 \bmod p^n$. Then the residue class of $a(1+p)$ has order $(p-1)p^{n-1}$.

(b) *For $p = 2$ and $n \geq 3$ the group $(\mathbb{Z}/2^n\mathbb{Z})^\times = \langle -\bar{1} \rangle \times \langle \bar{5} \rangle$ is the direct product of the cyclic subgroups generated by the residue classes of -1 and 5 (of order 2 and 2^{n-2} respectively).* (See §10.8(b) for the notion of the direct product of groups.)

9.12 Consider the abelian group $G = (\mathbb{Z}/m\mathbb{Z})^\times$, where $m > 1$. Describe the subgroup G_2 consisting of elements σ such that $\sigma^2 = 1$. Show, in particular, that if r is the number of odd primes in the prime factorization of m and e is the exponent of the factor 2, then G_2 has order 2^r, 2^{r+1}, or 2^{r+2}, depending on whether $e \leq 1$, $e = 2$, or $e \geq 3$, respectively. (In the case $m = 2^n$, with $n \geq 3$, write out the three elements of order 2.) For what values of m is G cyclic? (Answer: Only when $m = p^n$ or $m = 2p^n$, with p prime and $n \geq 1$, where moreover $n = 1$ if $p = 2$.)

9.13 What are the prime factorizations of $X^9 - X$ and $X^{27} - X$ in $\mathbb{F}_3[X]$?

9.14 Let K be the field with 729 elements. How many subfields does K possess? Prove that K has exactly 696 elements α with the property that $K = \mathbb{F}_3(\alpha)$, and that there are exactly 116 normalized prime polynomials of degree 6 over $\mathbb{F}_3$. Formulate more-general statements of this sort.

9.15 Let K be a finite field. Show that every element of K is a sum of two squares in K. More generally, if $a, b \in K^\times$, any $c \in K$ can be represented as $c = ax^2 + by^2$ in K. *Hint:* How many elements are in the sets $\{ax^2 \mid x \in K\}$ and $\{c - by^2 \mid y \in K\}$?

9.16 Prove that any root of unity contained in $\mathbb{Q}(\zeta_n)$ has the form $\pm\zeta_n^k$.

9.17 Take $n \in \mathbb{N}$ and a prime p not dividing n. Show that, for $x \in \mathbb{Z}$, we have $F_n(x) \equiv 0 \bmod p$ if and only if $\bar{x} \in \mathbb{F}_p^\times$ has order n. (*Hint:* $X^n - \bar{1}$ is a separable polynomial in $\mathbb{F}_p[X]$.)

9.18 Take $n \in \mathbb{N}$. Show that there are infinitely many primes p such that $p \equiv 1 \bmod n$. (More generally, the famous *Dirichlet Theorem* says that for any integer a relatively prime to n there are infinitely many primes $p \equiv a \bmod n$.) *Hint:* Let $P(F_n)$ be the set of all primes p such that the equation $F_n(X) \equiv 0 \bmod p$ has a solution in $\mathbb{Z}$. From §9.17 it follows, for every prime $p \nmid n$, that

$$p \in P(F_n) \iff p \equiv 1 \bmod n.$$

Now the assertion follows from the general result given in the next exercise.

9.19 Let $f(X) = a_n X^n + \cdots + a_1 X + a_0$ be a nonconstant polynomial in $\mathbb{Z}[X]$. Show that there are infinitely many primes p such that the congruence $f(X) \equiv 0 \bmod p$ has a solution in $\mathbb{Z}$. (*Hint:* If $a_0 = 1$, the result follows using the same sort of argument as in Euclid's proof that there are infinitely many primes. The general case can then be reduced to the case $a_0 = 1$.)

Note: If $P(f)$ denotes the set of primes p in question, it can actually be proved that

$$\sum_{p \in P(f)} \frac{1}{p} = \infty.$$

Unfortunately the author is not aware of an elementary proof of this fact, in spite of some stabs at the problem.

9.20 Why is $\sqrt[4]{5}$ not in $\mathbb{Q}(\zeta_{25})$? (*Hint:* $\mathbb{Q}(\sqrt[4]{5})/\mathbb{Q}$ is not normal.) Why is $\sqrt{15}$ not in $\mathbb{Q}(\zeta_{15})$? (*Hint:* $\sqrt{-3} \in \mathbb{Q}(\zeta_3)$, $\sqrt{5} \in \mathbb{Q}(\zeta_5)$.) Prove more precisely that $\mathbb{Q}(\zeta_{15})/\mathbb{Q}$ has exactly 6 proper intermediate fields, namely $\mathbb{Q}(\sqrt{-3})$, $\mathbb{Q}(\sqrt{5})$, $\mathbb{Q}(\sqrt{-15})$, $\mathbb{Q}(\sqrt{-3}, \sqrt{5})$, $\mathbb{Q}(\zeta_5)$, and $\mathbb{Q}(\zeta_{15} + \zeta_{15}^{-1})$.

9.21 Show that $\mathbb{Q}(\zeta_8)/\mathbb{Q}$ has exactly three intermediate fields, namely $\mathbb{Q}(\sqrt{2})$, $\mathbb{Q}(\sqrt{-1})$, and $\mathbb{Q}(\sqrt{-2})$.

9.22 Why does $\mathbb{Q}(\zeta_{41})/\mathbb{Q}$ have exactly 8 intermediate fields? Show, more generally, that if q is an odd prime and $p_1^{e_1} \ldots p_r^{e_r}$ is the prime factorization of $q-1$, then $\mathbb{Q}(\zeta_q)$ has exactly $(e_1+1) \ldots (e_r+1)$ subfields. (*Hint:* Theorem 3, Theorem 2, F6.)

9.23 Take $E_1 = K(\zeta_{16})$ and $E_2 = K(\zeta_{17})$, with $K = \mathbb{Q}$. Show that the degree of E_1/K is half as much as the degree of E_2/K, but E_1/K has more proper intermediate fields than E_2/K (in fact, twice as many).

9.24 Examine again the assertions of §3.8 in light of recently acquired knowledge and show how easily they can be proved now.

Chapter 10: Group Actions

10.1 Let G be a nontrivial group, i.e., a group having more than one element. Prove the equivalence between:

(i) G is finite of prime order.

(ii) G has no subgroups apart from itself and 1.

(iii) G is cyclic of prime order.

10.2 Let G be a nontrivial group. Prove the equivalence between:

(i) There is a proper subgroup of G containing all others.

(ii) G is a finite cyclic group and its order is a prime power.

(iii) G is finite and the set of subgroups of G is totally ordered.

10.3 Let k be a field with algebraic closure C. Let α be a element of $C \setminus k$. Prove that among the intermediate fields of C/k not containing α, there is a maximal one. Let K be such a maximal field and E/K a finite extension (inside C/k). Prove that if E/K is separable, E/K is Galois with a cyclic group of order equal to a prime power. (*Hint:* §10.2.) If E/K is not separable and $p = \text{char } K$, then E/K is purely inseparable, and the extension K/K^p has degree p. (*Hint:* See §8.7.)

10.4 Let G be a finite group and H a subgroup of G. Prove that if $G : H = 2$, then H is *normal* in G and lies in the center of G.

More generally, let p be the *smallest* prime dividing the order of G. Prove:

(a) If H is a normal subgroup of order p, then $H \subseteq ZG$.

(b) If H has index p, then H is a *normal subgroup* of G.

(*Hint:* (a) Inner automorphisms of G give rise to elements of the group Aut $H \simeq (\mathbb{Z}/p\mathbb{Z})^\times$. (b) Let M be the set of subgroups of G conjugate to H. By letting G act on M by conjugacy, one gets a homomorphism of G into the group $S(M)$ of permutations of M; look at its kernel.)

10.5 Let G be a finite p-group and H a proper subgroup of G. Show that H is a *proper* subgroup of its normalizer $N_G H$ in G. Consequences: If H is a *maximal* subgroup of G (that is, maximal among proper subgroups of G), then H is a *normal subgroup* of index p in G. Further: Every proper subgroup H of G lies in a normal subgroup N of index p in G. (*Hint:* $ZG \neq 1$. Work by induction on $|G|$. The case $ZG \nsubseteq H$ is trivial.)

10.6 Let G be a finite p-group and $N \neq 1$ a normal subgroup of G. Prove:

(a) $N \cap ZG \neq 1$.

(b) N contains a subgroup H normal *in G* and such that $N : H = p$.

(*Hint:* Use the orbit formula for the action of G on N by inner automorphisms. Part (b) follows from (a) by induction.)

10.7 Let N be a normal subgroup of a group G such that the quotient G/N is *cyclic*. Prove:

(a) If in addition N is contained in the *center* of G, then G is *abelian*. Consequence: Every group of order p^2 (for p prime) is *abelian*.

(b) If G is finite and G/N has order f, then G has a cyclic subgroup C of order f. If in addition f is relatively prime to the order of N, then $N \cap C = 1$ and $G = NC$.

10.8 (a) Let H be a subgroup of a group G. Denote by $p : G \to G/H$ the map defined by $\sigma \mapsto \sigma H$. Can a multiplication operation be defined on $\bar{G} := G/H$ in such a way that $p(\sigma) \cdot p(\tau) = p(\sigma\tau)$ for all $\sigma, \tau \in G$? If so, there is clearly only one way to do it; then G/H is a *group* with this operation, and p is a *group homomorphism*. Prove that the answer to the question is yes if and only if H is a *normal subgroup* of G.

(b) Let H_1, H_2 be subgroups of a group G and let $f : H_1 \times H_2 \to G$ be the map $(x_1, x_2) \mapsto x_1 x_2$. Show that f is a homomorphism if and only if $x_1 x_2 = x_2 x_1$ for all $x_1 \in H_1$, $x_2 \in H_2$. In this situation we say that G is the *direct product of H_1 and H_2* if f is an isomorphism. Show that this happens if and only if the following conditions are satisfied:

$$G = H_1 H_2, \qquad H_i \trianglelefteq G \text{ for } i = 1, 2, \qquad H_1 \cap H_2 = 1.$$

(*Hint:* Two elements a, b of a group commute if and only if their *commutator* $[a, b] := aba^{-1}b^{-1}$ equals 1.)

(c) Let $H_1, \ldots, H_n$ be subgroups of a group G, and let H be their intersection. Show that if every H_i has finite index in G, so does H. (*Hint:* The maps $p_i : G \to G/H_i$ defined by $p_i(\sigma) = \sigma H_i$ give rise to a well defined map $G/H \to G/H_1 \times \cdots \times G/H_n$.)

10.9 Let p be a prime number. Prove:

(a) Any group of order $p^a m$ with $m \leq p$ has a normal subgroup of order p^a. (*Hint:* The case $m = p$ is taken care of by F8. For $m < p$ see Sylow's Theorems.)

(b) If G is a group of order pq, where $q < p$ is also prime and does not divide $p - 1$, then G is cyclic.

(c) If q is a prime distinct form p, every group of order $p^2 q$ has a normal Sylow p-subgroup or a normal Sylow q-subgroup.

(d) Every group of order $2^n \cdot 3$ possesses a normal subgroup N with $G : N = 3$ or $G : N = 2$. (*Hint:* The action of G on $M = \mathrm{Syl}_2 G$ via inner automorphisms yields a homomorphism $G \to S(M)$.)

(e) Every group of order 45 is abelian.

(f) If q and r are primes with $p < q < r$ and G is a group of order pqr, then one of the nontrivial Sylow subgroups of G is normal in G. (*Hint:* Otherwise there would be too many elements of order p, q and r all combined.)

10.10 A subgroup H of a group G is called *characteristic* in G, and we write $H \blacktriangleleft G$, if $\varphi(H) = H$ for all automorphisms φ of G. A group $G \neq 1$ is called *simple* if it has no nontrivial proper subgroup H with $H \blacktriangleleft G$, and it is called *characteristic-simple* if it has no nontrivial proper subgroup H with $H \trianglelefteq G$. Prove:

(a) The condition $H \blacktriangleleft N \trianglelefteq G$ implies $H \trianglelefteq G$. In particular, any subgroup H of a *cyclic normal subgroup* N of G is normal in G.

(b) If G is a characteristic-simple finite group and N is a *minimal normal subgroup* of G (that is, minimal among nontrivial normal subgroups of G), then $G \simeq N \times \cdots \times N$ and N is simple. (*Hint:* $G = \langle \varphi(N) \mid \varphi \in \mathrm{Aut}\, G \rangle = \varphi_1(N) \ldots \varphi_r(N)$, and for r minimal the product is direct.)

(c) If $G \simeq N \times \cdots \times N$ and N is a *simple* group, G is characteristic-simple.

10.11 Among groups G of order $n < 60$ there are no simple groups apart from cyclic groups of prime order.

(*Hint:* By §10.9 the only cases that remain doubtful are $n = 36$, $n = 40$ and $n = 56$. For $n = 40$ a normal Sylow 5-group is available, for $n = 36$ you can argue as in §10.9(d), and for $n = 56$ you can count elements of order 7 and those of order a power of 2.)

10.12 Let E be a splitting field of $X^7 - 6$ over $\mathbb{Q}$. Prove:

(i) $E : \mathbb{Q} = 6 \cdot 7$.

(ii) There is exactly one intermediate field F of $E/\mathbb{Q}$ such that $E : F = 7$, namely $F = \mathbb{Q}(\zeta_7)$.

(iii) $\sqrt[7]{6} + \zeta_7$ is a primitive element of $E/\mathbb{Q}$.

(iv) The Galois group G of $E/\mathbb{Q}$ contains elements σ, τ such that ord $\sigma = 7$, ord $\tau = 6$ and $\tau \sigma \tau^{-1} = \sigma^3$.

Let's get an overview of *all* the intermediate fields of $E/\mathbb{Q}$. If $\alpha_1, \ldots, \alpha_7$ are the roots of $X^7 - 6$, then besides $\mathbb{Q}$ and E one necessarily has the following *pairwise distinct* intermediate fields: $\mathbb{Q}(\sqrt{-7})$, $\mathbb{Q}(\zeta_7 + \bar{\zeta}_7)$, $\mathbb{Q}(\zeta_7)$, $\mathbb{Q}(\alpha_k)$, $\mathbb{Q}(\alpha_k, \sqrt{-7})$, and $\mathbb{Q}(\alpha_k, \zeta_7 + \bar{\zeta}_7)$, with $1 \le k \le 7$ in each case. Show that there are no others. (*Hint:* The number of Sylow 3-groups is either 1 or 7, and since the $\mathbb{Q}(\alpha_k, \sqrt{-7})$ are distinct it must be 7. Note further that every subgroup of order 6 must be contained in one of these seven Sylow 3-groups. Correspondingly, a subgroup of order 14 or 21 contains the only Sylow 7-group of G. How much room is left now for the Sylow 2-groups? The number of elements of order 2 can also be read off from the structure of G.)

10.13 Let p^k be a prime power and m any natural number. For every group G of order mp^k, consider the set M of all subsets of G having p^k elements. G acts on M by translation; for $X \in M$ let $G_X = \{\sigma \in G \mid \sigma X = X\}$ be the corresponding stabilizer. Prove:

(a) For every $X \in M$ we have $|G_X| \le |X| = p^k$, and equality holds if and only if X is a coset of a subgroup H of order p^k.

(b) Denote by $n_G = n_G(p^k)$ the number of subgroups of order p^k in G. Then the orbit formula for the action of G on M yields the congruence

$$\binom{mp^k}{p^k} \equiv n_G \cdot m \bmod mp.$$

This holds for all groups G of order mp^k. In particular, if one takes G cyclic, we get

$$\binom{mp^k}{p^k} \equiv m \bmod mp.$$

Thus one reaches the following theorem of Frobenius: *In any group of order $p^k m$ the number of subgroups of order p^k is congruent to 1 modulo p.* In particular, any finite group has Sylow p-subgroups, and their number is congruent to 1 modulo p.

This chain of reasoning goes back to Wielandt (1959) and Miller (1915), but in those papers there is no reference to Frobenius's result.

10.14 Let P be a Sylow p-subgroup of a finite group G. Show, without using Sylow's Second Theorem, that if P is normal in G it is the only Sylow p-subgroup of G (and conversely). More generally: If N is an arbitrary normal subgroup of G, the intersection $N \cap P$ is a Sylow p-subgroup of N (and incidentally PN/N is a Sylow p-subgroup of G/N).

10.15 Let N be a normal subgroup of a finite group G. Show that, if P is a Sylow p-subgroup *of* N, then $N_G(P)N = G$.

10.16 Let P be a Sylow p-subgroup of a finite group G. Show that if H is a subgroup of G containing $N_G P$, then $N_G H = H$.

10.17 Let G be a finite group. Prove the existence of a Sylow p-subgroup of G by induction, using the *class formula* F6. (*Hint:* Why can one start by assuming that $|ZG| \not\equiv 0 \mod p$?)

10.18 Let G be a p-group of order p^n. Deduce from §10.13 that, for any $k \leq n$, the number of *normal subgroups* of order p^k in G is congruent to 1 modulo p.

10.19 A finite group G is called *nilpotent* if all Sylow subgroups of G are normal. Show that for a finite group G there is equivalence between:

(i) G is nilpotent.

(ii) G is the direct product of its Sylow subgroups.

(iii) Any proper subgroup H of G is a proper subgroup of $N_G H$.

(iv) Any maximal subgroup of G is normal (of prime index).

(v) Any nontrivial quotient group of G has nontrivial center.

(vi) Any two elements of G whose orders are relatively prime commute.

10.20 Let a finite group G act on a finite set M. For any $\sigma \in G$, denote by $i(\sigma)$ the number of *fixed points* of σ. Show that the average value of $i(\sigma)$ coincides with the number s of orbits:

$$s = \frac{1}{|G|} \sum_{\sigma \in G} i(\sigma).$$

Hint: Look at the subset $\{(\sigma, x) \mid \sigma x = x\}$ of $G \times M$ and count its elements.

10.21 Prove that a finite p-group $G \neq 1$ that has only one subgroup H_0 of index p must be *cyclic*. (*Hint:* Every proper subgroup of G lies in a maximal subgroup H of G. Now §10.19 implies $H = H_0$. The statement then follows using §10.2.)

10.22 Prove that a finite group G is never the union of the conjugates $\tau H \tau^{-1}$ of a proper subgroup $H \leq G$.

Simple and easy to prove as this fact is, it nonetheless plays a certain role on many different occasions. One can also stress its connection with §10.20 and recast it as follows: If a finite group G acts *transitively* and nontrivially on a set M, there exists σ in G that *leaves no point of M fixed*.

10.23 Let G be a group and take $a \in G$. For $m \in \mathbb{N}$, set $G_m(a) = \{x \in G \mid x^m = a\}$ and $G_m = G_m(1)$. Is G_m always a subgroup of G? Prove that G_2 is a subgroup of G if and only if $xy = yx$ for all $x, y \in G_2$. If G is abelian, every G_m is a subgroup of G.

Now suppose G finite of order n. Prove:

(a) If G is *cyclic*, then G_m has order (m, n).

(b) If G is *abelian*, the order of G_m is divisible by (m, n), and more generally

(∗) $|G_m(a)| \equiv 0 \mod (m, n)$.

(*Hint* for (b): One can assume $m \mid n$ and $m > 1$. Choose a prime factor p of m and use induction on m.)

Does (∗) hold when G is not abelian? In this connection consider the case $G = S_3$ and $m = 3$.

But by a theorem of *Frobenius* (see B. Huppert, *Finite groups* I), $|G_m(a)|$ is always divisible by $(m, n/c)$, where c is the number of elements of the *conjugacy class* of a in G. Thus one gets a statement of general validity if one replaces in (∗) the order n of G by the order n/c of the centralizer $Z_G(a)$ of a in G.

Chapter 11: Applications of Galois Theory to Cyclotomic Fields

11.1 Let α be an algebraic number such that $\mathbb{Q}(\alpha):\mathbb{Q} = 4$. Prove that, if the Galois group of the minimal polynomials f of α over $\mathbb{Q}$ has order greater than 8, then $\alpha \notin \triangle \mathbb{Q}$. (One example is $f(X) = X^4 + X + 1$; see §5.7.)

11.2 What is the smallest angle whose measure in degrees is a natural number and that can be constructed with ruler and compass from $\mathbb{Q}$?

11.3 Let p be a prime and $\gamma = g \bmod p$ a generator of $G := G(\mathbb{Q}(\zeta_p)/\mathbb{Q}) = (\mathbb{Z}/p\mathbb{Z})^\times$. Show that for any $t \in \mathcal{N}$ dividing $p - 1$ there exists a unique subfield $K = K_t$ of $\mathbb{Q}(\zeta_p)$ such that $\mathbb{Q}(\zeta_p):K = t$ and $K:\mathbb{Q} = (p-1)/t =: s$. If $H = H_t$ is the corresponding subgroup of G, we have $H = \{1, \gamma^s, \dots, \gamma^{s(t-1)}\}$. Set

(1)
$$\eta_i(t) = \sum_{j=0}^{t-1} \zeta_p^{g^{sj+i}} \quad \text{for } 0 \le i \le s-1.$$

Then $\eta_0(t)$ lies in K_t, and $\eta_0(t), \dots, \eta_{s-1}(t)$ are precisely the s distinct conjugates of $\eta_0(t)$ over $\mathbb{Q}$. It follows that $K_t = \mathbb{Q}(\eta_0(t)) = \mathbb{Q}(\eta_i(t))$. In this way one gets a description of all subfields of $\mathbb{Q}(\zeta_p)$ via the sums in (1), called *Gaussian periods*.

11.4 Consider the situation of §11.3 in the case where $p = 17$ and $\zeta = e^{i\varphi}$ with $\varphi = 2\pi/17$. Show that $g = 3$ can serve as a primitive root mod 17. Then we have, for instance,

$$\eta_0(2) = \zeta + \zeta^{-1} = 2\cos\varphi,$$
$$\eta_0(4) = \zeta + \zeta^4 + \zeta^{-1} + \zeta^{-4} = 2(\cos\varphi + \cos 4\varphi),$$
$$\eta_0(8) = \zeta^{-8} + \zeta^8 + \zeta^2 + \zeta^{-2} + \zeta^4 + \zeta^{-4} + \zeta^{-1} + \zeta.$$

Why is $\eta_0(8) + \eta_1(8)$ equal to -1? By computing $\eta_0(8)^2$, or in any other way, derive the equality

$$X^2 + X - 4 = \left(X - \eta_0(8)\right)\left(X - \eta_1(8)\right).$$

Because $\eta_0(8) = 2(\cos 8\varphi + \cos 4\varphi + \cos 2\varphi + \cos\varphi) > 0$ we then have

$$\eta_0(8) = \tfrac{1}{2}(-1 + \sqrt{17}), \quad \eta_1(8) = \tfrac{1}{2}(-1 - \sqrt{17})$$

(incidentally showing again that $\mathbb{Q}(\sqrt{17})$ is the quadratic subfield of $\mathbb{Q}(\zeta_{17})$). Now clearly $\eta_0(4) + \eta_2(4) = \eta_0(8)$ and you should show that $\eta_0(4)\eta_2(4) = -1$, so that

$$X^2 - \eta_0(8)X - 1 = \big(X - \eta_0(4)\big)\big(X - \eta_2(4)\big);$$

similarly,

$$X^2 - \eta_1(8)X - 1 = \big(X - \eta_1(4)\big)\big(X - \eta_3(4)\big).$$

It is easy to see that $\eta_0(4) > \eta_2(4)$ and $\eta_1(4) > \eta_3(4)$. This implies, for example, that

$$\eta_0(4) = \tfrac{1}{4}\big(-1 + \sqrt{17} + \sqrt{34 - 2\sqrt{17}}\big), \quad \eta_1(4) = \tfrac{1}{4}\big(-1 - \sqrt{17} + \sqrt{34 + 2\sqrt{17}}\big).$$

Finally, show that

$$X^2 - \eta_0(4)X + \eta_1(4) = \big(X - \eta_0(2)\big)\big(X - \eta_4(2)\big)$$

and observe that $\eta_0(2) > \eta_4(2)$. Putting it all together we get an explicit construction for $\eta_0(2) = 2\cos\varphi$ by successive adjunction of square roots, and hence an explicit ruler-and-compass construction for the 17-gon.

11.5 Prove:

(a) If p is a prime of the form $4k + 3$, the number $2p + 1$, if also a prime, divides $2^p - 1$. Give at least two examples of numbers of the form $2^p - 1$ (*Mersenne numbers*) that are not prime.

(b) Any prime divisor p of a Fermat number F_k, with $k \geq 2$, satisfies $p \equiv 1 \bmod 2^{k+2}$.

11.6 (a) Is 14993 a quadratic residue modulo 65537?

(b) Describe all primes p such that 7 is a quadratic residue modulo p.

11.7 Let a and b be natural numbers, with b odd. Prove:

(a) If $\left(\frac{x}{b}\right) = 1$ for all $x \in \mathbb{N}$ relatively prime to b, then b is a square. (*Hint:* Use the Chinese Remainder Theorem).

(b) There are infinitely many primes p such that $\left(\frac{a}{p}\right) = 1$. (Compare §9.19.)

(c) If $\left(\frac{a}{x}\right) = 1$ for all $x \in \mathbb{N}$ odd and relatively prime to a, then a is a square. (*Hint:* Use the quadratic reciprocity law and part (a).)

(d) If a is not a square, there exist infinitely many primes p not dividing a such that a is not a quadratic residue modulo p. (*Hint:* If $p_1, \ldots, p_n$ have already been found, there exists $y \in \mathbb{N}$ such that $y \equiv p_1 \bmod 4a$ and $y \equiv 1 \bmod p_1 \ldots p_n$.)

11.8 Prove that the congruence $(X^2 + 1)(X^4 - 4) \equiv 0 \bmod p$ has a solution in $\mathbb{Z}$ for every prime p, although the corresponding equation has no solution in $\mathbb{Z}$.

11.9 Let n be an odd natural number. Prove:

(a) If $x^{n-1} \equiv 1 \bmod n$ for all $x \in \mathbb{Z}$ relatively prime to n, then n is square-free (but not necessarily prime; the smallest counterexample is $n = 3 \cdot 11 \cdot 17$).

(b) If $\left(\frac{x}{n}\right) \equiv x^{(n-1)/2} \bmod n$ for all $x \in \mathbb{Z}$ relatively prime to n, then n is prime (this is a converse of sorts to Euler's criterion). *Hint:* Take $n = pm$ with $m > 1$. The assumption implies $x^{(m-1)/2} \equiv \left(\frac{x}{m}\right) \bmod p$. By §11.7 there exists some y with $\left(\frac{y}{m}\right) = -1$. Now take x such that $x \equiv y \bmod m$ and $x \equiv 1 \bmod p$.

11.10 An *algebraic number field* is an extension $K/\mathbb{Q}$ with $K:\mathbb{Q} < \infty$. Such a field is called *quadratic* if $K:\mathbb{Q} = 2$. It is called *cyclotomic* if it is a subfield of $\mathbb{Q}^{(m)} = \mathbb{Q}(\zeta_m)$ for some m. Prove that *every quadratic number field is cyclotomic*. More precisely, suppose $K = \mathbb{Q}(\sqrt{d})$, with $d \neq 1$ a square-free integer. Set $m = |d|$ if $d \equiv 1 \bmod 4$ and $m = 4|d|$ otherwise. Then K is contained in $\mathbb{Q}^{(m)}$ (and m is minimal with this property).

This is a good place to mention the famous *Kronecker–Weber Theorem:* Every *abelian* number field K (meaning that $K/\mathbb{Q}$ is Galois with a finite abelian Galois group) is cyclotomic. (Kronecker did not supply a complete proof; this was done by H. Weber.)

11.11 So that the example in §11.8 does not leave something of a false impression, let it be mentioned that *a nonconstant polynomial $f \in \mathbb{Z}[X]$ for which the congruence $f(X) \equiv 0 \bmod p$ has a solution in $\mathbb{Z}$ for almost all primes p cannot be prime over $\mathbb{Q}$*. The proof of this fairly deep theorem (see F. Lorenz, *Algebraische Zahlentheorie*, BI-Verlag, 1993, p. 293) is founded on an approach pioneered by Kronecker, which in 1896 allowed Frobenius to prove the following stronger result:

Suppose the Galois group $G(f)$ of a polynomial $f \in \mathbb{Z}[X]$ of degree $n > 1$ contains an element σ whose cycle decomposition comprises r cycles, of lengths $n_1, \ldots, n_r$ (summing up to n). Then there exist infinitely many primes p such that f factors in $\mathbb{F}_p[X]$ into r prime polynomials whose degrees are $n_1, \ldots, n_r$.

Now, if f is assumed *irreducible* over $\mathbb{Q}$, the group $G(f)$ contains some σ for which all the n_i exceed 1 (see §10.22); thus there are infinitely many values of p for which f has no linear factor in $\mathbb{F}_p[X]$. Compare also with Chapter 16, F13.

11.12 Factor $X^{p-1} - 1$ in $\mathbb{F}_p[X]$ and derive *Wilson's Theorem*:

$$(2) \qquad (p-1)! \equiv -1 \bmod p.$$

This was already known to Leibniz, as was the following "converse": *If $(n-1)! \equiv -1 \bmod n$ for $n \in \mathbb{N}$, then n is prime.*

For p an odd prime, set $t = \frac{1}{2}(p-1)$ and $H = \{1, 2, \ldots, t\}$. By taking together the factors a and $p - a$, for each $a \in H$, derive from (2) that

$$(3) \qquad (t!)^2 \equiv -(-1)^{(p-1)/2} \bmod p.$$

Thus, if $p \equiv 1 \bmod 4$, the number $t!$ is a solution to the congruence $X^2 \equiv -1 \bmod p$; whereas $t! \equiv \pm 1$ if $p \equiv 3 \bmod 4$. In this latter case show more precisely that

$$(4) \qquad t! \equiv (-1)^\nu \bmod p,$$

where ν is the number of elements of H that are *not* quadratic residues mod p. Let $\mu = t - \nu$ be the number of elements of H that *are* quadratic residues mod p. Prove that $\mu = \nu = \frac{1}{4}(p-1)$ if $p \equiv 1 \bmod 4$, but $\mu \neq \nu$ if $p \equiv 3 \bmod 4$. (Remarkably, in the latter case ν is always *less* than μ, but this is much harder to prove; see, for example, Z. I. Borevich and I. R. Shafarevich, *Number theory*, Academic Press, 1966. More precisely, let $p \equiv 3 \bmod 4$; if the so-called *class group* of $\mathbb{Q}(\sqrt{-p})$ has order h, then

(5) $$\mu - \nu = \begin{cases} h & \text{if } p = 3 \text{ or } p \equiv 7 \bmod 8, \\ 3h & \text{otherwise.} \end{cases}$$

Thus, since $\mu + \nu = \frac{1}{2}(p-1)$, we get

$$\nu = \frac{1}{4}(p-1) - \frac{1}{2}h \quad \text{or} \quad \nu = \frac{1}{4}(p-1) - \frac{3}{2}h,$$

as the case may be. Note that h is always *odd*.)

Chapter 12: Further Steps into Galois Theory

12.1 Let F/K be a *separable* field extension of finite degree. Derive again, using F3, the primitive element theorem (Chapter 8, Theorem 3), which says that F contains an element β such that $F = K(\beta)$. Can one require that $\mathrm{Tr}_{F/K}(\beta) = 1$?

12.2 Let ζ_n be a primitive root of unity in $\mathbb{C}$. When do the conjugates of ζ_n form a normal basis for $\mathbb{Q}(\zeta_n)/\mathbb{Q}$? (Answer: If and only if n is *square-free*; see also §13.7.)

12.3 In the finite field $E = \mathbb{F}_{3^3}$, find: (a) a primitive root of E whose conjugates do not form a normal basis of $E/\mathbb{F}_3$; (b) a normal basis that does not consist of primitive roots of E.
    For an arbitrary finite field E with prime field $\mathbb{F}_p$, the extension $E/\mathbb{F}_p$ does always have at least one normal basis consisting of primitive roots. This was proved by Carlitz for E large enough, and by Davenport for any E. Davenport's proof is elementary, subtle and long (*J. London Math. Soc.* **43** (1968), 21–39). The statement remains true when $\mathbb{F}_p$ is replaced by an arbitrary subfield K of E.

12.4 By giving $\mathbb{N}$ the divisibility partial order and by taking the natural homomorphisms $f_{mn} : \mathbb{Z}/n\mathbb{Z} \to \mathbb{Z}/m\mathbb{Z}$, for m dividing n, one makes $(\mathbb{Z}/n\mathbb{Z})_{n \in \mathbb{N}}$ into a projective system of groups (or rings). Set

$$\hat{\mathbb{Z}} = \varprojlim_{n} \mathbb{Z}/n\mathbb{Z}.$$

Prove:

(a) For every finite field $\mathbb{F}_p$ there is a canonical isomorphism $G(C/\mathbb{F}_q) \simeq \hat{\mathbb{Z}}$, where C denotes an algebraic closure of $\mathbb{F}_q$.

(b) The *open* subgroups of $\hat{\mathbb{Z}}$ are precisely the subgroups of the form $n\hat{\mathbb{Z}}$. The map $\mathbb{Z} \to \hat{\mathbb{Z}}$ gives rise to an isomorphism $\mathbb{Z}/n\mathbb{Z} \simeq \hat{\mathbb{Z}}/n\hat{\mathbb{Z}}$. There are nontrivial closed subgroups of $\hat{\mathbb{Z}}$ other than those of the form $n\hat{\mathbb{Z}}$.

(c) For each prime p, set $\mathbb{Z}_p = \varprojlim_i \mathbb{Z}/p^i\mathbb{Z}$. Show there is a natural isomorphism $\hat{\mathbb{Z}} \simeq \prod_p \mathbb{Z}_p$.

12.5 Let W be the group of all roots of unity in $\mathbb{C}$. Show that the following natural isomorphisms hold:

$$G(\mathbb{Q}(W)/\mathbb{Q}) \simeq \hat{\mathbb{Z}}^\times \simeq \varprojlim_n (\mathbb{Z}/n\mathbb{Z})^\times \simeq \prod_p \mathbb{Z}_p^\times.$$

12.6 Using Artin's result (Theorem 2′) prove again Theorem 4 of Chapter 8. *Hint:* Suppose, in contradiction with equation (8) of that chapter (page 79), that E/K contains n linearly independent vectors $b_1, \ldots, b_n$, with $n > |G|$. Now consider the system of linear equations

$$\sum_j \sigma^{-1}(b_j)x_j = 0 \quad \text{for } \sigma \in G,$$

and notice that $\mathrm{Tr}_G \neq 0$.

12.7 (Algebraic independence of field homomorphisms). It is possible to strengthen Theorem 2 under the additional assumption that the ground field K is *infinite*. In fact, in this case *a polynomial f in $C[X_1, \ldots, X_n]$ that satisfies*

(1) $\qquad f(\sigma_1(\beta), \ldots, \sigma_n(\beta)) = 0 \quad \text{for all } \beta \in E$

must be the zero polynomial.

To prove this, choose a basis $\beta_1, \ldots, \beta_n$ of E/K and consider the polynomial

(2) $\qquad g(X_1, \ldots, X_n) := f\left(\sum_{j=1}^n \sigma_1(\beta_j)X_j, \ldots, \sum_{j=1}^n \sigma_n(\beta_j)X_j \right).$

Because of (1) we have $g(x_1, \ldots, x_n) = 0$ for all $x_1, \ldots x_n \in K$. Since K is infinite, we have $g = 0$. But then $f = 0$ as well, for the following reason: By Theorem 2, the matrix $(\sigma_i(\beta_j))_{i,j}$ in $M_n(C)$ has an *inverse*, say $(a_{rs})_{r,s}$. By substitution in (2) we get

$$g\left(\sum_{k=1}^n a_{1k}X_k, \ldots, \sum_{k=1}^n a_{nk}X_k \right) = f(X_1, \ldots, X_n) = f.$$

12.8 The theorem asserting the existence of *normal bases* (Theorem 3) appears to have first been stated by *Emmy Noether*, in a 1932 paper in *Crelle's Journal*. Perhaps because the paper was essentially about number theory, or perhaps also because Noether was not sure the theorem was original, she merely outlined the proof of this purely algebraic result in a three-line footnote. As the nature of her argument makes clear, she was not including the case of a *finite* ground field. Was this all she meant when she remarked in a later work (*Gesammelte Abhandlungen*, p. 638) that her proof had "a gap"? In any case, her argument can be fleshed out into a proof of the theorem (for *infinite* ground fields) as follows:

Let E/K be a finite Galois extension with K infinite, and let $E(X_1,\ldots,X_n)$ be the field of rational functions in the variables $X_1,\ldots,X_n$ over E. Clearly the extension $E(X_1,\ldots,X_n)/K(X_1,\ldots,X_n)$ is Galois and its Galois group can be identified in a natural way with $G = G(E/K)$. Choose a basis $\beta_1,\ldots,\beta_n$ of E/K, form the element

$$
(3) \qquad\qquad u = \sum_{j=1}^{n} \beta_j X_j \in E[X_1,\ldots,X_n],
$$

and then the $n \times n$ matrix $(\tau^{-1}\sigma(u))_{\tau,\sigma}$ in $M_n(E[X_1,\ldots,X_n])$, where τ and σ run over the elements of G in some fixed order. Denote by $d = d(X_1,\ldots,X_n)$ the determinant of this matrix. We claim that d is not the zero polynomial. Assuming this, d cannot vanish everywhere in K^n, so there exist $a_1,\ldots,a_n$ in K such that $d(a_1,\ldots,a_n) \neq 0$. Now set $\alpha := \sum a_j \beta_j$; then

$$
(4) \qquad\qquad \det (\tau^{-1}\sigma(\alpha))_{\tau,\sigma} \neq 0,
$$

and this is enough to show that the elements $\sigma(\alpha)$, for $\sigma \in G$, are linearly independent over K.

To show that $d \neq 0$, set $Y_\sigma := \sigma(u) = \sum \sigma(\beta_j)X_j$ for every $\sigma \in G$; see (3). By Theorem 2 the $n \times n$ matrix $(\sigma(b_j))_{\sigma,j}$ is invertible. Thus $E[X_1,\ldots,X_n]$ equals the ring of polynomials over E **in the n variables** Y_σ, for $\sigma \in G$. By definition,

$$
(5) \qquad\qquad d = \det (Y_{\tau^{-1}\sigma})_{\tau,\sigma}.
$$

Following Frobenius, we call this the *group determinant* of G. It is nonzero for any finite group G and any field E, which can be seen as follows: On every row of the matrix $M := (Y_{\tau^{-1}\sigma})_{\tau,\sigma}$ the variable Y_1 appears exactly once, and always on the diagonal. Thus, by expanding $d = \det M$ according to Leibniz's rule, we get $d = Y_1^n + g$, where $g \in E[Y_\sigma \mid \sigma \in G]$ has degree less than n with respect to Y_1. Therefore d cannot be 0.

12.9 Let $(\sigma(\alpha))_{\sigma \in G}$ be a *normal basis* of a finite Galois extension E/K. If F is an intermediate field of E/K and we set $H = G(E/F)$, the elements $\tau(\alpha)$, for $\tau \in H$, may turn out to be *linearly dependent* over F, so one does *not* get a normal basis of E/F in this way. Confirm this with the example $E/K = \mathbb{Q}(\zeta_9)/\mathbb{Q}$, $\alpha = \zeta_9 + \zeta_3$, $F = \mathbb{Q}(\zeta_3)$.

Chapter 13: Norm and Trace

13.1 Let A be a K-algebra and let L be a subalgebra of A contained in the center of A (so that A can be viewed as an L-algebra as well). Assume further that both L as a K-module and A as an L-module are free and finitely generated. Then the same holds for A as a K-module. Prove that for any $\alpha \in A$ we have

$$N_{A/K}(a) = N_{L/K}(N_{A/L}(a)),$$
$$\mathrm{Tr}_{A/K}(a) = \mathrm{Tr}_{L/K}(\mathrm{Tr}_{A/L}(a)),$$
$$P_{A/K}(a; X) = N_{L[X]/K[X]}(P_{A/L}(a; X)).$$

(*Hint:* See LA II, p. 181, Aufgabe 63.)   From this we can get F5 as a particular case, and hence also equations (26) and (27).

13.2 Let E/K be a Galois extension with cyclic Galois group $\langle \sigma \rangle$ of order 4. Assume also that char $K \neq 2$. Prove that the quadratic intermediate field L of E/K must be of the form $L = K(\sqrt{a^2 + b^2})$, with $a, b \in K$.

   Hint: Write $L = K(\sqrt{d})$ and $E = L(w)$, with $w^2 \in L$. From $G(E/L) = \langle \sigma^2 \rangle$ deduce that $\sigma^2(w) = -w$ and then $w\sigma(w) \in L$. Moreover $N_{L/K}(w\sigma(w)) = -N_{L/K}(w^2)$, so $-1 \in N_{L/K}(L^\times)$. The result follows upon observing that if -1 is a square in K, *every* element of K is a sum of two squares (why?).

   Prove also the converse: If L/K is a quadratic extension of the form $L = K(\sqrt{a^2 + b^2})$, with $a, b \in K$, there is an extension E/L such that E/K is *cyclic of degree* 4.

13.3 A field K is called *pythagorean* if every sum of two squares in K is a square in K. Prove that a field K such that char $K \neq 2$ is pythagorean if and only if it has no cyclic extension of degree 4.

13.4 Let E/K be a finite Galois extension with Galois group G. Every $\alpha \in E^\times$ gives rise to a map $\varphi = \varphi_\alpha : G \to E^\times$, defined by

(1)
$$\varphi(\tau) = \frac{\alpha^\tau}{\alpha},$$

where $\alpha^\tau := \tau(\alpha)$ (accordingly, the notation for the group operation of G obeys $\alpha^{\sigma\tau} = (\alpha^\sigma)^\tau$). The function φ thus defined clearly satisfies the functional equation

(2)
$$\varphi(\sigma\tau) = \varphi(\sigma)^\tau \varphi(\tau).$$

Any map $\varphi : G \to E^\times$ with property (2) is called a *crossed homomorphism* from G to $E^\times$.

(a) Suppose G is *cyclic*, with generator σ. Show that if $\varphi : G \to E^\times$ is a crossed homomorphism, $\varphi(\sigma)$ has norm 1. Conversely, if γ is an element of $E^\times$ such that $N_{E/K}(\gamma) = 1$, there is a unique crossed homomorphism φ taking σ to γ.

   This equivalence reduces Hilbert's Theorem 90 (see F9) to a special case of the following theorem of A. Speiser, which holds for *any finite* Galois group: *Every crossed homomorphism $\varphi : G \to E^\times$ is split*, that is, has the form (1).

   This more general statement, too, is often called Hilbert's Theorem 90. (There is nothing objectionable about that, so long as the common practice of misattributing to E. Noether this generalization of the original Theorem 90 is avoided; in this connection see F. Lorenz, "Ein Scholion zum Satz 90 von Hilbert", *Abh. Math. Univ. Hamburg* **68** (1998), 347–362.)

(b) Prove the preceding Theorem of A. Speiser — as a candidate for α in (1), start with

$$\alpha = \sum_{\sigma \in G} a_\sigma \beta^\sigma \quad \text{where } a_\sigma, \beta \in E,$$

and take into account the linear independence of all the σ (Chapter 12, Theorem 2).

(c) Let $I_G E^\times$ denote the *subgroup* of $E^\times$ consisting of all finite products of elements of the form $\alpha^{\tau-1} := \alpha^\tau/\alpha$, with $\alpha \in E^\times$ and $\tau \in G$. Prove that $N_{E/K}(\gamma) = 1$ for any $\gamma \in I_G E^\times$. If G is *cyclic* with generator σ, we have $I_G E^\times = (E^\times)^{\sigma-1}$. Thus Hilbert's Theorem 90 says that *for G cyclic*, the norm homomorphism $N_{E/K} : E^\times \to K^\times$ has kernel $I_G E^\times$.

For G noncyclic it is possible to have $\ker N_{E/K} \neq I_G E^\times$. Establishing this looks at first like an easy algebraic exercise — take for instance $K = \mathbb{Q}$ and $E = K(\sqrt{a}, \sqrt{b})$ a *biquadratic* extension. But it's not as simple as that; see the article cited in part (a). Nonetheless, N. Zimmermann (at the time a graduate student in Münster) was able to prove explicitly, by relatively simple *number-theoretical* means, that in $E = \mathbb{Q}(\sqrt{2}, \sqrt{3})$, for example, the element $\gamma = 1 + \sqrt{2}$, which obviously satisfies $N_{E/K}(\gamma) = 1$, does not lie in $I_G E^\times$. Actually in this case it can be shown even that the quotient $\ker N_{E/K} / I_G E^\times$ has order 2, but this is a deeper result; see again the article cited in part (a).

13.5 Let f and g be distinct normalized irreducible polynomials in the polynomial ring $K[X]$ over the field K, of degrees m and n, respectively. In some extension E of K, let α and β be elements satisfying $f(\alpha) = 0$ and $g(\beta) = 0$. Prove the following *reciprocity result* concerning the norm:

$$N_{K(\alpha)/K}(g(\alpha)) \cdot N_{K(\beta)/K}(f(\beta))^{-1} = (-1)^{mn}.$$

13.6 A function $f : \mathbb{N} \to \mathbb{C}$ such that $f \neq 0$ is called a *number-theoretic function*; f is called *multiplicative* if $f(mn) = f(m)f(n)$ for all relatively prime m, n. Examples: the Euler totient function φ; the function ι such that $\iota(n) = 1$ for all n; the function ε such that $\varepsilon(1) = 1$ and $\varepsilon(n) = 0$ otherwise; the multiplicative function μ defined by $\mu(p) = -1$ and $\mu(p^e) = 0$, for p prime and $e > 1$. Given two number-theoretic functions f, g, define their product $f * g$ by

$$(f * g)(n) = \sum_{d|n} f(d)g(n/d).$$

Prove that $f * g = g * f$, $(f * g) * h = f * (g * h)$, $\varepsilon * f = f * \varepsilon = f$, and $\mu * \iota = \varepsilon$. Derive from this the *Möbius inversion formula*: $g = f * \iota$ if and only if $f = g * \mu$.

13.7 Given $n \in \mathbb{N}$, denote by ζ_n a primitive n-th root of unity in $\mathbb{C}$. Prove that

$$(3) \qquad \qquad \text{Tr}_{\mathbb{Q}(\zeta_n)/\mathbb{Q}}(\zeta_n) = \mu(n),$$

where μ is the *Möbius function* of §13.6. *Hint:* Define f via $f(n) = \text{Tr}_{\mathbb{Q}(\zeta_n)/\mathbb{Q}}(\zeta_n)$ and prove that $\sum_{d|n} f(d) = \varepsilon(n)$; this reduces the desired equality to the Möbius inversion formula.

For $n > 1$, let $s_2(n)$ denote the third coefficient of the n-th cyclotomic polynomial. Prove that

(4)
$$2s_2(n) = \operatorname{Tr}_{\mathbb{Q}^{(n)}\mathbb{Q}}(\zeta_n)^2 - \operatorname{Tr}_{\mathbb{Q}^{(n)}/\mathbb{Q}}(\zeta_n^2)$$

and deduce that $s_2(n)$ can only take the values $0, 1, -1$. The latter property is shared by the fourth coefficient $-s_3(n)$, as can be seen from the relation

(5)
$$3s_3(n) = \operatorname{Tr}_{\mathbb{Q}^{(n)}/\mathbb{Q}}(\zeta_n^3) - \mu(n)^3 + 3\mu(n)s_2(n)$$

(compare §15.24). What secrets lurk behind the distribution functions $s_k(n)$, for $0 \le k \le \varphi(n)$, can only be guessed at.

Chapter 14: Binomial Equations

14.1 Let K be a field containing a primitive n-th root of unity. Let E/K be a cyclic extension of degree n and suppose $E = K(\alpha)$, where α is an n-th root of some element a in $K^\times$. Take $a' \in K^\times$. Prove that E contains an n-th root α' of a' if and only if there is a natural number r for which a'/a^r is an n-th power in K. When is $K(\alpha') = E$ as well? (*Hint:* See the proof of F1. Also, these results can also be easily derived from *Kummer theory*.)

14.2 Let K be a field and q a prime distinct from char K.

(a) Prove that an element of $K^\times$ that is a q-th power in $K(\zeta_q)$ is already a q-th power in K (compare F2).

(b) Let $E = K(\alpha)$ and $E' = K(\alpha')$ be extensions of K containing, respectively, q-th roots α and α' of elements a and a' in $K^\times$. If $E/K \simeq E'/K$, there exists a number r relatively prime to q such that a'/a^r is a q-th power in K. If $E' : K = q$ the converse also holds. Generalizations? Counterexamples when q is not prime? (See also §14.5 and §14.6.) *Hint:* One may as well assume $E = E'$. Now see §14.1 and part (a).

14.3 Suppose char $K = p > 0$. Let $E = K(\alpha)$ and $E' = K(\alpha')$ be extensions of K, where α, α' satisfy $\wp(\alpha) = a$ and $\wp(\alpha') = a'$, with $a, a' \in K$ (for $\wp$ see Remark 1 after Theorem 3). Prove that E/K and E'/K are isomorphic if and only if there is a natural number $r < p$ such that $a' - ra \in \wp(K)$.

More generally, formulate a Kummer theory, analogous to Theorem 4, for abelian extensions of exponent p; in this version the multiplicative group of K is replaced by the additive group of K and the n-th power map is replaced by $\wp$. This topic is explored further in §14.15.

14.4 Let K be a field of characteristic $p > 0$.

(a) Let E/K be a *cyclic* extension of degree $p^e \ne 1$, with σ a generator of $G = G(E/K)$. Set $m = p^{e-1}$ and prove: There exists a unique intermediate field L of E/K such that $E : L = p$; this field is of the form $E = L(\alpha)$, where α

satisfies $\wp(\alpha) = a$ for some $a \in L$; moreover $\sigma^m \alpha = \alpha + 1$. Prove that, as a consequence, $E = K(\alpha)$. Prove also that $\beta := \sigma\alpha - \alpha$ lies in L and satisfies $\sigma a - a = \wp(\beta)$ and $S_{L/K}(\beta) = 1$. Is it the case that $L = K(\beta)$?

(b) Let L/K be a cyclic extension of degree $p^{e-1} \neq 1$. Prove there exists an extension E of L such that E/K is *cyclic* of degree p^e. (*Hint:* Use part (a) as a road map.)

14.5 Let n be a natural number not divisible by 4.

(a) Prove that if $X^n - a \in \mathbb{Q}[X]$ is irreducible, the Galois group G of $X^n - a$ over $\mathbb{Q}$ has order $\varphi(n)n$ or $\varphi(n)n/2$. Do both cases really occur? (*Hint:* Investigate whether $X^n - a$ is irreducible over $\mathbb{Q}(\zeta_n)$.)

(b) Suppose we're in the case $|G| = \varphi(n)n$ of part (a). Prove that G is isomorphic to the subgroup of matrices in $\mathrm{GL}(2, \mathbb{Z}/n\mathbb{Z})$ having the form $\left(\begin{smallmatrix} a & b \\ 0 & 1 \end{smallmatrix}\right)$.

14.6 Let q be a prime number and K a field of characteristic distinct from q; assume also that $\sqrt{-1} \in K$ if $q = 2$. Prove that any element of K that is a q^n-th power in $K(\zeta_{q^n})$ is already a q^n-th power in K. (*Hint:* Use induction on n, and be aware that some perseverance is required; apply Theorem 2 and the results from §9.11, §10.2 and §14.2.)

14.7 Let E/K be a finite field extension and M a subgroup of $E^\times$ such that $K^\times \subseteq M$. Assume the quotient $M/K^\times$ is finite. Explain why

$$[K(M):K] \leq M:K^\times$$

necessarily. Prove that $K(M)/K$ is separable if and only if the order of $M/K^\times$ is not divisible by char K.

14.8 In the situation of §14.7, assume that E/K is *separable* and that M satisfies the following conditions: (i) For any prime p, any p-th root of unity contained in M is already in K. (ii) If M contains a $1 + \zeta_4$, where ζ_4 is a primitive fourth root of unity, then $\zeta_4 \in K$. Prove in order:

(a) If $M/K^\times$ has prime order p, then $[K(M):K] = p$.

(b) If $M/K^\times$ is a p-group and $[K(M):K] = p$, then $M:K^\times = p$. (*Hint:* See Theorem 2 and §14.2.)

(c) If $M/K^\times$ is a group of order p^n, then $[K(M):K] = M:K^\times$. (*Hint:* Induction on the index $[K(M):K]$; from $M = \langle \alpha \rangle M'$ with $M':K^\times = p^{n-1}$ deduce that $X^p - \alpha^p$ is irreducible over $K(M')$.)

(d) $[K(M):K] = M:K^\times$. (*Hint:* Consider each Sylow group $P/K^\times$ of $M/K^\times$ and keep §14.7 in mind.)

14.9 Let E/K be a field extension and $\alpha_1, \ldots, \alpha_r$ elements of E. Suppose that $\alpha_i^{n_i} = a_i \in K^\times$, where $n_1, \ldots, n_r$ are natural numbers relatively prime to the characteristic of K. Prove:

(a) If $K(\alpha_1, \ldots, \alpha_r):K = n_1 n_2 \ldots n_r$, the following conditions are satisfied: (i) For any prime p, a product $\prod_{p|n_i} a_i^{t_i}$ cannot be a p-th power in K unless each

t_i is a multiple of p. (ii) A product $\prod_{2 \mid n_i} a_i^{t_i}$, with $n_i t_i \equiv 0 \bmod 4$, cannot be of the form $-4\lambda^4$ with $\lambda \in K$ unless all the t_i are multiples of 4.

(b) If conditions (i) and (ii) in part (a) are satisfied and M is the subgroup of $E^\times$ generated by $\alpha_1, \ldots, \alpha_r$ and $K^\times$, then $M/K^\times$ has order $n_1 n_2 \ldots n_r$.

(c) If conditions (i) and (ii) in part (a) are satisfied, then $K(\alpha_1, \ldots, \alpha_r) : K = n_1 n_2 \ldots n_r$. (*Hint:* By part (b), $M/K^\times$ has order $n_1 n_2 \ldots n_r$. Now check that M also satisfies conditions (i) and (ii) of §14.8.)

Apart from the assumptions made on the characteristic, these results are a direct generalization of *Capelli's Theorem* (Theorem 2 in Section 14.1).

14.10 Let G be any *abelian group* of finite order. Prove that G can be realized as a Galois group over the field $\mathbb{Q}$ (see the remarks after F5 in Section 15.3). *Hint:* Handle first the case where G is cyclic, by considering $\mathbb{Q}(\zeta_p)$ for primes $p \equiv 1 \bmod |G|$; see §9.18.

14.11 Let p be any prime number. Prove that the Galois group of the polynomial $X^{4p} + 4p^2$ over $\mathbb{Q}$ has order $\varphi(4p)\,p$. (*Hint:* Prove that $-4p^2$ is a fourth power in $\mathbb{Q}(\zeta_{4p})$.)

14.12 Let E/K be a *cyclic* extension of degree n, where K contains a primitive n-th root of unity ζ. Let σ be a generator of $G(E/K)$. The statement and proof of Theorem 1 (page 144) boil down to the existence of a nonzero $\alpha \in E$ such that $\sigma\alpha = \zeta\alpha$. Justify the existence of such an α using *linear algebra*.

Hint: Consider σ as an endomorphism of the K-vector space E, with minimal polynomial f. Since $\sigma^n = 1$, this polynomial divides $X^n - 1$. Moreover f splits into linear factors, because $\zeta \in K$, and it has only simple roots. Therefore σ is *diagonalizable* (see LA I, p. 215). Now let W be the set of eigenvalues of σ in K. Why is W a *group*?

The goal is to show that $\zeta \in W$. Suppose otherwise. Then $k := \operatorname{ord} W < n$. Applying σ^k to a basis of eigenvectors of σ we get $\sigma^k = 1$, a contradiction. (Note: Using the linear independence of $1, \sigma, \sigma^2, \ldots, \sigma^{n-1}$, we see right away from Theorem 2 in Chapter 12 that $f(X) = X^n - 1$, so ζ is an eigenvalue of σ.)

14.13 Let m be a natural number greater than 2. Prove that

$$\mathbb{Q}(\zeta_{2^m}, \sqrt[2^m]{2}) : \mathbb{Q} = 2^{2m-2} \quad \text{and} \quad \mathbb{Q}(\zeta_{2^m}) \cap \mathbb{Q}(\sqrt[2^m]{2}) = \mathbb{Q}(\sqrt{2}).$$

Hint: Set $n = 2^m$, $\alpha = \sqrt[n]{2}$, $K = \mathbb{Q}(\zeta_n)$. To determine $K(\alpha) : K$ use F1 after observing that $\sqrt{2} \in K$, $\sqrt[4]{2} \notin K$.

14.14 Let p be a prime number, K a field of characteristic distinct from p, ζ a primitive p-th root of unity (in an algebraic closure C of K), $L = K(\zeta)$, γ a generator of $G(L/K)$ and k a natural number such that $\zeta^\gamma := \gamma(\zeta) = \zeta^k$. Prove:

(a) If E/L is a cyclic extension of degree p, the extension E/K is *abelian* if and only if $E = L(\sqrt[p]{a})$ for some $a \in L^\times$ such that $a^{\gamma-k} \in L^{\times p}$.

(b) There is a bijection between abelian extensions F/K of exponent p (in C/K) and subgroups A of $L^\times$ such that $L^{\times p} \subseteq A$ and $A^{\gamma-k} \subseteq L^{\times p}$.

14.15 Let K be a field of prime characteristic p, C an algebraic closure of K and $\wp : C \to C$ the map given by $\wp(x) = x^p - x$. For each subset B of K, denote by $E_B = K(\wp^{-1}(B))$ the subfield of C obtained by adjoining to K the roots of all polynomials $X^p - X - b$ such that $b \in B$. Prove that *the map $B \mapsto E_B$ provides a one-to-one correspondence between additive subgroups B of K such that $\wp(K) \subseteq B$ and abelian extensions E/K of exponent p (in C/K); moreover E_B/K is finite if and only if $K/\wp(K)$ is finite, and in that case there are natural isomorphisms $G \simeq (B/\wp(K))^*$ and $G^* \simeq B/\wp(K)$, where $G = G(E_B/K)$.*

This is a generalization of the Artin–Schreier Theorem (Theorem 3), so it's sometimes called *Artin–Schreier theory*. It represents an analog of *Kummer theory*; not a full one, however, since it does not encompass abelian extensions of exponent p^n for $n > 1$. There is a theory for the latter as well, the key to which stems from the calculus of *Witt vectors*; see Volume II, Chapter 26, Theorem 6.

Chapter 15: Solvability of Equations

15.1 Let G be a solvable, characteristic-simple finite group (see §10.10). Prove that G is *elementary abelian*, that is, isomorphic to a product $G \simeq \mathbb{Z}/p\mathbb{Z} \times \cdots \times \mathbb{Z}/p\mathbb{Z}$ for p prime. In particular, every minimal normal subgroup of a solvable group is elementary abelian.

15.2 Let G be a finite group. A chain $G = H_0 \supseteq H_1 \supseteq \cdots \supseteq H_r = 1$ of subgroups H_i of G such that $H_i \trianglelefteq H_{i-1}$ is called a *normal series* of G. The groups H_{i-1}/H_i are called the *factors* of the normal series. If the H_i are all normal **in** G, the chain is called a *principal series* of G. The subgroup G' of G generated by all the *commutators* (elements of the form $aba^{-1}b^{-1}$ for $a, b \in G$) is called the *commutator subgroup* of G. Define $G^{(0)} = G$ and $G^{(i+1)} = G^{(i)\prime}$ by recursion. Prove:

(a) G' is the smallest normal subgroup of G with an abelian quotient group.

(b) There is equivalence between: (i) G is solvable; (ii) G has a normal series whose factors are abelian. (iii) There exists n such that $G^{(n)} = 1$. (iv) G has a *principal series* whose factors are abelian. (v) G has a *principal series* whose factors are elementary abelian (§15.1).

15.3 Let K be a field of characteristic 0. Prove that for every n the Galois group of a polynomial of the form $X^{4n} + aX^{3n} + bX^{2n} + cX^n + d \in K[X]$ is solvable.

15.4 Suppose a cubic polynomial $f \in \mathbb{Q}[X]$ is irreducible and has three real roots. Prove that nevertheless there is no radical extension $F/\mathbb{Q}$ such that $F \subseteq \mathbb{R}$ and that F contains the splitting field of f over $\mathbb{Q}$.

15.5 Prove:

(a) Every group of order $n < 60$ is solvable (use §10.11).

(b) A_5 is a simple group of order 60. (*Hint:* Otherwise A_5 would be solvable, because of part (a); but this contradicts F6.)

15.6 Prove:

(a) If a *simple* group G has a strict subgroup H of index $n \leq 5$ and G is not cyclic, then $n = 5$ and $G \simeq A_5$. (*Hint:* G acts on G/H, and is therefore isomorphic to a subgroup G^* of S_n; since sgn must be trivial on G^*, we have $G^* \subseteq A_n$.)

(b) *Every simple group G of order 60 is isomorphic to A_5.* Hint: Otherwise, by part (a), G has no strict subgroup H such that $G : H \leq 5$. Hence there must be fifteen 2-Sylow groups, ten 3-Sylow groups and six 5-Sylow groups. This means that there exists some $x \neq 1$ lying in two different 2-Sylow groups P and P'; since both are abelian, x lies in the center of $H = \langle P, P' \rangle = G$, a contradiction. (Only the count of 2-Sylow groups is needed for the argument if one uses Remark (b') following Sylow's Third Theorem, page 100.)

15.7 Prove: *For every maximal subgroup H of a solvable group G, the index $n = G : H$ is a prime power.* Hint: By induction one can restrict oneself to the case where H contains no nontrivial normal subgroups of G. Moreover, under the action of G on G/H, every normal subgroup $N \neq 1$ acts transitively on G/H. Now consider a minimal normal subgroup N of G and use §15.1.

15.8 Let a group G act on a set M with n elements. Prove that there exists a unique homomorphism

$$\text{sgn} = \text{sgn}_{(G,M)} : G \to \{+1, -1\}$$

with the following property: If $\varphi : \{1, 2, \ldots, n\} \to M$ is any bijection and we denote by $\sigma^\varphi = \varphi^{-1} \circ \sigma \circ \varphi$ the element of S_n corresponding to $\sigma \in G$ via φ, then

$$\text{sgn}(\sigma) = \text{sgn}_n(\sigma^\varphi) \quad \text{for all } \sigma \in G,$$

where $\text{sgn}_n : S_n \to \{+1, -1\}$ is the signature map.

15.9 Let $f \in K[X]$ be a separable polynomial over a field K and let E be a splitting field of f over K. The Galois group G of E/K acts on the set $M = \{\alpha_1, \ldots, \alpha_n\}$ of roots of f in E; let $\text{sgn} : G \to \{+1, -1\}$ be the corresponding signature map (see §15.8). Prove that the element $\Delta = \prod_{i<j}(\alpha_i - \alpha_j)$ satisfies $\sigma(\Delta) = \text{sgn}(\sigma)\Delta$ for all $\sigma \in G$.

Let $H = \{\sigma \in G \mid \text{sgn}(\sigma) = 1\}$ be the subgroup of *even* permutations σ of G. Assuming char $K \neq 2$, the fixed field of H in E is $K(\sqrt{D(f)})$; in particular, $D(f)$ is a square in K if and only if G contains only even permutations. (For the definition of $D(f)$ see Chapter 8, Definition 4; one may as well assume also that f is normalized.)

15.10 Consider the natural action of S_n on the rational function field $k(X_1, \ldots, X_n)$, whose fixed field is $K = k(s_1, \ldots, s_n)$ (see F3). Set $\Delta = \prod_{i<j}(X_i - X_j)$ and show, under the assumption char $k \neq 2$, that $K(\Delta)$ is the fixed field of the alternating group A_n. As we saw in Section 15.3, the intermediate field $k(s_1, \ldots, s_n)$ of $k(X_1, \ldots, X_n)$ is a *field of rational functions in n variables over k*. It is natural to ask whether this is the case also for the intermediate field $K(\Delta) = k(s_1, \ldots, s_n, \Delta)$, that is, whether there are in this field *variables* $t_1, \ldots, t_n$ over k such that $k(s_1, \ldots, s_n, \Delta) = k(t_1, \ldots, t_n)$. The answer to this question is unknown.

15.11 Why does A_4 have no subgroup of order 6? Why are the solvable groups S_3, A_4 and S_4 not nilpotent?

15.12 Prove:

(a) The conjugacy class of a permutation $\sigma \in S_n$ of type $c_1, \ldots, c_n$ has cardinality $n!/(c_1!1^{c_1} c_2!2^{c_2} \ldots c_n!n^{c_n})$; equivalently, exactly $c_1!1^{c_1} c_2!2^{c_2} \ldots c_n!n^{c_n}$ elements of S_n commute with σ.

(b) An element $\sigma \in S_n$ of type $c_1, \ldots, c_n$ belongs to A_n if and only if the number $\sum_j (j-1)c_j$ is even. Let $k(\sigma)$ be the conjugacy class of σ *in* S_n. Given $\sigma \in A_n$, either $k(\sigma)$ is also a conjugacy class *in* A_n or the disjoint union of two conjugacy classes in A_n, and the choice hinges on whether there exists an odd permutation that commutes with σ. The former case happens if and only if $c_j \geq 1$ for some even j or $c_j \geq 2$ for some odd j.

15.13 Using §15.12, show that A_5 is *simple*. (*Hint:* Any normal subgroup N of A_5 consists of full conjugacy classes in A_5. But the equation $1+20x+15y+12z = |N|$ has no solution in integers $x, y, z \geq 0$ when $|N|$ is a proper divisor of 60.)

15.14 Given a finite group G, consider the action of G on itself by left translations. Prove that the corresponding signature map is nontrivial if and only if G has *even* order and the 2-Sylow groups of G are *cyclic*.

15.15 Let n be an *odd* natural number. Prove:

(a) Any group G of order $2n$ contains a normal subgroup of index 2. (*Hint:* There exists $\tau \in G$ of order 2. Consider the cycle decomposition of τ as a permutation of G, and look at the image of τ under $\text{sgn} : G \to \{+1, -1\}$.)

(b) If G is a group of order $2^k n$ containing an element of order 2^k, then G has a normal subgroup N with cyclic quotient group of order 2^k. (*Hint:* Argue by induction on k.)

15.16 Let G be a group acting on a set M. We say that the action is k-transitive if any k-tuple of *distinct* elements of M is mapped to any other such k-tuple by some $\sigma \in G$.

(a) Let G be a subgroup of the full permutation group $S(M)$ of an n-element set M, acting 2-transitively on M. Prove that every normal subgroup $N \neq 1$ of G acts transitively on M.

(b) Prove that for $n \geq 3$ the action of A_n on $\{1, 2, \ldots, n\}$ is $(n-2)$-transitive.

(c) Using induction on n, prove that A_n is *simple* if $n \geq 5$. *Hint:* To start the induction apply §15.13 or §15.5. Now let $n > 5$ and regard A_{n-1} as a subgroup of A_n. Suppose $1 \neq N \triangleleft A_n$. Since N acts transitively, we have $A_n = NA_{n-1}$. By the induction assumption, either $A_{n-1} \cap N = A_{n-1}$ or $A_{n-1} \cap N = 1$. In the second case, take some $\sigma' \in N$ such that $\sigma'(n) = 1$. Prove that there exists $\sigma \in A_{n-1}$ such that $\sigma\sigma'\sigma^{-1} \neq \sigma'$ but $\sigma\sigma'\sigma^{-1}(n) = 1$. This contradicts the fact that $A_{n-1} \cap N = 1$.

15.17 For $n \geq 3$, determine all nontrivial normal subgroups of S_n. (Answer: For $n \neq 4$ the only one is A_n and for $n = 4$ there is also the four-group V_4; see §15.16.)

15.18 Derive from §15.17 *Bertrand's Theorem:* For $n \neq 4$ the group S_n has no subgroup H of index m, where $2 < m < n$. (*Hint:* S_n acts transitively on G/H; now look at $S_n \to S(G/H)$.) As an interesting historical footnote, when *Bertrand* proved his theorem in 1845 he had to make the assumption that for every natural number $n > 1$ there exists a prime p such that $n < p < 2n$; this became known as *Bertrand's Postulate* and was later proved by *Chebyshev*.

As to subgroups of index n in S_n, it can be proved (with more effort) that — assuming $n \neq 6$ — the only ones are those that leave invariant a particular element of the set $\{1, \ldots, n\}$; otherwise put, every automorphism of S_n is inner. The case $n = 6$ is exceptional: then there is also another type.

15.19 Let $p \geq 5$ be a prime number. (a) Prove that there exists $n_0 \in \mathbb{N}$, depending on p, such that for any natural number $n > n_0$ the polynomial

$$f(X) = (X - p)(X - 2p) \ldots (X - (p-2)p)(X^2 + np) - p$$

over $\mathbb{Q}$ is not solvable by radicals. (b) Same question, proving that f in fact has Galois group isomorphic to S_p. (*Hint:* For part (a) see F12, and for part (b) the remarks thereafter.)

15.20 Let $f \in K[X]$ be irreducible of prime degree p. Suppose the splitting field L of f over K contains roots $\alpha \neq \beta$ of f with $\beta \in K[\alpha]$. Prove that L/K is *cyclic* of degree p. (*Hint:* Using F1(b) in Chapter 6, prove that $K(\alpha)/K$ is Galois.)

15.21 Let the subgroup G of S_n act *transitively* on $\{1, 2, \ldots, n\}$. Prove that, if G contains a cycle of length $n - 1$ and also a transposition, then $G = S_n$. (*Hint:* If $(1 \, 2 \ldots n{-}1) \in G$, show that G contains all transpositions of the form $(i \, n)$.)

15.22 (a) Find the Galois group of the polynomial $f(X) = X^5 + X + 1$ over $\mathbb{Q}$. (b) Let $\alpha_1, \alpha_2, \alpha_3$ be the roots of $X^3 - X^2 + 1$ in $\mathbb{C}$. Prove that the splitting field of f has exactly the following distinct nontrivial subfields: $\mathbb{Q}(\sqrt{-3})$, $\mathbb{Q}(\sqrt{-23})$, $\mathbb{Q}(\sqrt{69})$, $\mathbb{Q}(\sqrt{-3}, \sqrt{69})$, $\mathbb{Q}(\alpha_1, \alpha_2, \alpha_3)$, $\mathbb{Q}(\alpha_k)$, $\mathbb{Q}(\alpha_k, \sqrt{-3})$, $\mathbb{Q}(\alpha_k, \sqrt{69})$, with $1 \leq k \leq 3$ in the last three cases.

15.23 Let K be a subfield of $\mathbb{R}$ and let $f \in K[X]$ be a *solvable* prime polynomial of prime degree p. Prove that $D(f) > 0$ if $p \equiv 1 \bmod 4$. For $p \equiv 3 \bmod 4$ the inequality $D(f) < 0$ is equivalent to f having a single real root.

15.24 In the polynomial ring $\mathbb{Z}[X_1, \ldots, X_n]$, consider for $i \geq 0$ the polynomials $p_i = X_1^i + X_2^i + \cdots + X_n^i$ (sums of powers) and prove *Newton's formulas:*

$$p_r - s_1 p_{r-1} + s_2 p_{r-2} - \cdots + (-1)^n s_n p_{r-n} = 0 \quad \text{for } n \leq r,$$
$$p_r - s_1 p_{r-1} + s_2 p_{r-2} - \cdots + (-1)^r s_r r \quad\ = 0 \quad \text{for } n > r \geq 1.$$

Hint: With f as in F3, set $g = f^* = X^n f(X^{-1}) = \sum (-1)^i s_i X^i$. Then the quotient $X g'(X)/g(X)$ can be expanded in power series as $n - \sum_{k=0}^{\infty} p_k X^k$. The assertion follows by comparing coefficients for $X g'(X)$.

Incidentally, $\log f^*(X)$ has the power series expansion $- \sum_{k=1}^{\infty} (1/k) p_k X^k$, from which it likewise follows that each of the p_k's is a polynomial in $s_1, \ldots, s_k$ with integer coefficients (where $s_k = 0$ for $k > n$).

15.25 Let G be the Galois group of the polynomial $f(X) = X^5 - X - 1$ over $\mathbb{Q}$ (see page 189). Use Theorem 7 to prove that $G = S_5$.

Hint: By §16.10 below the discriminant D of f equals $2869 = 19 \cdot 151$. Therefore G is not contained in A_5, by §15.9. Suppose $G \neq S_5$. Then $G \cap A_5$ is a strict subgroup of A_5, and therefore *solvable* (§15.5). But then so is G. Theorem 7 and its proof then imply that $G = NH$, with $N \triangleleft G$ and $H \leq G$, where N is generated by a cycle of length 5 and H is isomorphic to a subgroup of $\mathbb{F}_5^\times$. Since G is not contained in A_5, there must exist $\tau \in H$ such that $\operatorname{sgn} \tau = -1$. Since $G \neq S_5$, this τ cannot be a transposition. Deduce that τ is a cycle of length 4. Letting F be the fixed field of N, the extension $F/\mathbb{Q}$ is then cyclic of degree 4, with $\sqrt{D} \in F$. Now §13.2 leads to a contradiction, because $D = 19 \cdot 151$ is not of the form $a^2 + b^2$ with $a, b \in \mathbb{Q}$. (To see this apply §9.3).

Chapter 16: Integral Ring Extensions with Applications to Galois Theory

16.1 Let A/R be an *integral* ring extension. Prove that, if A is an integral domain and R is a field, A is a field.

16.2 Let A/R be an *integral* ring extension. Let $\mathfrak{P}$ be a prime ideal of A and $\mathfrak{p}$ a prime ideal of R such that $\mathfrak{P}$ lies over $\mathfrak{p}$, meaning that $\mathfrak{P} \cap R = \mathfrak{p}$. Prove:

(a) If $\mathfrak{p}$ is a maximal ideal in R, then $\mathfrak{P}$ is a maximal ideal in A. (*Hint:* Go over to $A/\mathfrak{P}$.)

(b) If $\mathfrak{P}_0$ is a second prime ideal of A lying over $\mathfrak{p}$ and it satisfies $\mathfrak{P}_0 \subseteq \mathfrak{P}$, then $\mathfrak{P} = \mathfrak{P}_0$. (*Hint:* By going over to $A/\mathfrak{P}_0$ one can assume that A is an integral domain and $\mathfrak{P} \cap R = 0$.)

16.3 Let A/R be an extension of commutative rings and let S be a *multiplicative* subset of R (hence also of A); see §3.12. The ring $S^{-1}R$ can be regarded as a subring of $S^{-1}A$. Prove that if A/R is integral, the ring extension $S^{-1}A/S^{-1}R$ is also integral.

16.4 Let A/R be an *integral* ring extension, $\mathfrak{q}$ an ideal of R and $\mathfrak{q}A$ the ideal of A generated by $\mathfrak{q}$. Prove that every $\alpha \in \mathfrak{q}A$ satisfies an integrality equation $f(\alpha) = 0$ whose coefficients, apart from that of the leading term, lie in $\mathfrak{q}$. (*Hint:* Suppose $\alpha = a_1 \alpha_1 + \cdots + a_n \alpha_n$, with $a_i \in \mathfrak{q}$ and $\alpha_i \in A$. The subring $A' = R[\alpha_1, \ldots, \alpha_n]$ is finitely generated as a module over R (see F3), and $\alpha A' \subseteq \mathfrak{q}A'$. Now adapt the proof of F1 to derive the assertion.)

16.5 In the situation of §16.4, assume further that A is an *integral domain*, R is *integrally closed* in its fraction field K and $\mathfrak{q}$ is a *prime ideal*. Prove that if $\alpha \in \mathfrak{q}A$, the *minimal polynomial* g of α over K has all its coefficients in $\mathfrak{q}$, apart from that of the leading term. (*Hint:* Take f as in §16.4 and consider $f(X) = g(X)h(X)$ modulo $\mathfrak{q}$.)

16.6 Let B/R be an extension of commutative rings and $\mathfrak{q}$ a prime ideal of R. If $\mathfrak{q}B \cap R = \mathfrak{q}$, there exists a prime ideal $\mathfrak{Q}$ of B such $\mathfrak{Q} \cap R = \mathfrak{q}$ (and vice versa).

Hint: Set $S = R \smallsetminus \mathfrak{q}$; then $\mathfrak{q}B \cap S = \varnothing$. Thus $\mathfrak{q}S^{-1}B$ is a proper ideal of $S^{-1}B$ (see §4.12) and thus lies in a maximal ideal $\mathfrak{M}$ of $S^{-1}B$. Now set $\mathfrak{Q} = \mathfrak{M} \cap B$, in the sense of §4.12.

16.7 The *fundamental theorem on symmetric functions* in Chapter 15 was only given for the case of polynomials over a field, since that was what the context demanded. Now, with the methods of Chapter 16, prove:

Let R be a commutative ring with unity. Then every symmetric polynomial h in $R[X_1,\dots,X_n]$ has a unique representation $h = g(s_1,\dots,s_n)$ with some polynomial $g \in R[X_1,\dots,X_n]$. Hence the ring of symmetric polynomials of $R[X_1,\dots,X_n]$ coincides with the subring $R[s_1,\dots,s_n]$ of $R[X_1,\dots,X_n]$, and it is even a polynomial ring in $s_1,\dots,s_n$ as variables over R.

Hint: Since the question is one of existence and uniqueness involving finitely many coefficients only, it suffices to prove the theorem for the case of a polynomial ring $R = \mathbb{Z}[Y_1,\dots,Y_r]$ over $\mathbb{Z}$. Let $k = \operatorname{Frac} R$. From Chapter 15, F4 it follows that there exists a unique g in $k(X_1,\dots,X_n)$ such that $h = g(s_1,\dots,s_n)$. There remains to show that g in fact lies in $R[X_1,\dots,X_n]$. This follows from the fact that $h(X_1,\dots,X_n)$ is integral over $R[s_1,\dots,s_n]$ and the latter ring is a UFD, hence integrally closed.

16.8 Derive from §16.7 the existence of a certain n-variable polynomial d_n over $\mathbb{Z}$ with the following property: If $f = X^n + a_1 X^{n-1} + \cdots + a_n$ is any normalized polynomial of degree n over any field K, the discriminant of f is given by $D(f) = d_n(a_1,\dots,a_n)$.

16.9 In the situation of Section 16.3, prove:

(i) $D(f) \in \mathbb{Z}$ (or $D(f) \in R$ as the case may be).

(ii) $\bar{f}$ has no multiple roots if and only if $D(f) \not\equiv 0 \bmod p$ (or $D(f) \not\equiv 0 \bmod \mathfrak{p}$).

16.10 (a) Let E/K be a field extension of degree n and α a primitive element of E/K, with minimal polynomial f. Prove that
$$D(f) = (-1)^{n(n-1)/2} N_{E/K}(f'(\alpha)).$$

(b) Let K be a field. Prove that a polynomial $f \in K[X]$ of the form $f(X) = X^n + bX + c$ has discriminant
$$D(f) = (-1)^{n(n-1)/2}\left(n^n c^{n-1} + (-1)^{n-1}(n-1)^{n-1}b^n\right)$$

Hint: By §16.8, you can assume that f is irreducible and that $\operatorname{char} K = 0$. Suppose $f(\alpha) = 0$; set $\beta = f'(\alpha)$ and show that α can be expressed in terms of β by means of a simple formula. Now compute the minimal polynomial of β and so also $N_{K(\alpha)/K}(\beta)$.

16.11 Prove that the two statements in F12* are indeed equivalent.

16.12 Let A/R be an extension of commutative rings and let $\mathfrak{p}$ be a prime ideal of R. Prove that, if A/R is integral, there exists a prime ideal $\mathfrak{P}$ of A such that $\mathfrak{P} \cap R = \mathfrak{p}$. (This *going up theorem* goes back to *Krull*, but is often named after *Cohen* and *Seidenberg*).

Hint: Consider first the case where R is a *local ring* (see §4.13) and $\mathfrak{p}$ its maximal ideal. Then it's actually the case that *every* maximal ideal $\mathfrak{P}$ of A satisfies $\mathfrak{P} \cap R = \mathfrak{p}$, as can be easily seen by applying F11 to the rings $A/\mathfrak{P}$ and $R/\mathfrak{P} \cap R$. The general case can be reduced to the local case by examining the ring extension $S^{-1}A/S^{-1}R$, where $S = R \setminus \mathfrak{p}$; see §4.13, §4.12 and §16.3.

16.13 Let E/K be a finite *separable* field extension of degree n, and let $\sigma_1, \ldots, \sigma_n$ be the distinct K-homomorphisms from E into the algebraic closure C of E. Fix a basis $\beta_1, \ldots, \beta_n$ of E/K and consider the matrix

(1) $$M = (\sigma_i(\beta_j))_{i,j}.$$

Its determinant Δ is nonzero (Chapter 12, Theorem 2). The square

$$D_{E/K}(\beta_1, \ldots, \beta_n) := \Delta^2$$

is also nonzero. But the square must lie K, as can be seen for instance from the equality $\Delta^2 = \det{}^t\!M \det M$ and its consequence

(2) $$D_{E/K}(\beta_1, \ldots, \beta_n) = \det\big((S_{E/K}(\beta_i\beta_j))_{i,j}\big).$$

The element $D_{E/K}(\beta_1, \ldots, \beta_n) \in K$ is called the *discriminant of the basis* $\beta_1, \ldots, \beta_n$ of E/K. If $E = K(\alpha)$ and $f = \mathrm{MiPo}_K(\alpha)$, the discriminant of the basis $1, \alpha, \ldots, \alpha^{n-1}$ agrees with the *discriminant of the polynomial* f:

(3) $$D_{E/K}(1, \alpha, \ldots, \alpha^{n-1}) = D(f) = (-1)^{n(n-1)/2} N_{E/K}(f'(\alpha));$$

see (11) in Chapter 12 and (17) in Chapter 8.

Now assume that K is the fraction field of an *integrally closed* domain R, and let A be the integral closure of R in E. Suppose a basis $\beta_1, \ldots, \beta_n$ of E/K consists only of elements of A; such a basis exists by F7. Prove that

(4) $$A \subseteq \frac{1}{D}(R\beta_1 + \cdots + R\beta_n),$$

with $D = D_{E/K}(\beta_1, \ldots, \beta_n) \in R$. If we now choose $\alpha \in A$ with $E = K(\alpha)$, then

(5) $$A \subseteq \frac{1}{D}R[\alpha].$$

Hint: Any $\beta \in A$ has first of all a representation $\beta = \sum_j x_j\beta_j$, with each x_j in K. Applying σ_i one gets

$$\sigma_i(\beta) = \sum_j x_j\sigma_i(\beta_j) \quad \text{for } 1 \le i \le n.$$

The assertion follows using Cramér's Rule (see for example LA II, p. 12).

16.14 Let $E = \mathbb{Q}(\sqrt{d})$, where $d \in \mathbb{Z}$ is square-free and distinct from 1. Find the integral closure A of $\mathbb{Z}$ in E. *Hint:* An element $\alpha \in E$ is an algebraic integer if and only if its trace an norm are integers. Use this to show that $A = \mathbb{Z}[\sqrt{d}]$ for $d \equiv 2 \bmod 4$ and for $d \equiv 3 \bmod 4$, whereas $A = \mathbb{Z}[\frac{1}{2}(1 + \sqrt{d})]$ for $d \equiv 1 \bmod 4$.

16.15 Let K be an *algebraic number field*, that is, an extension of $\mathbb{Q}$ such that $K : \mathbb{Q} < \infty$. Denote by $\mathcal{O}_K$ the integral closure of $\mathbb{Z}$ in K. Prove that $\mathcal{O}_K$ possesses a $\mathbb{Z}$-basis with $n = K : \mathbb{Q}$. Such a basis is called an *integral basis* of K. (*Hint:* Using equation (4) in §16.13 above, the assertion follows easily from Section 14.3; see the proof of Theorem 6 there.)

Prove further that all integral bases of K have the same discriminant; the integer D_K thus defined is called the *discriminant of the algebraic number field K*. Find the discriminant of the quadratic number field $K = \mathbb{Q}(\sqrt{d})$ (see §16.14).

16.16 Let E/K be a extension of algebraic number fields, of degree d, and suppose that $E = K(\zeta_m)$, where ζ_m is a primitive m-th root of unity. Prove that

$$m^d \mathcal{O}_E \subseteq \mathcal{O}_K[\zeta_m].$$

Hint: $f = \mathrm{MiPo}_K(\zeta_m)$ divides $X^m - 1$. Therefore $f'(\zeta_m)$ is a divisor of m in $\mathcal{O}_E$. Now use equations (3) and (5) from §16.13.

16.17 (a) Let p^r be a prime power and $\zeta = \zeta_{p^r}$ a primitive p^r-th root of unity in $\mathbb{C}$. Prove that the integral closure of $\mathbb{Z}$ in the field $F = \mathbb{Q}(\zeta)$ is $\mathcal{O}_F = \mathbb{Z}[\zeta]$.

    Hint: Set $\pi := 1 - \zeta$. In the ring $\mathbb{Z}[\zeta] = \mathbb{Z}[\pi]$ we have $(p) = (\pi^e)$ with $e = F : \mathbb{Q}$; see §9.7. Now use this to show that

$$p\mathcal{O}_F \cap \mathbb{Z}[\zeta] = p\mathbb{Z}[\zeta].$$

At the same time, by §16.16, there is a power p^a such that $p^a \mathcal{O}_F \subseteq \mathbb{Z}[\zeta]$.

(b) Now let n be any natural number and ζ_n a primitive n-th root of unity. Prove that $\mathbb{Z}[\zeta_n]$ is the ring of algebraic integers of $\mathbb{Q}(\zeta_n)$. *Hint:* Use induction founded on §16.16(a).

16.18 Let n be a natural number and $c > 0$ a real number. Prove the following simple but remarkable observation of *Kronecker*: There are only finitely many *algebraic integers* α whose conjugates have absolute value at most c and such that $\mathbb{Q}(\alpha) : \mathbb{Q}$ is at most n. Hence a given algebraic number field K (see §16.15) contains only a finite number of algebraic integers whose conjugates all have absolute value at most c. (*Hint:* The coefficients of the minimal polynomial of an algebraic number α can be expressed in terms of the conjugates of α.)

16.19 Deduce from §16.18 that any *algebraic integer* α whose conjugates all have absolute value 1 must be a *root of unity*. Is this also true when α is only assumed to be *algebraic*?

16.20 Let K be a subfield of $\mathbb{C}$ and let $K_0 = K \cap \mathbb{R}$ be its maximal real subfield. We say that K is a *complex multiplication field*, or *CM-field* in short, if two conditions are satisfied: (i) $K:K_0 = 2$, and (ii) K_0 is *totally real*, that is, every homomorphism $\sigma : K_0 \to \mathbb{C}$ maps K_0 into $\mathbb{R}$. An important example of a CM-field is $K = \mathbb{Q}(\zeta_n)$. Assume that K is a CM-field and prove:

(a) If $\rho : \mathbb{C} \to \mathbb{C}$ is complex conjugacy and $\sigma : K \to \mathbb{C}$ is any homomorphism, the composition $\sigma^{-1}\rho\sigma$ is defined and coincides with ρ on K.

(b) If $\alpha \in K$ has absolute value 1, so do all conjugates of α.

(c) If ε is a *unit* of K, that is, a unit in the ring $\mathbb{O}_K$ of algebraic integers in K, then $\varepsilon/\bar{\varepsilon}$ is a root of unity in K (see §16.19).

(d) The map $d : \alpha \to \alpha/\bar{\alpha}$ gives rise to an exact sequence

$$1 \to W_K E_{K_0} \to E_K \xrightarrow{d} W_K/W_K^2,$$

where W_K is the group of roots of unity of K and E_K, E_{K_0} are the groups of units of K, K_0. In particular we have $E_K : W_K E_{K_0} \leq 2$.

16.21 (a) Prove that $K = \mathbb{Q}(\sqrt{2}, \sqrt{1+i})$ is not a CM-field (see §16.20). Find the maximal real subfield K_0 of K.

(b) Find an algebraic integer of absolute value 1 whose conjugates don't all have absolute value 1 (but compare §16.20(b)). *Hint:* Consider $K(\sqrt[4]{2})$ with K as in (a).

(c) Let α be an algebraic number such that $\bar{\alpha} \neq \alpha$ and $\alpha\bar{\alpha} \in \mathbb{Q}$. Prove that if all conjugates of α have the same absolute value, then $\mathbb{Q}(\alpha)$ is a CM-field.

16.22 (a) Let p^r be an odd prime power and $\zeta = \zeta_{p^r}$ a primitive p^r-th root of unity in $\mathbb{C}$. Prove *Kummer's Lemma*: Every unit ε in $\mathbb{Z}[\zeta]$ has the form $\varepsilon = \xi\eta$ for some p^r-th root of unity ξ and some *real* unit η.

(b) Does the same hold in the case $p = 2$?

(c) Let n be a natural number and ζ_n a primitive n-th root of unity. Assume also that $n \not\equiv 2 \bmod 4$. Set $E = \mathbb{Z}[\zeta]^\times$, $E_0 = E \cap \mathbb{R}$ and $W = \langle -\zeta_n \rangle$. Prove that $E \neq WE_0$ if n is not a prime power. (*Hint:* See §16.19 and §16.20, and take §9.8(vi) into account.)

16.23 (a) Take $K = \mathbb{Q}(\alpha)$, where $\alpha^3 + \alpha + 1 = 0$. Justify why $D_K = d(1, \alpha, \alpha^2)$ and hence why $1, \alpha, \alpha^2$ is an integral basis of K.

(b) Take $K = \mathbb{Q}(\alpha)$, where $\alpha^5 - \alpha + 1 = 0$. Show that $\mathbb{O}_K = \mathbb{Z}[\alpha]$.

(c) Take $K = \mathbb{Q}(\alpha)$, where $\alpha^3 + \alpha^2 - 2\alpha + 8 = 0$. Prove that $\mathbb{O}_K \neq \mathbb{Z}[\alpha]$. (*Hint:* Consider $4/\alpha$.)

16.24 Let $K = \mathbb{Q}(\zeta_n)$, where ζ_n is a primitive n-th root of unity, with $n > 2$. Prove that

$$D_K = (-1)^{\varphi(n)/2} \frac{n^{\varphi(n)}}{\prod_{p|n} p^{\varphi(n)/p-1}}.$$

Hint: Consider first the case $n = p^r$; then the n-th cyclotomic polynomial $f = F_n$ satisfies $X^{p^r} - 1 = (X^{p^{r-1}} - 1) f(X)$. Taking the derivative, plugging in ζ_n and taking the norm then leads to the desired assertion. For other values of n work by induction; use the fact that if A is an $r \times r$ matrix and B is an $s \times s$ matrix, the tensor product matrix $A \otimes B = (a_{ij} B)$ has determinant $\det(A)^s \det(B)^r$ (see LA II, p. 181, problem 63).

16.25 Let m, n be odd integers, with n not divisible by 5. Prove that the Galois group of the polynomial

$$f(X) = X^5 + 5mX + 5n$$

over $\mathbb{Q}$ is isomorphic to S_5. (*Hint:* Use Theorem 1 with $p = 2$.)

16.26 Prove that all the roots of the polynomial

$$f(X) = X^5 - 20X^3 + 9X + 1$$

are *real*, and that the Galois group of f over $\mathbb{Q}$ is isomorphic to S_5. (*Hint:* Use Theorem 3 of Chapter 14 with $p = 5$ and F13 of Chapter 16 with $p = 2$.)

16.27 Prove the irreducibility of the polynomial

$$f(X) = X^5 - 5X + 12$$

over $\mathbb{Q}$, convince yourself that f has only one real root and that $D(f)$ is a square, and deduce therefrom that the Galois group G of f over $\mathbb{Q}$ must be isomorphic either to A_5 or to a subgroup of order 10 of A_5 (see Section 15.6). (Note: It turns out that G is *not* isomorphic to A_5, but this is apparently not to be proved without a good deal more trouble.) Prove that

$$f(X) = X^5 + 20X - 16$$

has Galois group over $\mathbb{Q}$ isomorphic to A_5. *Hint:* Again infer from §16.10 that $D(f)$ is a square. Use F13 with $p = 3$ and $p = 7$ to show that f is irreducible and $G(f)$ contains a 3-cycle.

Chapter 18: Fundamentals of Transcendental Field Extensions

18.1 Let E/K be a field extension. Prove the equivalence of: (i) $E : K(B)$ is finite for any transcendence basis B of E/K. (ii) $\mathrm{TrDeg}(E/K) < \infty$, and $E : K(B)$ is finite for any transcendence basis B of E/K. (iii) There exists a finite transcendence basis B of E/K such that $E : K(B)$ is finite. (iv) E/K is finitely generated, that is, $E = K(x_1, \ldots, x_n)$ for some $x_1, \ldots, x_n$ in E.

18.2 Let E/K be a field extension. Prove:

(a) If $M \subseteq E$ is algebraically independent over K and F is an intermediate field of E/K such that F/K is algebraic, then M is also algebraically independent over F.

(b) If E/K is purely transcendental, K is algebraically closed in E.

(c) Let B be a transcendence basis of E/K and M a subset of E such that $E/K(M)$ is algebraic. Then B can be written as a union $B = \bigcup_{\alpha \in M} B_\alpha$, where the B_α are finite sets. Thus for B infinite the cardinality of B does not exceed that of M.

18.3 Prove:

(i) $\mathrm{TrDeg}(\mathbb{R}/\mathbb{Q}) = \mathrm{TrDeg}(\mathbb{C}/\mathbb{Q}) = \mathrm{Card}(\mathbb{R})$.

(ii) $\mathrm{Aut}(\mathbb{C}) = \mathrm{Aut}(\mathbb{C}/\mathbb{Q})$ is *uncountable* (contrast with §8.20).

(iii) If C is the algebraic closure of $\mathbb{Q}$ in $\mathbb{C}$ and B is a transcendence basis of $\mathbb{C}/C$, then $\mathbb{C}/C(B)$ is never finite and hence $\mathbb{C}/C$ is not purely transcendental. (*Hint:* Consider $\sqrt{B}$.)

(iv) There exist subfields K of $\mathbb{C}$ such that $\mathbb{C}:K = 2$ and $K \neq \mathbb{R}$.

18.4 In the polynomial ring $K[X_1, \ldots, X_n]$ over the field K, consider the power sum polynomials $p_1, \ldots, p_n$ defined in §15.24. Prove that if K has characteristic zero, $p_1, \ldots, p_n$ are algebraically independent over K. *Hint:* §15.24 implies that $K[p_1, \ldots, p_n] = K[s_1, \ldots, s_n]$.

18.5 Let k be a field of characteristic other than 2. For any $a, b \in k^\times$, consider the polynomial

(1) $$f(X,Y) = aX^2 + bY^2 - 1 \in k[X,Y]$$

and the corresponding homogeneous polynomial $\tilde{f} = aX^2 + bY^2 - Z^2$ in $k[X, Y, Z]$. Let C be an algebraic closure of $k(X)$ and y an element in C such that $f(X, y) = 0$. Set $x := X$ and $F := k(x, y)$. Prove:

(a) f is absolutely irreducible, that is, irreducible over the algebraic closure C_0 of k (in C).

(b) $\mathrm{TrDeg}(F/k) = 1$, $F : k(x) = 2$, and k is algebraically closed in F.

18.6 Let the setup be as in §18.5. We say that *the projective curve over k defined by* (1) *has a rational point* if there is a point $(\alpha, \beta, \gamma) \neq 0$ in k^3 such that

$$\tilde{f}(\alpha, \beta, \gamma) = a\alpha^2 + b\beta^2 - \gamma^2 = 0.$$

Prove that the following statements are equivalent:

(i) The projective curve over k defined by (1) has a rational point.

(ii) b is a norm in the extension $k(\sqrt{a})/k$.

(iii) F/k is *purely transcendental*, that is, there exists $t \in F$ such that $F = k(t)$.

Hint: Suppose $a\alpha^2 + b - \beta^2 = 0$, with $\alpha, \beta \in k$. Parametrize the line joining the point (α, β) to the point $(x/y, 1/y)$ of the curve by its slope; that is, set

$$t := \frac{1/y - \beta}{x/y - \alpha} = \frac{1 - \beta y}{x - \alpha y} \quad \text{and} \quad s := \frac{1 + \beta y}{x + \alpha y}.$$

An easy calculation shows that $st = a$. Then one can find x, y by solving a system of linear equations with coefficients in $k(s,t)$ via Cramér's rule; the result is

$$x = \frac{2\beta - (s+t)\alpha}{\beta(s+t) - 2a\alpha}, \quad y = \frac{s-t}{\beta(s+t) - 2a\alpha},$$

with a nonzero denominator.

These formulas become, in the case of the equation $x^2 + y^2 - 1 = 0$ of a circle,

$$(2) \qquad\qquad x = \frac{2t}{1+t^2}, \quad y = \frac{1-t^2}{1+t^2}$$

— a *rational parametrization of the circle*! Discuss also the curve $x^2 + y^2 + 1 = 0$, which has no rational points over $\mathbb{Q}$, over the field $k = \mathbb{Q}(i)$.

18.7 Using (2), parametrize all the *rational* solutions of the equation $x^2 + y^2 = 1$. As an application, prove that the set of solutions (a,b,c) of $X^2 + Y^2 = Z^2$ in natural numbers a, b, c (such triples are called *Pythagorean*) with a, b, c relatively prime and b even is parametrized by

$$a = m^2 - n^2, \quad b = 2mn, \quad c = m^2 + n^2,$$

where (m, n) runs over all pairs of relatively prime natural numbers such that $m > n$ and mn is even. (Obviously any Pythagorean triple can be reduced to one satisfying the stated restrictions.)

18.8 Let k be a field of characteristic distinct from 2. Lest the special examples of curves in §18.5 to §18.7 leave something of a false impression, it should be said that for an arbitrary curve defined over k the situation for k-rational points may be radically different. For example, replace the polynomial (1) of §18.5 by a polynomial of the form

$$(3) \qquad\qquad f(X, Y) = Y^2 - d(X),$$

where $d(X)$ is any *separable* polynomial over k of degree at least 3, and let $\tilde{f}$ be the corresponding homogeneous polynomial. A (projective) curve defined over k by a such a polynomial f is called *hyperelliptic* (beware: often this term is restricted to the case $\deg d \geq 5$).

If $k = \mathbb{Q}$ and $\deg d \geq 5$, the curve defined by (3) *always has at most finitely many rational points.* This follows from a general result of *Gerd Faltings*, one of the most significant advances in mathematical research from the last couple of decades.

We also say something about the case where d has degree 3; the projective curve determined by (3) is then called an *elliptic curve* over k, and it must contain at least one rational point over k, since $\tilde{f}(0, 1, 0) = 0$. (When $\deg d = 4$, too, one sometimes talks of the curve as being elliptic, but such a curve need not contain regular rational points over k. If such a point does exist, the curve can be reduced to the case of degree 3 by a change of coordinates.)

Again with $k = \mathbb{Q}$, the elliptic curve defined by $y^2 = x^3 + 7x$ has the rational points $(0, 1, 0)$ and $(0, 0, 1)$, and it can be shown with somewhat more advanced techniques that these are the only ones. By contrast, the elliptic curve defined by $y^2 = x^3 - 2$ has infinitely many rational points (which moreover form a *cyclic* group); this is all the more remarkable because only two of these points have *integer coordinates* (see §4.5). More broadly, the question whether a given elliptic curve over $\mathbb{Q}$ possesses finitely or infinitely many rational points illustrates a challenging and interesting research area in contemporary mathematics, likely to remain a topic of lively investigation for a long time to come.

18.9 Let $K(X_1, \ldots, X_n)/K$ be a purely transcendental extension and E an intermediate field thereof. *Lüroth's Theorem* (see §5.9) says that in the case $n = 1$ the extension E/K is *purely transcendental* as well. Is is natural to ask what can be said in the case $n > 1$, if one assumes additionally that $\mathrm{TrDeg}(E/K) = n$, or equivalently that $K(X_1, \ldots, X_n)/E$ is finite.

For $n = 2$ we have a theorem of *Castelnuovo*: If K is *algebraically closed* and $K(X_1, X_2)/E$ is finite and *separable*, then E/K is also purely transcendental. (The separability assumption cannot be lifted.)

The parallel statement for $n = 3$ is no longer valid (for all this see references in Hartshorne, *Algebraic Geometry*, GTM 52, Springer, New York, 1977). Also *Ischebeck* and others were able to show that even for $n = 2$ there are counterexamples when instead of assuming K algebraically closed one takes $K = \mathbb{R}$ as the ground field. Compare also the remarks at the end of Chapter 16.

Chapter 19: Hilbert's Nullstellensatz

19.1 Let A be an affine K-algebra, that is, a K-algebra $A = K[x_1, \ldots, x_n]$ generated by finitely many elements, and let C be an algebraically closed extension of K. Prove that *there exists a homomorphism of K-algebras* $K[x_1, \ldots, x_n] \to C$.

(*Hint:* Consider the kernel $\mathfrak{a}$ of $K[X_1, \ldots, X_n] \to K[x_1, \ldots, x_n]$. Then the statement can be most easily deduced from Theorem 2; in fact it can be regarded as yet another version of the *Hilbert Nullstellensatz*.)

Can one still demand in the case of an *integral domain* $K[x_1, \ldots, x_n]$ that among specified elements $f_1, \ldots, f_r \in K[x_1, \ldots, x_n]$, none should map to 0? (*Hint:* Consider $K[x_1, \ldots, x_n, 1/f_1, \ldots, 1/f_r]$.)

19.2 Let C be a fixed algebraic closure of K. Prove that the maximal ideals of $K[X_1, \ldots, X_n]$ are exactly the ideals of the form $\mathfrak{m}(\alpha) = \{f \in K[X_1, \ldots, X_n] \mid f(\alpha) = 0\}$ with $\alpha = (\alpha_1, \ldots, \alpha_n) \in C^n$. If K is algebraically closed, these ideals are more explicitly described in the form $\mathfrak{m}(\alpha) = (X_1 - \alpha_1, X_2 - \alpha_2, \ldots, X_n - \alpha_n)$.

19.3 Prove that the *nilradical* $\sqrt{0}$ of an affine K-algebra $A = K[x_1, x_2, \ldots, x_n]$ coincides with the *Jacobson radical* of A, which is defined as the intersection of all the maximal ideals of A. This is a weak form of the *Hilbert Nullstellensatz*.

(*Hint:* Set $A = K[X_1, \ldots, X_n]/\mathfrak{a}$ and take $f \in K[X_1, \ldots, X_n]$. If $\bar{f}$ is not nilpotent, there exists $(\alpha_1, \ldots, \alpha_n) \in \mathcal{N}_C(\mathfrak{a})$ such that $f(\alpha_1, \ldots, \alpha_n) \neq 0$.)

19.4 Prove that if every prime ideal in a commutative ring R with unity is finitely generated, R is *Noetherian*. *Hint:* Let $\mathfrak{a}$ be maximal among all nonfinitely generated ideals of R. For $b \notin \mathfrak{a}$, consider the ideal $\mathfrak{a} : bR = \{x \in R \mid xb \in \mathfrak{a}\}$.

19.5 Let A/R be an *integral* extension of *integral domains*, and let R be *integrally closed* (in its fraction field K). Also let $\mathfrak{P}$ be a prime ideal of A and set $\mathfrak{p} = \mathfrak{P} \cap R$. Prove that every prime ideal $\mathfrak{q}$ of R contained in $\mathfrak{p}$ satisfies

$$\mathfrak{q} A_{\mathfrak{P}} \cap R = \mathfrak{q}.$$

Hint: Any $\beta \in \mathfrak{q} A_{\mathfrak{P}}$ has the form $\beta = \alpha/s$ with $\alpha \in \mathfrak{q}A$ and $s \in S = A \smallsetminus \mathfrak{P}$. The minimal equation of $\alpha \in \mathfrak{q}A$ over K then has the form

$$\alpha^n + a_{n-1}\alpha^{n-1} + \cdots + a_0 = 0 \quad \text{with } a_i \in \mathfrak{q},$$

by §16.5. Therefore, if one further assumes that $\beta \in R$, the *minimal equation* of $s = \alpha/\beta$ over K is

$$s^n + \frac{a_{n-1}}{\beta} s^{n-1} + \cdots + \frac{a_0}{\beta^n} = 0.$$

But s is integral over R, so all the a_{n-i}/β^i lie in R. If β were not in $\mathfrak{q}$, we would have $s^n \in \mathfrak{q}A \subseteq \mathfrak{p}A \subseteq \mathfrak{P}$.

19.6 Prove F6*, *Krull's Descent Lemma*. *Hint:* It suffices to show that there is a prime ideal Q in $A_{\mathfrak{P}}$ such that $Q \cap R = \mathfrak{q}$. Indeed, by §4.12, Q is of the form $\mathfrak{Q}A_{\mathfrak{P}}$, for $\mathfrak{Q}$ a prime ideal of A satisfying $\mathfrak{Q} \subseteq \mathfrak{P}$ and $Q \cap A = \mathfrak{Q}$; there follows $\mathfrak{q} = Q \cap R = \mathfrak{Q} \cap R$. Now, in order to show that such a Q exists, we can't just use the going up theorem of §16.12, since $A_{\mathfrak{P}}$ is generally not integral over R. But we can use the result of §19.5, which suffices because of §16.6.

19.7 Prove (using the *Hilbert Basis Theorem*) that the content of Hilbert's Nullstellensatz (Theorem 1) can easily be quantitatively strengthened as follows: For every ideal $\mathfrak{a}$ of $K[X_1, \ldots, X_n]$ there is a natural number m such that $f^m \in \mathfrak{a}$ whenever $f \in K[X_1, \ldots, X_n]$ vanishes on the zero set of $\mathfrak{a}$. (Actually finding such an m for a specified $\mathfrak{a}$, described say by a set of generators, is a different matter.)

19.8 *Over an algebraically closed field, a system of homogeneous algebraic equations in more unknowns than there are equations always has a nontrivial solution.* Prove this using the road map below, following *H.-J. Nastold*:

Let K be an algebraically closed field, and suppose $f_1, \ldots, f_r$ are polynomials in $K[X_1, \ldots, X_n]$ such that $\mathcal{N}(f_1, \ldots, f_r) = \{0\}$. Then:

(a) There exists $q \in \mathbb{N}$ such that $X_i^q \in (f_1, \ldots, f_r)$ for all i.

(b) Take q as in (a). If the f_i are homogeneous, there exists for each i certain homogeneous polynomials $h_1, \ldots, h_r$ in $K[X_1, \ldots, X_n]$ such that

$$X_i^q = h_1 f_1 + \cdots + h_r f_r \quad \text{and} \quad \deg h_j < q.$$

(c) Set $A = K[X_1, \ldots, X_n]$ and $R = K[f_1, \ldots, f_r]$. If the f_i are homogeneous, the ring extension A/R is *finite*. (*Hint:* By (b), any monomial of degree $\geq nq$ can be generated by monomials of lower degree.)

(d) If the f_i are homogeneous, then $r \geq n$.

At the end of Chapter 27 we shall see what is needed in order to derive this theorem by means of a simple dimensionality argument.

19.9 (a) Let $\mathfrak{a}$ be any ideal of a commutative ring R with unity, and let $\mathfrak{p}_1, \mathfrak{p}_2, \ldots, \mathfrak{p}_r$ be prime ideals of R such that

$$\mathfrak{a} \subseteq \mathfrak{p}_1 \cup \mathfrak{p}_2 \cup \cdots \cup \mathfrak{p}_r.$$

Prove that there exists i such that $\mathfrak{a} \subseteq \mathfrak{p}_i$.

Hint: Let r be minimal with the property that $\mathfrak{a}$ is contained in a union as above. For $1 \leq j \leq r$, there exists $a_j \in \mathfrak{a} \cap \mathfrak{p}_j$ such that $a_j \notin \mathfrak{p}_i$ for all $i \neq j$. Now, if $r > 1$, consider the element $a_1 a_2 \ldots a_{r-1} + a_r$.

(b) Deduce from this that, if $V, W_1, \ldots, W_r$ are algebraic K-sets in C^n with the W_i irreducible and none of them contained in V, there is f in $K[X_1, \ldots, X_n]$ such that f vanishes on V but not on any of the W_i.

19.10 Let V be any algebraic K-set of C^n, where C is assumed algebraically closed. Prove that, for each $1 \leq i \leq n+1$, there exist polynomials $f_1, \ldots, f_i$ and algebraic K-sets $Z_1, \ldots, Z_i$ of C^n such that

(1) $$\mathcal{N}(f_1) \cap \cdots \cap \mathcal{N}(f_i) = V \cup Z_i, \quad \text{with } \dim Z_i \leq n - i.$$

Hint: We can start by assuming that there is some nonzero f_1 that vanishes on V. Then $\mathcal{N}(f_1) = V \cup Z_1$, where Z_1 is the union of all the K-components of $\mathcal{N}(f_1)$ that don't already lie in V. If $Z_1 \neq \varnothing$, there is by §19.9 some f_2 that vanishes on V but not on any K-component of Z_1. Then $\mathcal{N}(f_1) \cap \mathcal{N}(f_2) = V \cup (Z_1 \cap \mathcal{N}(f_2))$, and by (30) we have $\dim(Z_1 \cap \mathcal{N}(f_2)) \leq \dim Z_1 - 1 \leq (n-1) - 1 = n - 2$. Continuing in this way leads to the assertion.

In the case $i = n + 1$, equation (1) amounts to *Kronecker's* result given in F9.

19.11 Round off §19.2 by proving that every *maximal* ideal $\mathfrak{a}$ of $K[X_1, \ldots, X_n]$ can be generated by n irreducible polynomials $f_1, \ldots, f_n$. (As remarked at the end of Chapter 19, it is even true that every *reduced* ideal of $K[X_1, \ldots, X_n]$ can be generated by n polynomials, but the proof cannot be supplied so readily as for the special case of this exercise.)

Hint: By §19.2, such an $\mathfrak{a}$ is the vanishing ideal of a point $\alpha = (\alpha_1, \ldots, \alpha_n)$ in C^n. Now choose f_n in $K[X_1, \ldots, X_{n-1}][X_n]$ such that $f_n(\alpha_1, \ldots, \alpha_{n-1}, X_n)$ is the minimal polynomial of α_n over the field $K[\alpha_1, \ldots, \alpha_{n-1}]$. Take $f \in \mathfrak{a}$. Division with rest over $K[X_1, \ldots, X_{n-1}]$ yields $f = q f_n + r$, with some polynomial r all of whose coefficients lie in the kernel $\mathfrak{a}_{n-1}$ of $K[X_1, \ldots, X_{n-1}] \rightarrow k[\alpha_1, \ldots, \alpha_{n-1}]$. By induction we can assume that $\mathfrak{a}_{n-1}$ is generated by $f_1, \ldots, f_{n-1}$. Then f lies in $(f_1, \ldots, f_{n-1}, f_n)$.

19.12 Prove the following generalization of F4: Every *reduced* ideal $\mathfrak{a}$ in a *Noetherian* ring is an intersection of finitely many prime ideals; if we insist that none of these ideals be contained in one another, they are uniquely determined up to order.

Hint: Use the proof of F4 for guidance. Note that $\sqrt{\mathfrak{b}\mathfrak{c}} = \sqrt{\mathfrak{b}} \cap \mathfrak{c}$.

Index of Notation

$\mathbb{N} = \{1, 2, 3, \ldots\}$ set of natural numbers

$\mathbb{Z}$ ring of integers

$\mathbb{Q}$ field of rational numbers

$\mathbb{R}$ field of real numbers

$\mathbb{C} = \mathbb{R}^2$ complex plane, field of complex numbers

$|M|$ cardinality of a set M

Index

Abel, Niels Henrik (1802–1829)   154, 166, 178, 244
abelian   *see also under* Galois group
  extension   151
  field extension   152
  group   149
    elementary   268
    finite   154, 160
    free   161
    number field   259
adjunction   6
  of a square root   7
affine
  K-algebra   224
  algebraic set   217
  coordinate ring   224
  subgroup of $S(\mathbb{F}_p)$ 187
  variety   222
Alfes, Rainer   ii
algebra over K 11
algebraic
  K-set   217
  closure
    absolute   57
    in an extension   19
  extension   17
  geometry   217
  independence of field homomorphisms   261
  number   15
  number field   259, 275
  number theory   113
  over K 15

algebraically (in)dependent   209
algebraically closed   56, 234
alternating group   182
angle   *see* trisection
Artin, Emil (1898–1962)   55, 113, 118, 120, 147, 261, 268
associated element   34
automorphism
  group of a field extension   65, 76, 246
  inner   96
axiom of choice   39

Bertrand, Joseph Louis François (1822–1900)   271
Bertrand's Theorem and Postulate   271
Bézout, Etienne (1730–1783)   235
Bézout ring   235
BI-Wissenschaftsverlag   ii
bilinear form   134
bilinear map   150
binomial equation   143
Borevich, Zenon I. (1924–1995)   260
Brandis, Albrecht   232

Capelli, Alfredo (1855–1910)   267
Cardano, Geronimo (1501–1576)   165, 185
Carlitz, Leonard (1907–1999)   260
Castelnuovo, Guido (1865–1952)   280
casus irreducibilis   186
Cauchy, Augustin (1789–1857)   101, 154
Cayley, Arthur (1821–1895)   94
center of a group   97
central subring   191
centralizer   97